CAMBRIDGE TRACTS IN MATHEMATICS

122 Duality in Analytic Number Theory

P. D. T. A. Elliott
University of Colorado

Duality in Analytic Number Theory

CAMBRIDGE UNIVERSITY PRESS
Cambridge, New York, Melbourne, Madrid, Cape Town, Singapore, São Paulo

Cambridge University Press
The Edinburgh Building, Cambridge CB2 8RU, UK

Published in the United States of America by Cambridge University Press, New York

www.cambridge.org
Information on this title: www.cambridge.org/9780521560887

First published 1997
This digitally printed version 2008

A catalogue record for this publication is available from the British Library

Library of Congress Cataloguing in Publication data

Elliott, P. D. T. A. (Peter D. T. A.)
Duality in analytic number theory / P.D.T.A. Elliott.
p. cm. – (Cambridge tracts in mathematics ; 122)
Includes bibliographical references and indexes.
ISBN 0 521 56088 8 (hardcover)
1. Number theory. 2. Duality theory (Mathematics) I. Title. II. Series.
QA274.73.B48 1996
512′.73 – dc20 95-48214 CIP

ISBN 978-0-521-56088-7 hardback
ISBN 978-0-521-05808-7 paperback

Table of Contents

Acknowledgements

This book was largely written in my free time. I thank the American Mathematical Society and Cambridge University Press for allowing me to adapt Chapters two and three of my AMS monograph *On the Correlation of Multiplicative and the Sum of Additive Arithmetic Functions*. Partial summer support on National Science Foundation contract DMS 9300551 is reflected in some two hundred and seventy exercises that elaborate the text.

I thank David Tranah, editor of the Cambridge University Press, for his challenging advice.

I particularly thank Liz Stimmel who typed the whole project with virtuoso handling of TeX and its distant relatives.

The yellow sea cycled into a grey swell, lifting bergs of printout, feet deep. For the third time Jean heard the cry of the Dutchman, saw him break through the mist ....

Preface

In this book I have two aims. My first is to give a coherent account of a general method in analytic number theory, and to develop that method sufficiently far that it solves problems otherwise beyond reach. The method applies the simplest notions from functional analysis, and has its roots in geometry.

My second aim, bound to the first, and to me of equal interest, is a light discussion of the creation of the method as a raising of the underlying philosophical motivation into consciousness. In particular, this offers a paradigm for the application of the method itself.

I wrote the present work and my memoir: *The Correlation of Multiplicative and the Sum of Additive Arithmetic Functions* together. To facilitate a bridge between the two works I have elaborated the treatment of approximate functional equations given in Chapters 2 and 3 of the monograph. In particular, I preserve the same notation. For permission to do this I thank both the American Mathematical Society and Cambridge University Press.

The memoir applies the method to a problem not treated in this book. Background details in the construction of the method are omitted. Consideration of the problem to hand remains paramount. A large number of auxiliary results are required.

The present work is quite different in nature. The method itself is the object of study. Essential inequalities are derived in detail. In the body of the text background results either illustrate a mathematical idea, and can be omitted at first reading, or they are chosen to simplify the presentation of a proof, and can often be replaced by something weaker and more easily accessible. Otherwise only the basics of functional analysis on vector spaces of finite dimension, some elementary number theory, a little Fourier analysis and a familiarity with Cauchy's theorem in complex analysis are assumed.

Notation

A function is *arithmetic* if it is defined on the positive integers. Unless otherwise stated, arithmetic functions will be complex valued. The arithmetic function **1** is identically 1.

n, p generally denote a positive integer and a positive prime, respectively.

q often denotes a prime power, q_0 the prime of which it is a power.

$g(n)$ is *multiplicative* if $g(ab) = g(a)g(b)$ whenever (a, b), the highest common factor of a and b, is 1.

$f(n)$ is *additive* if $f(ab) = f(a) + f(b)$ whenever $(a, b) = 1$. A strongly additive function satisfies $f(p^m) = f(p)$, $m = 1, 2, \ldots$.

Additive and multiplicative functions are determined by their values on the prime powers. For completely multiplicative or additive functions we may suppress the condition of coprimality.

I employ the following standard arithmetic functions, the first three of which are multiplicative:

$\phi(n)$ of Euler, the order of the group of reduced residue classes $(\operatorname{mod} n)$,

$\tau(n)$, Dirichlet's function, the number of positive divisors of n,

$\mu(n)$ of Möbius, one at 1, otherwise zero when n has a squared factor, $(-1)^k$ when n is the product of k distinct primes,

$\Lambda(n)$ of von Mangoldt, $\log p$ when $n = p^m$, $m = 1, 2, \ldots$, zero otherwise,

$\pi(x)$, the number of primes not exceeding x.

An initial account of these and related functions may be found in Hardy and Wright [96].

In the Introduction, Chapters 19 and 34, $\tau(n)$ will also denote Ramanujan's modular coefficient function.

$\pi(x, D, \ell)$ denotes the number of primes, not exceeding x, which lie in the residue class $\ell(\operatorname{mod} D)$.

$s = \sigma + i\tau$, $\sigma = \operatorname{Re}(s)$, is a complex variable.

$L(s,\chi)$ denotes the standard Dirichlet series formed with the Dirichlet character χ, $\zeta(s)$ the Riemann zeta function. Classical results concerning these and related functions may be found in Davenport [19], Prachar [134].

$f(x) = O(g(x))$, $f(x) \ll g(x)$ both denote that $|f(x)| \le Ag(x)$, for some constant A, holds uniformly on a specified set of x-values. When qualified 'as $x \to \infty$', the set is a half-line $[x_0, \infty)$ with an undisclosed x_0.

$O(g(x))$ denotes a function f that satisfies $f(x) \ll g(x)$.

$f(x) = o(g(x))$ as $x \to \infty$, means that $\lim_{x\to\infty} f(x)g(x)^{-1} = 0$.

$o(g(x))$ denotes a function f that satisfies $f(x) = o(g(x))$.

Various natural generalisations of these notions are self-evident.

I shall many times employ the elementary bounds

$$\sum_{n\le x} \frac{1}{n} = \log x + \gamma + O\left(\frac{1}{x}\right), \quad x \ge 1,$$

with Euler's constant γ, and

$$\sum_{p\le x} \frac{\log p}{p} = \log x + O(1), \quad \sum_{p\le x} \frac{1}{p} = \log\log x + c + O\left(\frac{1}{\log x}\right), \quad x \ge 2,$$

where c is constant. These estimates are largely established in Hardy and Wright [96], Chapter XXII, in part employing the bound $\pi(x) \ll x/\log x$, $x \ge 2$, of Chebyshev. See, also, Chapter 15 of the present work.

$\nu_x(n; \cdots)$ denotes the frequency $[x]^{-1}N$, where $[x]$ counts the positive integers up to x, N the number of such integers n for which property $\cdots$ holds.

On the space of functions $f : S \to \mathbb{C}$, measurable with respect to μ,

$$\|f\|_\alpha = \left(\int_S |f|^\alpha d\mu\right)^{1/\alpha}, \quad \alpha \ge 1,$$

as usual. $1/\alpha' = 1 - 1/\alpha$, suitably interpreted when $\alpha = 1$.

Families of basic spaces, together with their duals, are introduced as follows:

$\mathbb{C}^s$, $L^\alpha(\mathbb{C}^s)$, $L^\alpha \cap L^\beta(\mathbb{C}^s)$, $\mathbb{C}^t$, $M^\alpha(\mathbb{C}^t)$, in Chapter 3,

$J^\alpha(\mathbb{C}^u)$, $J^\alpha \cap J^\beta(\mathbb{C}^u)$, $K^\alpha(\mathbb{C}^v)$, in Chapter 25, pp. 211–212,

$\lambda^\alpha(F)$, $\lambda^\alpha \cap \lambda^\beta(F)$, $\mu^\alpha(G)$, in Chapter 25, pp. 222-223.

E denotes expectation in the sense of probability theory in Chapters 25, 26, 28.

E denotes a shift operator in the Introduction and Chapters 23, 30, 31; and (with a slightly changed definition) again a shift operator in Chapters 32, 33.

In a composition TS of mappings

$$\xrightarrow{S} B \xrightarrow{T}$$

it is assumed only that the range of S is contained in the domain of definition B of T.

Introduction

Du Doppeltgänger! du bleicher Geselle! H. Heine.

In this volume I give a unified account of a method in the analytic theory of numbers: *the method of the stable dual.* The method is particularly effective in the study of arithmetic functions possessing algebraic structure.

The reader may check details of selected applications of the method; turn to the continuing developments discussed in the penultimate chapter; press on to the new. However, works in analysis, especially in analytic number theory, can seem formless. The leading thread, as Hadamard would have called it, [91] p. 105, becomes obscured by a mass of detail. All too often the conclusion will appear atop a pyramid of small steps, each step apparently insignificant. Sometimes this is due to the nature of the subject; sometimes it is not.

As footnote 2 on page 136 of his book [114], Lakatos makes the following remark:

> *Rationalists doubt that there are methodological discoveries at all. They think that method is unchanging, eternal. Indeed methodological discoverers are very badly treated. Before their method is accepted, it is treated like a cranky theory; after it is treated as a trivial commonplace.*

Experience has convinced me of the validity of this statement. Analytic number theory is a dynamic subject, with partial results as moments of clarity. There rarely comes a static final result. Indeed, analytic number theory is largely concerned with method. An heterogeneous nature, allowing opportunities to exhibit frailties of human psychology, gives it a forbidding aspect.

Accordingly, from time to time I discuss not only the details of a proof, but also an intent of its argument, or the manifestation of a background philosophy. I set these comments in historical context, account the occasional false step, and indicate unsolved problems. In short I give not only an account of these results, but also of the construction of their proofs. I believe that most readers will want not only to understand, but also to create.

The inequality of the Large Sieve has its origins in a paper of Linnik, [120]. He applied Fourier analysis in a manner derived from the Hardy–Littlewood circle method. Subsequent papers: Rényi [142], Roth [145], Bombieri [6], Davenport and Halberstam [20], Gallagher [89], are important references, and the list is by no means complete. In this introduction I am concerned only with the fact that all of these treatments were Fourier analytic. Under this umbrella I include Rényi's use of the theory of probability.

In the Fall of 1962 I entered Trinity College, Cambridge, as a graduate student of H. Davenport. For my first semester I was supervised by A. E. Ingham. Almost at once I became interested in the inequality of the Large Sieve. Although I had already noticed it, Ingham brought to my attention Rényi's paper on the representation of an even integer as the sum of a prime and an almost prime, [141]. There Rényi applies a version of the Large Sieve together with Fourier analysis on the complex plane in the manner of classical analytic number theory. So matters stood.

Things intervened. My thesis for the doctorate in large part concerned integral zeros of cubic forms. However, I continued to apply the inequality of the Large Sieve, and began again to ponder its nature. I obtained a connection between the Large Sieve and the Turán–Kubilius inequality, [27], and corresponded with Rényi himself.

During the academic year, Fall 1969–Summer 1970, I was a visitor to the University of Colorado, Boulder, and as a visitor gave a Colloquium in January of 1970, on the Large Sieve. I summarised my interpretation of the Large Sieve at that time: *Inequalities of Large Sieve type come in pairs, the inequality and its dual* (*or conjugate*); *to establish such an inequality is to determine the spectral radius of a self-adjoint operator*. An account appears in [30].

Once seen in functional analytic terms, as an application of the notion of duality, the inequality of the Large Sieve becomes redolent of the general scheme of analytic number theory otherwise begun by Euler and developed by Dirichlet: Study the integers through Fourier analysis on an appropriate group.

I was more interested in the notion of duality itself, and the precise logical rôle that it played or might play in the proofs of arithmetical propositions. Familiar with projective geometry, where the duality is between point and line, I knew that Desargues' theorem, which is of course an axiom, has its converse as dual. Since application of inequalities of Turán–Kubilius form developed information on the integers from information on the primes, so the dual of the Turán–Kubilius inequality, directly related to the Large Sieve, might yield properties of the primes from those of the integers. Mathematically expressed, the directions *primes* $\to$ *integers*, *integers* $\to$ *primes* were dual. Moreover, notions and arguments might be similarly paired. For example: *operator corresponds to sufficiency, dual operator corresponds to necessity.*

Readers familiar with functional analysis will have observed that the second dual of an operator is often essentially the operator itself. According to the scheme of the example, proposition and dual proposition would then be equivalent; the combined result would have a finished form. This is indeed manifest in certain applications of the method of the stable dual. A proposition concerning an arithmetic function on the integers is valid if and only if another proposition concerning that function holds on the prime powers. Moreover, successive applications of the background philosophy allow loosely formulated initial propositions to be made increasingly precise, as well as suggesting their proof. The argument moves in tightening loops.

In this manifestation the method of the stable dual recalls *Analysis* from the mathematical thought of ancient Greece, a method briefly summarised by Euclid in his *Data.* At home in geometry, that method would here be augmented by the systematic application of the notion of duality, and realised in functional analytic terms. But is an operator not a tangent?

By the seventeenth century the Greek method of Analysis had been abandoned; it was insufficiently effectual. The quest for certainty came to prevail over the quest for finality; Lakatos, [114] footnote to p. 64. Besides this, unswerving fealty to the notion of finality also allows interesting and non-trivial partial results to go unappreciated. Even in projective geometry, my experience is that without further argument the application of duality in pursuit of a converse proposition may not suffice. The method of the stable dual can be enhanced by employing results derived under other aesthetics. Indeed, its form suggests this.

Chapters 1 to 3, 5 and 6 of the present work contain an initial account of the method of the stable dual. Chapter 3 introduces the Banach spaces $L^\alpha(\mathbb{C}^s)$, $M^\alpha(\mathbb{C}^t)$, $\alpha > 1$ defined on intervals of the prime powers, and the integers, respectively. Beginning with these spaces I construct others on which the Turán–Kubilius inequality and various generalisations of it are interpreted as bounds for operator norms. The formulation of the inequalities and the construction of the spaces go hand in hand.

The various spaces depend upon at least one real valued parameter. Moving this parameter I introduce the notion of stability.

Applications of the method follow.

Spaces $\mathcal{L}^\alpha$ of arithmetic functions are introduced in Chapter 8. These spaces embrace many functions of active interest in number theory.

In Chapter 8 I characterise the additive functions which belong to a given space $\mathcal{L}^\alpha$.

In Chapters 9, 10 and 11 the multiplicative functions which lie in $\mathcal{L}^\alpha$ and possess a non-zero (limiting) mean value are characterised. To formulate an appropriate generalisation of the classical result of Delange, [21], is part of the problem.

The results of these four chapters are in a sense final; equivalent propositions are obtained.

Chapter 13 contains short proofs of theorems of Wirsing, [170], and Halász, [92], concerning multiplicative functions with values in the real interval $[-1, 1]$, and in the complex unit disc, respectively, and having asymptotic mean value zero.

The second of these theorems is applied in Chapter 16 to obtain Erdős' characterisation of finitely distributed additive functions. The proof uses Fourier analysis.

Chapter 17 employs the method of the stable dual together with the results of Chapter 16 to characterise those multiplicative functions which belong to $\mathcal{L}^\alpha$ and do not have mean value zero. The theorems of this chapter are also largely final. They may be applied to give necessary and sufficient conditions for a multiplicative function in $\mathcal{L}^\alpha$ to have (asymptotic) mean value zero. Aesthetically this result is not quite as satisfactory as the theorem obtained for non-zero values. In studying the value distribution of an arithmetic function the fact that it has asymptotic mean value zero is not particularly helpful. The function might be identically zero.

Chapter 19 is an anecdote which applies results of Chapter 17 to Ramanujan's modular function coefficient $\tau(n)$.

The space $L^2(\mathbb{C}^s)$ introduced in Chapter 3 can be given an inner product; it becomes a Hilbert space. The operator A_2, there attached to the classical Turán–Kubilius inequality, has an adjoint, A_2^*. The spectrum of the self-adjoint operator I(dentity) – $A_2^*A_2$ can be largely determined. As I show in Chapters 20 and 21, this allows a localised version of the theorem of Chapter 8 to be obtained that is in a sense final. Underlying the classical Turán–Kubilius inequality and certain of its generalisations are approximate isometries. Within the aesthetic of probability, the particular case $\alpha = 2$ was first obtained by Ruzsa, [150], using complex Fourier analysis.

The characterisation of additive functions f for which the difference (discrete derivative) $f(n) - f(n-1)$ belongs to $\mathcal{L}^\alpha$, with its implicit connection to the correlation of multiplicative functions, apparently lies deeper than that for a plain additive function. Motives for such an aim together with ideas for its attainment are given in Chapter 23. The application of an inequality of Large Sieve type is suggested, and a short discussion given to exemplify possible procedures.

The case that $f(n) - f(n-1)$ belongs to $\mathcal{L}^\alpha$ for some α in the range $1 < \alpha \le 2$ is particularly interesting. In the notation of Chapter 3, underlying the study of additive functions in such $\mathcal{L}^\alpha$ is the chain of operators

$$(L^2 \cap L^{\alpha'}(\mathbb{C}^s))' \xrightarrow{A_\alpha} M^\alpha(\mathbb{C}^t) \simeq (M^{\alpha'}(\mathbb{C}^t))' \xrightarrow{A'_{\alpha'}} (L^2 \cap L^{\alpha'}(\mathbb{C}^s))'.$$

In the case $\alpha = 2$, $A_2'A_2$ may be replaced by the self-adjoint $A_2^*A_2$. In order to study the difference $f(n) - f(n-1)$ of additive functions in a similar manner it is necessary to interpose a difference operator between A_α and $A'_{\alpha'}$:

$$M^\alpha(\mathbb{C}^t) \overset{(I-E^{-1})}{\longrightarrow} M^\alpha(\mathbb{C}^t) \simeq (M^{\alpha'}(\mathbb{C}^t))'$$

where E^{-1} largely coincides with the classical forward shift of sequences. The operator $A_2^*(I - E^{-1})A_2$ is no longer self adjoint. Ideally, the effect of the operator $A'_{\alpha'}E^{-1}A_\alpha$ would be negligible. Although there is no initial reason for this to be true, an appropriate conjecture and evidence in its favour are given in Chapter 30. In the case $\alpha = 2$, the conjecture would be a particular yet more general inequality related to the theorem of Bombieri and Vinogradov, [6], and the conjecture of Elliott and Halberstam, [80], concerning primes in arithmetic progression. It would be an abstract norm inequality.

A partial validation of the conjecture, sufficient for many needs, is given in Chapters 27, 28. Satisfactorily small bounds are established for the norms of operators $PA'_{\alpha'}E^{-1}A_\alpha$, where the projections P preserve almost all of the space $(L^2 \cap L^{\alpha'}(\mathbb{C}^s))'$, so-to-speak. Additive functions are regarded as the convolution of two functions, one simple, the other supported on the prime powers. A Mellin transformation (in the complex plane) is used to strip off the simple function and an inequality of Large Sieve type applied to treat the function on the prime powers.

Fractional power Large Sieve inequalities appropriate for this purpose are developed in Chapter 25. It is difficult to see how these inequalities might be formulated or established without the notion of duality. Each desired inequality is the dual of an inequality involving high powers in mean, requiring in part a Riesz–Thorin interpolation. To facilitate a rapid proof I appeal to an inequality of Rosenthal, from the theory of probability. As its form suggests, this inequality can be readily established by adapting the methods of Chapter 2. I demonstrate so in a short series of exercises in Chapter 26.

In part, the treatment of differences of additive functions follows argument for a plain additive function, lifted onto larger spaces.

Chapter 31 contains analogues of the results in Chapters 20 and 21. The additive function $f(n)$ is replaced by its difference $f(n) - f(n-1)$. The inequalities obtained have the desired form but are not quite as sharply localised as their (earlier) models. They much improve an earlier result of Wirsing, [171]. Apparently the case $\alpha = \infty$ was conjectured by Ruzsa.

In Chapter 32 I employ the results of Chapter 31, together with a polynomial ring of shift operators, to characterise those additive functions f for which $\lambda_1 f(n + a_1) + \cdots + \lambda_k f(n + a_k)$ belongs to $\mathcal{L}^\alpha$, where $a_1, \ldots, a_k$

are distinct integers and the λ_j complex. This bears upon a conjecture of Kátai, [108]. The results of this chapter also have a final form; equivalent propositions are obtained. They may be compared with those of Chapter 8.

Apart from changes of detail, the results of Chapters 25, 27, 28, and 30 to 32, motivated by the ideas of Chapter 23, I had already established in 1987, [59], [63]. Owing to exigencies of publication, they first appear here.

In Chapter 34 I discuss a variety of theorems derived using the method of the stable dual, and offer exercises and comments related to parts of the text. It should perhaps be emphasised that whilst the results of Chapters 8 to 11, and 13 can also be obtained without application of the method of the stable dual, at present the general results of Chapters 31, 32 and many of the results in Chapter 34, cannot.

At appropriate locations in the text I have placed chapters that contain only exercises and comments. Not including Chapter 34, there are ten such chapters, of a more or less elaborate nature, together containing 250 exercises. At the expense of an occasional repetition I have rendered the main body of the text independent of these chapters. However, to omit them is to omit much. There I gloss problems and ideas considered in the text, discuss approaching difficulties and indicate connections with classical or modern technique in analytic number theory.

Exercises face variously from analysis towards number theory or from number theory to analysis; many are linked in a chain around a specific object. Thus according to Chapter 14, experience in the method of the stable dual allows a proof of Wirsing's theorem different from that given in Chapter 13, and which does not apply the prime number theorem. With this as catalyst, Chapter 15 deploys a series of exercises to telegraph a commentary upon the classical approaches to the prime number theorem.

Taken together with the main body of the text, the exercise chapters introduce a wide collection of notions that have been of advantage to me in an ongoing study of analytic number theory.

The present volume might be used as a text or for private study. In particular, I provide at once a chapter on the qualitative nature of duality and Fourier analysis. In a short course it would suffice to read section **5**, concerning the duality principle.

0

Duality and Fourier analysis

The notion of duality and its action in analytic number theory informs this entire work. Emphasis is given to the interplay between the arithmetic and analytic meaning of inequalities. The following remarks place ideas employed in the present work within a broader framework.

1. Conics. By duality the notion of a point conic gives rise to the notion of a line conic. The members of the line conic comprise the tangents to the point conic. Slightly surrealistically we may regard a conic to be a geometric object, defined from the inside by a point locus, and from the outside by a line envelope.

2. Dual spaces. Let V be a finite dimensional vector space over a field F. The dual of V is the vector space of linear maps of V into F. The space V and its dual, V', are isomorphic.

To every linear map $T : V \to W$ between spaces, there corresponds a dual map $T' : W' \to V'$. In standard notation, the action $f(x)$ of a function f upon x is written $\langle x, f\rangle$. The dual map T' is defined by $\langle Tx, y'\rangle = \langle x, T'y'\rangle$ where x, y' denote typical elements of V, W' respectively.

Let $V = F^n$, $W = F^m$. We may identify W' with the set of maps $W \to F$ given by $k \mapsto k^t y'$, where y' is a vector in W, t denotes transposition. If we employ a similar identification for V', then T is represented by an m-by-n matrix with entries in F, T' by the transpose of the same matrix. The defining relation of the dual map may be expressed in the form $(Ha)^t b = a^t(H^t b)$, where $H = (h_{ij})$, $1 \le i \le m$, $1 \le j \le n$, represents T. In other terms

$$\text{(i)} \qquad \sum_{i=1}^{m}\sum_{j=1}^{n} h_{ij}a_j b_i = \sum_{j=1}^{n} a_j \sum_{i=1}^{m} h_{ij} b_i.$$

If we set $a_i = 1 = b_j$ for every i, j, then we may view the interchange of two summations as a 'computation through the dual'.

A non-zero vector x is an eigenvector of an operator $T : V \to V$, and λ is the corresponding (scalar) eigenvalue, if $Tx = \lambda x$. The set of all such x corresponding to a fixed eigenvalue comprises an eigenspace of T.

3. Spectral decomposition. On a finite dimensional vector space V over the complex numbers an inner product $(\, , \,)$ can be defined. Riesz' representation theorem asserts that each linear map of V into $\mathbb{C}$ has the form $x \mapsto (x, w)$ for a unique vector w in V. To each linear operator T from V to itself there corresponds an adjoint operator T^*, also from V into itself, and defined by $(Tx, y) = (x, T^*y)$ for all x, y in V.

The operator T is self-adjoint with respect to the inner product if $T = T^*$. The spectral theorem then asserts the existence of a basis for V, orthogonal with respect to the inner product and comprised of eigenvectors of T.

An operator S which commutes with T takes each eigenspace of T into itself. If S is self-adjoint with respect to the same inner product, then we may spectrally decompose each eigenspace by S. Continuing in this manner we see that the basis vectors in the spectral decomposition of V by T may be chosen simultaneous eigenvectors of any (mutually) commuting family of self-adjoint operators which contains T.

The notion of an eigenvector has a meaning for any finite dimensional vector space over a field F, but complications occur. Let $T : F^n \to F^n$ be represented by the matrix H, and let I denote the n-by-n unit matrix. Then λ is (formally) an eigenvalue of T if and only if $\det(H - \lambda I) = 0$. It is clear that the solutions λ to this polynomial equation need not belong to the ground field F. In order to employ all the eigenfunctions, it is apparently necessary to work in a space on which the associated operator is not initially defined. The simplest groundfield analogue of $\mathbb{C}$ would be an algebraically closed field; but that abandons structural advantages attached to fields that are not algebraically closed. For example, over the reals a linear combination of self-adjoint operators is again self-adjoint. Over the complex numbers it need not be.

The Riesz representation allows an inner product space V to be identified with its dual. The identification is not quite a linear map since for a scalar λ, $(x, \lambda w) = \bar{\lambda}(x, w)$; a complex conjugation appears. However, with an appropriate interpolation of this identification and its inverse, to each linear map $T : V \to W$ between inner product spaces, there corresponds an adjoint linear map $T^* : W \to V$, formally defined by $(Tx, y) = (x, T^*y)$, the inner products evaluated in their respective spaces.

On $\mathbb{C}^n$, the standard inner product is given by $(u, v) = \sum_{j=1}^{n} u_j \bar{v}_j$. If the operator $T : \mathbb{C}^n \to \mathbb{C}^m$ is represented by the matrix H, then T^* is represented by $\bar{H}^t$, the complex conjugate transpose of H. There is an analogue of the identity (i), with every b_i replaced by $\bar{b}_i$, and we may correspondingly speak of 'computation through the adjoint'.

4. Banach spaces. The dual of a Banach space over $\mathbb{C}$ is the space of bounded linear operators into $\mathbb{C}$. Two Banach spaces are isometric if there is an isomorphism between them in which the norms of corresponding elements have the same value. Whilst a Banach space is isometric to a subspace of its second dual, neither the first nor second dual of a Banach space need be isometric to the original space. The formalism of **2** carries over, moreover $\|T\| = \|T'\|$.

Hilbert spaces are the Banach spaces whose norm is induced by an inner product. The definition of the adjoint operator between Hilbert spaces slightly extends that for finite dimensional vector spaces, and $\|T\| = \|T^*\|$. For each self-adjoint operator there is a corresponding spectral decomposition of the space on which it acts. However, the spectrum of an operator may contain a continuous component, and the corresponding decomposition of an arbitrary vector in the Hilbert space will then employ projection valued measures.

I assume the spectral decomposition theorem for Hilbert spaces only in the background discussion of Chapter 21. However, the interpretation of finite dimensional vector spaces as Banach spaces with respect to various norms is a notion central to the present work. Moreover, estimates of the spectra of operators on an appropriate inner product space are explicitly or implicitly important. The methods are elementary, the viewpoint perhaps sophisticated.

5. Duality principle. Let h_{ij}, $1 \le i \le m$, $1 \le j \le n$, be mn complex numbers. Narrowly interpreted, the *principle of duality*, as it is often called, asserts that

$$\text{(ii)} \qquad \sum_{i=1}^{m} \left| \sum_{j=1}^{n} a_j h_{ij} \right|^2 \le \lambda \sum_{j=1}^{n} |a_j|^2$$

is valid for all complex a_j, if and only if

$$\text{(iii)} \qquad \sum_{j=1}^{n} \left| \sum_{i=1}^{m} b_i h_{ij} \right|^2 \le \lambda \sum_{i=1}^{m} |b_i|^2$$

is valid for all complex b_i. A simple proof of this particular assertion is given in Chapter 3. The principle may be employed to give a unifying interpretation of inequalities of Large Sieve type, [30].

Let H denote the matrix (h_{ij}). Inequality (ii) asserts that the operator $T : \mathbb{C}^n \to \mathbb{C}^m$, given by $a \mapsto Ha$, in terms of the standard euclidean norms on $\mathbb{C}^n$ and $\mathbb{C}^m$ satisfies $\|T\| \le \lambda^{1/2}$. The companion inequality asserts that the dual operator $T' : \mathbb{C}^m \to \mathbb{C}^n$ given by $b \mapsto H^t b$ satisfies $\|T'\| \le \lambda^{1/2}$. The principle of duality follows from the assertion $\|T\| = \|T'\|$ of **4**.

In terms of norms

$$|a|_\alpha = \left(\sum_{j=1}^{n} |a_j|^\alpha \right)^{1/\alpha}, \quad |b|_{\alpha'} = \left(\sum_{i=1}^{m} |b_i|^{\alpha'} \right)^{1/\alpha'}, \quad \frac{1}{\alpha} + \frac{1}{\alpha'} = 1,$$

a similar application of functional analysis shows that

$$\sup_{a \ne 0} |a|_\alpha^{-1} \left(\sum_{i=1}^{m} \left| \sum_{j=1}^{n} a_j h_{ij} \right|^{\alpha} \right)^{1/\alpha}$$

$$\sup_{b \ne 0} |b|_{\alpha'}^{-1} \left(\sum_{j=1}^{n} \left| \sum_{i=1}^{m} b_i h_{ij} \right|^{\alpha'} \right)^{1/\alpha'}$$

have identical values. Applications of Hölder's inequality show that each of these expressions has the same value as

$$\sup_{\substack{a,b \\ \text{non null}}} \left(|a|_\alpha |b|_{\alpha'} \right)^{-1} \left| \sum_{i=1}^{m} \sum_{j=1}^{n} b_i a_j h_{ij} \right| .$$

From the case $\alpha = \alpha' = 2$, we see that in this setting the duality principle is equivalent to estimating a bilinear form universally.

The left hand side of (ii) has the representations $(Ha, Ha) = (a, \bar{H}^t Ha)$ by the standard inner product on $\mathbb{C}^n$. The best value of λ is the spectral radius of the self-adjoint operator on $\mathbb{C}^n$ represented by $\bar{H}^t H$. This same value is the spectral radius of the self-adjoint operator on $\mathbb{C}^m$ represented by $H\bar{H}^t$.

More generally, the duality principle asserts that certain inequalities come in pairs, and that each pair is equivalent to an estimate for the (equal) spectral radii of a corresponding pair of self-adjoint operators, [30].

Over the last twenty years the duality principle has been increasingly applied in analytic number theory. A main difficulty is the conceptualisation of the problem to hand in terms which allow interpretations of a universal nature. For example, many problems may be reduced to the estimation of sums

$$\sum_{i=1}^{m} \left| \sum_{j=1}^{n} a_j h_{ij} \right|^2$$

or similar. Typically, the a_j are known only in some weak sense, their individual values difficult to obtain. In such circumstances it is often advantageous to allow the a_j to vary freely and seek estimates of the type (ii) through the agency of an appropriate dual, such as (iii). To aid this process we may modify the h_{ij}, for example, by introducing convenience factors, so that the spectral radius of the underlying operator may be readily approached. Amongst many objects, the unlikely may serve as a variable, for example a chain of characters defined on differing groups but possessing a common property.

Variations on the duality principle are employed throughout the present volume. When the h_{ij} are values of characters, this principle has something of the nature of a reciprocity theorem. An illustrative example in the estimation of a character sum using duality is given at the end of Chapter 23. An example applied to the study of the global distribution of the class number of imaginary quadratic fields occurs in [41] Chapter 22.

Several applications of the duality principle, one involving the use of prime numbers as a support set, are vital to the characterisation of differences of additive functions made in [39]. A typical application of the principle to the estimation of character sums appears in Heath-Brown, [97], Lemma 11.1, p. 319, in the course of his proof that for all sufficiently large moduli q, every reduced residue class (mod q) contains a prime not exceeding $q^{11/2}$. I indicate other examples as this volume proceeds.

6. Fourier analysis (mod 1). Fourier analysis arose historically from the practice of constructing general solutions to partial differential equations important in physics by using the superposition of simpler solutions. Dirichlet was the first to provide a wide class of functions of finite period which are equal to convergent sums of their corresponding formal Fourier series. Such complex valued functions f, of period 1, have an expansion

$$f(x) = \sum_{k=-\infty}^{\infty} a_k e^{2\pi i k x},$$

where

$$a_k = \int_0^1 f(x) e^{-2\pi i k x} dx.$$

Fourier series have been embraced within general disciplines.

7. Locally compact groups. The dual of a locally compact abelian group is the group of its continuous homomorphisms into the unit circle in the complex plane with topology induced by the standard topology on $\mathbb{C}^2$. These homomorphisms are called the characters of the group. G will denote a typical group, $\widehat{G}$ its dual and U the unit circle in $\mathbb{C}^2$. Pontryagin and van

Kampen showed that with a suitable topology $\widehat{G}$ becomes locally compact, and $\widehat{\widehat{G}}$ is isomorphic to G.

G has a translation invariant Haar measure, unique up to renormalisation. If $d\mu$ denotes this measure, and χ a typical character on G, then a formal Fourier transform of a function $f : G \to \mathbb{C}$ is the function $\hat{f} : \widehat{G} \to \mathbb{C}$ given by

$$\chi \mapsto \int_G f\bar{\chi}d\mu.$$

Under favourable circumstances there is an inverse transform relating f to the function

$$g \mapsto \int_{\widehat{G}} \hat{f}(\chi)\chi(g)d\nu,$$

where g denotes a typical element of G, and $d\nu$ the Haar measure on $\widehat{G}$.

G is discrete if and only if its dual group is compact. We may then renormalise the Haar measure to give the dual space measure 1, and regard the elements of G as random variables on $\widehat{G}$. To effect this view in general we may condition the measure on $\widehat{G}$.

To an extent, analytic number theory is the study of certain discrete groups by means of their (compact) duals.

8. Fourier analysis (mod 1) revisited. In remark **6**, f is defined on the quotient of additive groups $\mathbb{R}/\mathbb{Z}$. The characters on this group are $x \mapsto e^{2\pi ikx}$, one for each integer k. $\mathbb{R}/\mathbb{Z}$ has dual group $\mathbb{Z}$.

We can define an inner product on the Lebesgue class $L^2(0,1)$ and so on $L^2(\mathbb{R}/\mathbb{Z})$ by

$$(f,g) = \int_0^1 f(x)\overline{g(x)}dx.$$

With this definition $L^2(\mathbb{R}/\mathbb{Z})$ becomes a Hilbert space. In an abuse of notation, the classical orthogonality of exponentials may be expressed:

$$(e^{2\pi ikx}, e^{2\pi imx}) = 0 \quad \text{if} \quad k \neq m.$$

The exponentials $e^{2\pi ikx}$ are eigenfunctions of the differential operator $\frac{d^2}{dx^2}$. It is tempting to try and derive a Fourier expansion from the spectral decomposition theorem. However, this differential operator, although formally self-adjoint on the space of L^2 functions, does not take $L^2(\mathbb{R}/\mathbb{Z})$ into itself. If we cut $L^2(\mathbb{R}/\mathbb{Z})$ down to $\mathbb{C}^\infty(\mathbb{R}/\mathbb{Z})$, the subspace of functions infinitely differentiable, then $\frac{d^2}{dx^2}$ is indeed self-adjoint; but the space $C^\infty(\mathbb{R}/\mathbb{Z})$ is not complete with respect to the norm induced by the inner product. This difficulty is the topological analogue of the algebraic field-of-definition problem attached to the eigenvalues of operators on a finite dimensional space.

Again we wish to extend an operator to a more comprehensive space. Since $\mathbb{C}^\infty(\mathbb{R}/\mathbb{Z})$ is dense in $L^2(\mathbb{R}/\mathbb{Z})$ and $(-g'', g) = (g', g') \geq 0$ there, a theorem of Friedrichs ensures a self-adjoint extension of $-\frac{d^2}{dx^2}$ to $L^2(\mathbb{R}/\mathbb{Z})$. The rôle of $\mathbb{C}^\infty(\mathbb{R}/\mathbb{Z})$ may also be played by the polynomials in $\exp(2\pi i x)$. The subtleties of Fourier analysis begin to reveal themselves.

From the orthogonality of the exponential characters we derive Parseval's relation:

$$\int_0^1 \left| \sum_{|k|\leq n} a_k e^{2\pi i k x} \right|^2 dx = \sum_{|k|\leq n} |a_k|^2,$$

valid for all a_k in $\mathbb{C}$. If we regard this equality as a bound for a mean square norm, then by duality the functions f in $L^2(\mathbb{R}/\mathbb{Z})$ satisfy

$$\sum_{|k|\leq n} \left| \int_0^1 f(x) e^{2\pi i k x} dx \right|^2 \leq \int_0^1 |f(x)|^2 dx,$$

Bessel's inequality. Moreover, there is a function f, not essentially zero, which gives equality. The Hardy–Littlewood circle method amounts to estimating asymptotically the Fourier coefficients of a given periodic function $f(x)$. The interval $[0, 1)$ is covered by smaller intervals around rational numbers a/q, $1 \leq a \leq q$, $(a, q) = 1$, for varying q, and in part $f(x)$ is treated by reduction to $f(a/q)$.

With the reversal of this method Linnik devised (his inequality of) the Large Sieve, [120].

9. Fourier analysis on $\mathbb{R}$. A typical character of the additive group of reals is given by $x \mapsto e^{itx}$, for some real t. $\widehat{\mathbb{R}}$ is isomorphic to $\mathbb{R}$. The Haar measures on $\mathbb{R}, \widehat{\mathbb{R}}$ are renormalised Lebesgue measure. The choice of renormalisation may vary to effect elegance in the presentation of results.

Control of a function around the origin is equivalent to control of its Fourier transform at infinity (far from the origin). That a function and its Fourier transform cannot both be small at infinity was pointed out by Wiener. As a severe example, let the measurable f and its transform $\hat{f}$ vanish outside a compact set I. Then

$$\hat{f}(t) = \frac{1}{\sqrt{2\pi}} \int_I f(x) e^{-itx} dx$$

defines an everywhere analytic function of complex t. Since $\hat{f}(t)$ vanishes on a half-line $\operatorname{Re}(t) > t_0$, $\operatorname{Im}(t) = 0$, by analytic continuation $\hat{f} = 0$ identically. For functions belonging to $L^2(\mathbb{R})$, Plancherel's relation asserts that

$$\int_{\mathbb{R}} |\hat{f}(t)|^2 dt = \int_{\mathbb{R}} |f(x)|^2 dx;$$

and here $f = 0$ almost surely.

The property of changing ends through Fourier analysis is much exploited in the theory of probability. To each distribution function $F(u)$ on the line corresponds the Fourier–Stieltjes transform

$$\phi(t) = \int_{\mathbb{R}} e^{itu} dF(u),$$

otherwise known as the characteristic function. Asymptotic properties of the tail $1 - F(z) + F(-z)$, as $|z| \to \infty$, are then equivalent to properties of $\phi(t)$ as $t \to 0$.

Wiener's phenomenon may be compared with the reciprocating action of duality.

Applications of Fourier analysis on $\mathbb{R}$, with and without integration, appear in Chapters 14, 15, 16. Notions from the theory of probability, including that of independent random variables, are explicitly utilised in Chapters 25, 26 and 28.

For functions f supported on a half-line, it is often possible to view $\sqrt{2\pi}\ \hat{f}(-is)$ as a function of the complex variable s. This function is then called the Laplace transform of f. A series of exercises demonstrating the Laplace transform is given in Chapter 7.

10. Poisson summation. In Remark **9** the Fourier transform on $\mathbb{R}$ was renormalised to identify Plancherel's identity within an isometry between the space $L^2(\mathbb{R})$ and its dual. In the present remark it is appropriate to define

$$\hat{f}(t) = \int_{\mathbb{R}} f(x) e^{-2\pi i t x} dx, \quad t \text{ in } \mathbb{R}.$$

$\mathbb{Z}$ is a subgroup of $\mathbb{R}$. Under favourable circumstances the function f, defined on $\mathbb{R}$, may be summed over the representatives of a coset to generate a function $\sum_{n=-\infty}^{\infty} f(n+x)$ defined on $\mathbb{R}/\mathbb{Z}$. Expanding this as a Fourier series and reassembling formally gives

$$\sum_{n=-\infty}^{\infty} f(n+x) = \sum_{k=-\infty}^{\infty} e^{2\pi i k x} \hat{f}(n).$$

In particular,

$$\sum_{n=-\infty}^{\infty} f(n) = \sum_{n=-\infty}^{\infty} \hat{f}(n).$$

This last is Poisson summation. The choice of Fourier transform $\hat{f}$ on $\mathbb{R}$ ensures consistency between the Haar measures on $\mathbb{R}$ $(= \widehat{\mathbb{R}})$ and $\mathbb{R}/\mathbb{Z}$ $(= \widehat{\mathbb{Z}})$.

An example is furnished by $f(x) = \exp(-\pi x^2 y)$, $y > 0$. Then $\hat{f}(t) = y^{-1/2}\exp(-\pi t^2 y^{-1})$ and

$$\sum_{n=-\infty}^{\infty} e^{-\pi n^2 y} = \frac{1}{\sqrt{y}} \sum_{n=-\infty}^{\infty} e^{-\pi n^2/y}.$$

The left side representation gives precise asymptotic behaviour as $y \to \infty$, the right as $y \to 0+$. We can change ends.

Once the formula is established, we may extend it to hold for $y = -iz$, z complex, with the complex z-plane cut to guarantee a single valued meaning to $(-iz)^{1/2}$. Evaluating the resulting expressions at suitable rational points, Cauchy could give a proof of the quadratic reciprocity law through the agency of Gauss sums.

A derivation of standard Large Sieve inequalities as an application of the duality principle, implemented by a Poisson summation, is given in [56], Chapter 6, and in the exercises of the present Chapter 4. A sophisticated application of this methodology may be found in a recent paper of Duke and Iwaniec, [24].

11. Dirichlet characters. If we give a finite abelian group the discrete topology, then the dual group is isomorphic to the original group. This process was begun by Dirichlet, who explicitly constructed the characters of the groups of reduced residue classes $(\text{mod}\, m)$: $(\mathbb{Z}/m\mathbb{Z})^*$, in the course of his celebrated proof that each such class contains infinitely many primes.

12. Characters on Q^*. Dirichlet characters are readily extended to the multiplicative positive rationals prime to m. We endow Q^*, the multiplicative group of all positive rationals, with the discrete topology. Let Γ_m denote the subgroup of Q^* generated by the primes dividing m. Then there is a canonical homomorphism $Q^* \to Q^*/\Gamma_m$, and

$$Q^* \to Q^*/\Gamma_m \to (\mathbb{Z}/m\mathbb{Z})^* \to U$$

gives a character on Q^*. This is not quite the generalisation favoured in Analytic Number Theory, which extends a Dirichlet character to be zero on the integers that have a prime in common with m.

Since Q^* is a direct sum of the cyclic groups generated by the prime numbers, the dual of Q^* is isomorphic to the direct product of denumerably many copies of $\mathbb{R}/\mathbb{Z}$. A study of the characters of Q^* obtained by factoring by a suitable subgroup,

$$Q^* \to Q^*/\Gamma \to U,$$

can be brought to bear upon the problem of representing rationals as products of given rationals, [48], [51], [52], [56], [78]. To this end the method of the stable dual, including the principle of duality, may in particular be applied. This may be viewed as the beginning of a systematic study of Harmonic Analysis on Q^*.

13. Fourier analysis on $\mathbb{R}^*$. Let $\mathbb{R}^*$ denote the multiplicative group of positive reals. The map $u \mapsto \log u$ renders $\mathbb{R}^*$ isomorphic to the additive group of reals, and induces a topology on $\mathbb{R}^*$ from that on $\mathbb{R}$. Likewise the Haar measure on $\mathbb{R}^*$ is induced by Lebesgue measure on $\mathbb{R}$. The characters on $\mathbb{R}^*$ have the form $x \mapsto x^{it}$, t real. The dual of $\mathbb{R}^*$ is isomorphic to $\mathbb{R}$. Corresponding to the Fourier transform is the Mellin transform. The inversion of a Mellin transform takes place naturally along a line in the complex plane. Using substitution, appropriate formulae may be formally derived from those for Fourier analysis on $\mathbb{R}$.

It was Riemann, in his celebrated paper of 1860, who introduced Mellin transforms into analytic number theory, and with them the accompanying problem of providing analytic continuation for certain Dirichlet series.

Inspired by a method of Euler, Dirichlet based his investigation of the distribution of primes in residue classes upon the study of the L-series $\sum_{n=1}^{\infty} \chi(n)n^{-s}$ associated with the characters χ on $(\mathbb{Z}/m\mathbb{Z})^*$. His s is real, $s > 1$. Both the characters and the series of the form $\sum_{n=1}^{\infty} a_n n^{-s}$ are now named after him. It is clear that a Dirichlet series is a Mellin–Stieltjes transform

$$\int_{1-}^{\infty} y^{-s} d\left(\sum_{n \le y} a_n\right).$$

In its half-plane of absolute convergence, a Dirichlet series has the representation

$$\sum_{n=1}^{\infty} a_n n^{-s} = s\int_{1}^{\infty} y^{-s} \sum_{n \le y} a_n y^{-1} dy,$$

with $y^{-1}dy$ the Haar measure on $\mathbb{R}^*$. Under the property of changing ends, to investigate the asymptotic distribution of prime numbers requires knowledge of appropriate Dirichlet series as s approaches 1, their abscissa of absolute convergence. Note that for $\operatorname{Re}(s) > 1$, $d\left(\sum_{n \le x} a_n n^{-s}\right)$ assigns a finite measure to $\mathbb{R}$. As s approaches 1, $y^{-(s-1)}$ approaches the identity (origin) of the group $(\mathbb{R}^*)^\wedge$.

Plancherel's relation and a Mellin analogue occur in Chapters 14 and 15. In Chapter 28 I employ the Mellin transform of a Dirichlet series to estimate the norm of a composition of operators.

The function $G(s) = \sum_{n=1}^{\infty} g(n)n^{-s}$ formed with a completely multiplicative function g taking values in U may be viewed as a generalisation to Q^* of a Dirichlet L-series. In practice it is convenient to allow g to lie in the complex unit disc. The functions $G(s,\chi) = \sum_{n=1}^{\infty} g(n)\chi(n)n^{-s}$ attached to Q^*, in which g is braided with a standard Dirichlet character, play a vital rôle in the study of the correlation of multiplicative functions made in [77]. In [58] it was shown that for a given g, at most one $G(s,\chi)$ can approximate a pole as $\mathrm{Re}\,(s) \to 1+$; a result reminiscent of the Deuring–Heilbronn phenomenon in the study of the class number of imaginary quadratic number fields. Otherwise interpreted, a character on Q^* can be near to only one Dirichlet character.

14. Automorphic functions. The notion of an automorphic function is somewhat diffuse. Let H be a topological space upon which a group T acts. A function f defined on H is automorphic with respect to T if it has sufficient analytic properties and is largely invariant under T. I furnish two examples.

Example 1. Let $\mathrm{T} = SL_2(\mathbb{Z})$ be the group of 2-by-2 matrices with rational integer entries, and determinant 1. Let H be the upper complex half-plane $\mathrm{Im}\,(z) > 0$.

Let the function $f(z)$ satisfy $f(\gamma z) = (cz+d)^k f(z)$ for some even integer k, all $\gamma = \begin{pmatrix} \cdot & \cdot \\ c & d \end{pmatrix}$ in T and all z in H. Since $\begin{pmatrix} 1 & 1 \\ 0 & 1 \end{pmatrix}$ belongs to T, $f(z+1) = f(z)$. In a sense f generalises the notion of a function periodic (mod 1).

The group $\mathbb{Z}$ acts on $\mathbb{R}$ by $x \mapsto x+1$, so that a function defined on $\mathbb{R}$ and periodic (mod 1) can be viewed as a function defined on the interval $[0,1)$. A typical element $\begin{pmatrix} a & b \\ c & d \end{pmatrix}$ of Γ acts on H by $z \mapsto (az+b)/(cz+d)$. We may view $f(z)$ as defined on the complete set of representatives $-1/2 < \mathrm{Re}\,(z) \le 1/2$, $|z| \ge 1$. The (locally compact) region D : $-1/2 \le \mathrm{Re}\,(z) \le 1/2$, $|z| \ge 1$ is known as a fundamental domain for Γ. If we identify the half-lines $\mathrm{Re}\,(z) = 1/2$, $|z| \ge 1$ and $\mathrm{Re}\,(z) = -1/2$, $|z| \ge 1$, then we see that topologically the fundamental domain represents part of an upper infinite organ pipe.

The map $q = e^{2\pi i z}$ takes the semi-infinite box $-1/2 < \mathrm{Re}\,(z) \le 1/2$, $\mathrm{Im}\,(z) > 0$, one-to-one onto the punctured disc $0 < |q| < 1$ in the complex q-plane. We may compactify the fundamental domain by adding to it a point at infinity, corresponding to the disc centre $q = 0$.

Suppose now that f is meromorphic on H. We say that f is holomorphic at infinity if the function $g(q)$, given by $g(e^{2\pi i z}) = f(z)$, is holomorphic at $q = 0$. This is equivalent to requiring an expansion $g(q) = \sum_{n=0}^{\infty} a_n q^n$ in some neighbourhood of the origin. In other terms, there is an expansion of Fourier type,

$$f(z) = \sum_{n=0}^{\infty} a_n e^{2\pi i z n},$$

valid for all $\operatorname{Im}(z)$ sufficiently large.

The functions f, holomorphic on H and satisfying the above 'condition at infinity' are said to be (holomorphic) modular forms of weight k with respect to Γ. Further, they are cusp forms if the first coefficient a_0 in the Fourier expansion (at the cusp) is zero.

It is not *a priori* clear that there are any non-trivial modular forms.

The celebrated arithmetic function $\tau(n)$ of Ramanujan can be formally defined by

$$x \prod_{k=1}^{\infty} (1 - x^k)^{24} = \sum_{n=1}^{\infty} \tau(n) x^n.$$

If we set $x = e^{2\pi i z}$, then the left hand product is a cusp form of weight 12 for Γ, the right hand sum is its Fourier expansion at infinity. That $\tau(n)$ is a multiplicative function of n was conjectured by Ramanujan. It was established by Mordell, prefiguring the algebraic notion of a Hecke operator.

The space of cusp forms of (even) weight k for Γ bears the Petersson inner product

$$(f, g) = \int_D f(z)\overline{g(z)} y^{k-2} dx dy$$

with respect to which the members of a commutative algebra of Hecke operators are self-adjoint. This begins a mechanism for studying (holomorphic) modular forms.

It is characteristic of modular forms that their Fourier coefficients carry arithmetic information but are difficult to estimate. The circle method has a main root in the evaluation by Hardy and Ramanujan of the partition function $p(n)$, defined by

$$\prod_{j=1}^{\infty} (1 - x^j)^{-1} = 1 + \sum_{n=1}^{\infty} p(n) x^n.$$

The shape of the generating function suggests a connection with a modular function.

Example 2. By means of the Petersson inner product, $L^2(D)$ becomes a Hilbert space. The hyperbolic Laplacian $y^2 \left(\frac{\partial^2}{\partial x^2} + \frac{\partial^2}{\partial y^2} \right)$ has an extension which is self-adjoint with respect to this inner product. That such an extension exists is not immediate, but after the difficulties encountered with the differential operator $\frac{d^2}{dx^2}$ in $L^2(\mathbb{R}/\mathbb{Z})$ we are not surprised. There is now the theoretical guarantee that a typical function in $L^2(D)$ has a spectral decomposition via the Laplacian. A practical determination of such a decomposition is no straightforward matter. The papers of Maass [121] and Selberg [156], [157] are fundamental; Bruggeman [9], [10] and Kuznetsov, [113], are also significant. For the continuous spectrum, Eisenstein series come into play. For the discrete spectrum, the Γ-invariant, square summable eigenfunctions of the hyperbolic Laplacian are important. These are known as Maass forms.

The space of (holomorphic) modular forms of weight k with respect to Γ has finite dimension over $\mathbb{C}$. Holomorphic forms are not plentiful, but they are rich in explicit number theoretical connections. There is a sufficient supply of Maass forms, but their arithmetic nature is less clear. There is again a commutative algebra of Hecke operators, each of which commutes with the Laplacian.

The above notions lend themselves to much generalisation. Suitably defined, holomorphic modular forms of orders other than an even integer exist. The theory is of number theoretical interest when Γ is replaced by certain subgroups of itself. $SL_2(\mathbb{Z})$ may be replaced by other groups, which act on objects other than the upper half-plane. As the spaces corresponding to $L^2(D)$ enlarge, further differential operators may be needed.

Like that for Fourier analysis on $\mathbb{R}/\mathbb{Z}$, the literature on automorphic functions is extensive. Applications of modular functions and of spectral decomposition via the hyperbolic Laplacian, in particular to questions in analytic number theory, appear in the books of Sarnak, [151], and Venkov, [167].

Ramanujan's function $\tau(n)$ is featured in Chapter 19 of the present work to illustrate what can be achieved for a largely unknown multiplicative function by using general principles. Otherwise I do not employ the theory of automorphic functions.

A comprehensive account of functional analysis may be found in Yosida, [175]. The Riesz–Thorin interpolation theorem, which I employ in Chapters 23, 25, 28 and 29, is proved in Bergh and Löfström, [5]. Haar measures in the present volume are either on finite sets, or may be derived from Lebesgue measure on $\mathbb{R}$. A leisurely treatment of Haar measure on (otherwise arbitrary) locally compact groups is given in Nachbin, [133]; the beginnings of Fourier analysis on abelian locally compact groups in Rudin, [146]. An introduction to the theory of complex variables with emphasis on Dirichlet series and Mellin transformation, sympathetic to analytic number theory, may be found in Titchmarsh, [163]. However, the treatment of analytic continuation given there might be elaborated, at least to that in Conway, [13], proceeding up to and including the monodromy theorem. The relevance of such studies to the Riemann zeta function is demonstrated in Titchmarsh, [165]; see also Prachar [134], particularly the appendix.

A sufficient treatment of the Fourier transform on $\mathbb{R}$ may be found in Titchmarsh, [164]. As practical matters, if $f : \mathbb{R} \to \mathbb{C}$ belongs to $L(\mathbb{R})$, then

$$\hat{f}(y) = \int_{\mathbb{R}} f(x)e^{-ityx}\,dx$$

defines a continuous function $\mathbb{R} \to \mathbb{C}$. For points x contained in an open interval on the closure of which f has bounded variation, this transformation may be inverted:

$$\frac{1}{2}(f(x+) + f(x-)) = \lim_{T\to\infty} \frac{1}{2\pi} \int_{-T}^{T} e^{itxy}\hat{f}(y)dy,$$

where

$$f(x+) = \lim_{\varepsilon\to 0} f(x+\varepsilon^2), \quad f(x-) = \lim_{\varepsilon\to 0} f(x-\varepsilon^2).$$

It is straightforward to derive a result for the Laplace transform. Suppose that $e^{-cx}f(x)$ belongs to $L(0,\infty)$ for some real c. Then the Laplace transform

$$\tilde{f}(s) = \int_0^\infty f(x)e^{-sx}dx$$

exists for $\operatorname{Re}(s) \geq c$, in fact is analytic in the half-plane $\operatorname{Re}(s) > c$. It may be inverted:

$$\frac{1}{2}(f(x+) + f(x-)) = \lim_{T\to\infty} \frac{1}{2\pi i} \int_{c-iT}^{c+iT} e^{xs}\tilde{f}(s)ds,$$

integrating upwards along the line $\operatorname{Re}(s) = c$, under the same conditions upon f as for the Fourier transformation. We may increase c in this integral by any positive amount.

The evaluation of a Mellin inversion appropriate to the study of Dirichlet series:

$$\lim_{T\to\infty} \frac{1}{2\pi i} \int_{c-iT}^{c+iT} \frac{x^s}{s} ds = \begin{cases} 1 & \text{if } \ x > 1, \\ \dfrac{1}{2} & \text{if } \ x = 1, \\ 0 & \text{if } \ 0 < x < 1, \end{cases}$$

valid whenever $c > 0$, follows rapidly. It may also be accomplished by simple contour integration.

For functions f, g in $L(\mathbb{R})$, the convolution $f * g$, given at x by

$$\int_{\mathbb{R}} f(x-u)g(u)du,$$

also belongs to $L(\mathbb{R})$. Moreover, $(f * g)^{\wedge}$ is the product $\hat{f}\hat{g}$. Besides their own interest, convolutions serve to accelerate the convergence of Fourier integrals.

In the present text technique from specialised areas of Fourier analysis usually serves in an auxiliary capacity. Individual steps in the various arguments are largely elementary, subservient to the overview. A main purpose is a pursuit of the general notion of duality, and its implications for the methodology of constructing and establishing propositions in analytic number theory. In part the work is a study in the clarification of form. Nonetheless, the solution of otherwise intractable problems is reached.

The arithmetic implication of analytic inequalities interested me from the outset, and the number theoretical meaning attached to inequalities gives the present work some of its flavour. A typical inequality derived from functional analysis, even a simple one such as the dual Turán–Kubilius, can be so variously applied that arithmetic implications are almost completely obscured. Their full extent can only be imagined. Of the study of the arithmetic implications of analytic inequalities, particularly those obtained locally through duality, I here account only a beginning.

The interested reader may perhaps like to read the introduction to the present volume for a second time.

1

Background philosophy

The method of the stable dual is in two parts: the derivation of one or more approximate functional equations, by means of the notion of duality; and the solution of these equations, interpreted as a problem concerning the stability of functions defined by functional equations. Only widely applicable basic principles are employed.

Under the philosophy: *If operator corresponds to sufficiency, then dual operator corresponds to necessity* the method begins with an inequality of arithmetic nature derived by duality from an inequality of functional nature, such as a bound for an operator norm.

A standard variant of the Turán–Kubilius inequality asserts that

$$\sum_{n \le x} \left| f(n) - A(x) \right|^2 \ll x B(x)^2 \tag{1.1}$$

with

$$A(x) = \sum_{q \le x} \frac{f(q)}{q} \left(1 - \frac{1}{q_0}\right), \quad B(x) = \left(\sum_{q \le x} \frac{|f(q)|^2}{q} \right)^{1/2} \ge 0,$$

holds uniformly for all complex valued additive functions f, and real $x \ge 2$. Here q denotes a prime power, and q_0 the prime of which it is a power. For the probabilist this may be viewed as an analogue of the well-known inequality of Chebyshev, and reflects the fact that whether an integer is divisible by one prime is independent of its being divisible by another. For the algebraist a completely additive function f may be regarded as the restriction of a homomorphism from the multiplicative group of rationals into the additive group of complex numbers, and the Turán–Kubilius inequality as a quantitative representation of the freedom of Q, with the prime numbers as generators.

An important rôle in my presentation of the method is played by inequalities of the type

$$\sum_{q\le x} q \left| \sum_{\substack{n\le x\\ q\| n}} d_n - \frac{1}{q}\left(1-\frac{1}{q_0}\right)\sum_{n\le x} d_n \right|^2 \ll x \sum_{n\le x} |d_n|^2. \tag{1.2}$$

The inequality (1.2), which is valid for all complex numbers d_n, is a dual of the Turán–Kubilius inequality, Elliott, [41] Chapter 4. In the next two chapters I obtain estimates which much generalise these inequalities, and lend themselves very well to the method of the stable dual. I continue to use q to denote a prime power, q_0 the prime of which it is a power.

2

Operator norm inequalities

Theorem 2.1. *If $\alpha \geq 2$, then*

$$\left(x^{-1} \sum_{n \leq x} \left| f(n) - \sum_{q \leq x} \frac{f(q)}{q} \left(1 - \frac{1}{q_0}\right) \right|^{\alpha} \right)^{1/\alpha}$$

$$\ll \left(\sum_{q \leq x} \frac{|f(q)|^2}{q} \right)^{1/2} + \left(\sum_{q \leq x} \frac{|f(q)|^{\alpha}}{q} \right)^{1/\alpha}$$

uniformly for complex additive functions f, and for $x \geq 1$.

If $0 < \alpha < 2$, then a similar inequality holds with the upper bound replaced by

$$\inf_{\eta > 0} \left\{ \left(\sum_{\substack{q \leq x \\ |f(q)| \leq \eta}} \frac{|f(q)|^2}{q} \right)^{1/2} + \left(\sum_{\substack{q \leq x \\ |f(q)| > \eta}} \frac{|f(q)|^{\alpha}}{q} \right)^{1/\alpha} \right\}.$$

The implied constant may be taken an absolute multiple of α when $\alpha \geq 2$, and absolute when $1 < \alpha < 2$.

For use in the proof of Theorem 2.1 define

$$D = \left(\sum_{q \leq x} \frac{|f(q)|^2}{q} \right)^{1/2} \geq 0.$$

Define additive functions f_j by

$$f_1(q) = \begin{cases} f(q) & \text{if } |f(q)| \leq D, \\ 0 & \text{otherwise,} \end{cases} \qquad f_2(q) = \begin{cases} f(q) & \text{if } |f(q)| > D, \\ 0 & \text{otherwise,} \end{cases}$$

so that $f = f_1 + f_2$. The proof falls into two parts according to the size of $f(q)$. Such a division is suggested by experience in the theory of probability. Without loss of generality we may assume $D > 0$.

(i) Small values of $f(q)$

Lemma 2.2. *The estimate*

$$\sum_{2\le n\le x} g(n) \le \left(\frac{x}{\log x} + \frac{10x}{(\log x)^2}\right)\Delta \sum_{n\le x}\frac{g(n)}{n}$$

with

$$\Delta = \sup_{1\le y\le x} y^{-1}\sum_{q\le y} g(q)\log q$$

holds uniformly for all real non-negative multiplicative functions g, and all $x \ge 2$. Moreover,

$$\frac{1}{\log x}\sum_{n\le x}\frac{g(n)}{n} \ll \exp\left(\sum_{q\le x}\frac{g(q)-1}{q}\right)$$

with an absolute implied constant.

In particular, if $g(q) \le M$ on the prime powers, then

$$x^{-1}\sum_{2\le n\le x} g(n) \ll M\exp\left(\sum_{q\le x}\frac{1}{q}(g(q)-1)\right)$$

Proof. Employing the estimate $\log n = \sum_{q\|n}\log q$ we can write

$$S(x) = \sum_{n\le x} g(n)\log n = \sum_{q\le x}\log q \sum_{\substack{m\le x/q\\(m,q)=1}} g(qm).$$

Since g is multiplicative, the innermost sum does not exceed $g(q)\sum g(m)$, $m \le x/q$, so that after a change of summation

$$S(x) \le \sum_{m\le x} g(m)\sum_{q\le x/m} g(q)\log q \le \Delta x\sum_{m\le x}\frac{g(m)}{m}$$

Integrating by parts,

$$\sum_{2\le n\le x} g(n) = \frac{S(x)}{\log x} + \int_2^x \frac{S(t)}{t(\log t)^2}\,dt,$$

which together with the upper bound for $S(x)$ justifies the first assertion of the lemma.

Since $1+t \le e^t$ for all real non-negative t,

$$\sum_{n\le x} \frac{g(n)}{n} \le \prod_{p\le x} \left(1 + \sum_{k\le \log x/\log p} \frac{g(p^k)}{p^k}\right) \le \exp\left(\sum_{q\le x} \frac{g(q)}{q}\right),$$

and the second assertion follows from the elementary estimate $\sum_{q\le x} q^{-1} = \log\log x + O(1)$.

The Chebyshev bound $\pi(x) \ll x/\log x$ shows that in the particular case $\Delta \ll M$.

For complex z, w, define $h(n) = zf_1(n) + w\overline{f_1(n)}$,

$$\psi(z,w) = x^{-1} \sum_{n\le x} \exp\left(D^{-1}\left(h(n) - E(h)\right)\right),$$

where

$$E(f) = \sum_{q\le x} \frac{f(q)}{q}\left(1 - \frac{1}{q_0}\right).$$

Note that $E(f) = \sum_{q\le x} q^{-1} f(q) + O(D)$.

Lemma 2.3. *The function $\psi(z,w)$ is bounded uniformly on $|z| \le 1$, $|w| \le 1$.*

Proof. The function $g(n) = \exp(D^{-1}\mathrm{Re}(h(n)))$ is multiplicative, and satisfies $0 \le g(q) \le e^2$ on the product of the unit discs. By Lemma 2.2

$$|\psi(z,w)| \le \exp\left(-\mathrm{Re}(D^{-1}E(h))\right) x^{-1} \sum_{n\le x} g(n)$$

$$\ll \exp\left(\sum_{q\le x} \frac{1}{q}\left\{g(q) - 1 - \mathrm{Re}(D^{-1}h(q))\right\}\right)$$

Applying, for each q, the inequality $|e^t - 1 - t| \le \frac{1}{2}\,|t|^2 e^{|t|}$ which is valid for all complex t, we see that for $|z| \le 1$, $|w| \le 1$ the sum over prime powers in this exponential is

$$\ll \frac{1}{D^2} \sum_{q\le x} \frac{|f_1(q)|^2}{q} \ll 1.$$

Lemma 2.4. *For any complex β, w*

$$\frac{1}{2\pi i}\int_{|z|=1} e^{\beta z} e^{\overline{\beta z} w} z^{-1} dz = \sum_{k=0}^{\infty} \frac{|\beta|^{2k}}{k!^2} w^k.$$

Proof. We may expand the exponential functions and integrate term by term.

Lemma 2.5. *For $\alpha \geq 0$,*

$$\left(\frac{1}{x}\sum_{n\leq x} |f_1(n) - E(f_1)|^{\alpha}\right)^{1/\alpha} \ll D.$$

Proof. By Hölder's inequality it will suffice to establish this estimate for integral values $2k$ of α. In view of Lemma 2.4

$$\frac{1}{D^{2k}x}\sum_{n\leq x} |f_1(n) - E(f_1)|^{2k} = -\frac{(k!)^2}{4\pi^2}\int_{|w|=1}\int_{|z|=1} \psi\left(z, \frac{w}{z}\right) \frac{dzdw}{zw^{k+1}},$$

and so by Lemma 2.3, bounded.

(ii) Large values of $f(q)$

Lemma 2.6. *If the real valued additive function $t(n)$ is at most 1 on the prime powers, then*

$$\sum_{n\leq x} t(n)^k \leq x\left(k + \sum_{q\leq x}\frac{t(q)}{q}\right)^k$$

uniformly in $x \geq 1$, integers $k \geq 0$.

Proof. I give a proof by induction on $k \geq 0$. For $k = 0$ the result is clear. Assume that it is valid for exponents up to $r-1$, $r \geq 1$, and all functions t. Then the sum to be estimated has the representation

$$A = \sum_{n\leq x} t(n)^{r-1}\sum_{q\|n} t(q) = \sum_{q\leq x} t(q) \sum_{\substack{m\leq x/q \\ (m,q)=1}} t(qm)^{r-1}$$
$$\leq \sum_{q\leq x} t(q) \sum_{\substack{m\leq x/q \\ (m,q)=1}} (1 + t(m))^{r-1}.$$

A typical inner sum may be expressed in the form

$$\sum_{j=0}^{r-1} \binom{r-1}{j} \sum_{\substack{m \le x/q \\ (m,q)=1}} t(m)^j$$

which by the inductive hypothesis is at most

$$\sum_{j=0}^{r-1} \binom{r-1}{j} \frac{x}{q} \left(j + \sum_{\ell \le x/q} \frac{t(\ell)}{\ell} \right)^j,$$

ℓ denoting a prime power. Hence

$$A \le x \sum_{q \le x} \frac{t(q)}{q} \sum_{j=0}^{r-1} \binom{r-1}{j} \left(r - 1 + \sum_{\ell \le x} \frac{t(\ell)}{\ell} \right)^j,$$

from which the desired inequality immediately follows.

Remark. For each fixed $\alpha \ge 0$, an appropriate application of Hölder's inequality shows that

$$\sum_{n \le x} t(n)^\alpha \ll x \left(1 + \sum_{q \le x} \frac{t(q)}{q} \right)^\alpha.$$

Proof of Theorem 2.1 when $\alpha \ge 2$. After Lemma 2.5 it will suffice to consider the case $f = f_2$. We have

$$\sum_{\substack{q \le x \\ |f(q)| > D}} \frac{1}{q} \le \frac{1}{D^2} \sum_{q \le x} \frac{|f(q)|^2}{q} = 1,$$

so that the prime powers for which $|f(q)| > D$ are few in number. Let P denote that set. Let $t(n)$ count the members of P which exactly divide n.

We apply Hölder's inequality:

$$|f_2(n)|^\alpha \le t(n)^{\alpha-1} \sum_{q \| n} |f_2(q)|^\alpha.$$

With a change in the order of summation this gives

$$\sum_{n \le x} |f_2(n)|^\alpha \le \sum_{q \le x} |f_2(q)|^\alpha \sum_{\substack{n \le x \\ q \| n}} t(n)^{\alpha-1}.$$

Here a typical inner sum does not exceed

$$\sum_{m \le x/q} (1 + t(m))^{\alpha-1}$$

which is $\ll x/q$ from applications of Lemma 2.6, bearing in mind the size of the set P. Hence

$$\left(x^{-1} \sum_{n \le x} |f_2(n)|^\alpha \right)^{1/\alpha} \ll \left(\sum_{q \le x} q^{-1} |f_2(q)|^\alpha \right)^{1/\alpha}.$$

A similar upper bound is readily derived for $E(f_2)$ by means of Hölder's inequality, again using the small size of P.

This gives Theorem 2.1 when $\alpha \ge 2$.

The range $0 < \alpha \le 2$ is dealt with by means of the following result.

Lemma 2.7. *Let $0 < \alpha \le 2$. Let $c_j > 0, a_j$, $j = 1, \dots, n$, be real numbers. For $y \ge 0$ define*

$$h(y) = \left(\sum_{|a_j| \le y} c_j a_j^2 \right)^{1/2} + \left(\sum_{|a_j| > y} c_j |a_j|^\alpha \right)^{1/\alpha}.$$

If $h(y) \le w$ for some y, then $h(w) \le 8^{1/\alpha} w$. In particular, $h(h(y)) \le 8^{1/\alpha} h(y)$.

The difficulty with this last inequality is to believe in it, cf. [61]. A more general version may be obtained involving an arbitrary pair of exponents in place of $2, \alpha$.

Proof. Assume first that $y \ge w$. Then

$$h(w) \le 2^{1/\alpha} h(y) + \left(2 \sum_{w < |a_j| \le y} c_j |a_j|^\alpha \right)^{1/\alpha}.$$

The sum here does not exceed

$$w^{\alpha-2} \sum_{w < |a_j| \le y} c_j |a_j|^2 \le w^{\alpha-2} h(y)^2,$$

giving the desired inequality.

If $y < w$, then

$$h(w) \leq w\sqrt{2} + \left(2 \sum_{y < |a_j| \leq w} c_j a_j^2 \right)^{1/2},$$

and

$$\sum_{y < |a_j| \leq w} c_j a_j^2 \leq w^{2-\alpha} \sum_{y < |a_j| \leq w} c_j |a_j|^\alpha \leq w^{2-\alpha} h(y)^\alpha,$$

completing the proof of the lemma.

Remark. It is useful to note that if $\gamma > 0$, then by a similar argument one may establish the bound $h(\gamma w) \ll w$, the implied constant depending only upon γ, α.

Proof of Theorem 2.1 when $0 < \alpha < 2$. For $\eta > 0$ we define

$$\delta = \left(\sum_{\substack{q \leq x \\ |f(q)| \leq \eta}} \frac{|f(q)|^2}{q} \right)^{1/2} + \left(\sum_{\substack{q \leq x \\ |f(q)| > \eta}} \frac{|f(q)|^\alpha}{q} \right)^{1/\alpha},$$

and replace D in the treatment of the cases $\alpha \geq 2$, by δ. Note that by Lemma 2.7

$$\frac{1}{\delta^2} \sum_{\substack{q \leq x \\ |f(q)| \leq \delta}} \frac{|f(q)|^2}{q} \leq 2^{6/\alpha}, \qquad \sum_{\substack{q \leq x \\ |f(q)| > \delta}} \frac{1}{q} \leq 8.$$

The proof then proceeds as before.

Much of the difficulty in obtaining inequalities of the type given in Theorem 2.1 lies in making an appropriate formulation. These results can be naturally viewed within the format of functional analysis, and since that is how I found them, I shall discuss how this may be done. This will in part illustrate the philosophy underlying the method of the stable dual. It will also show that the above formulations are appropriate, and yield dual inequalities as a by-product.

3

Dual norm inequalities

For $\alpha > 1$ the inequalities of Theorem 2.1 have dual forms. To simplify the exposition I give them in the opposite order.

Theorem 3.1. *If $\alpha \geq 2$, then the inequality*

$$\left(\sum_{q\leq x} q^{\alpha-1}\left|\frac{1}{x}\sum_{\substack{n\leq x\\ n\cong 0(\mathrm{mod}\, q)}} a_n - \frac{1}{q}\left(1-\frac{1}{q_0}\right)\frac{1}{x}\sum_{n\leq x} a_n\right|^{\alpha}\right)^{\frac{1}{\alpha}} +$$

$$\left(\sum_{q\leq x} q\left|\frac{1}{x}\sum_{\substack{n\leq x\\ n\cong 0(\mathrm{mod}\, q)}} a_n - \frac{1}{q}\left(1-\frac{1}{q_0}\right)\frac{1}{x}\sum_{n\leq x} a_n\right|^{2}\right)^{\frac{1}{2}} \ll \left(\frac{1}{x}\sum_{n\leq x}|a_n|^{\alpha}\right)^{\frac{1}{\alpha}}$$

holds for all complex a_n, $1 \leq n \leq x$.

If $1 < \alpha < 2$, then there is a similar inequality but with the outer summation in the first of the norm terms over only those prime powers q for which

$$\frac{q}{x}\left|\sum_{\substack{n\leq x\\ n\cong 0(\mathrm{mod}\, q)}} a_n - \frac{1}{q}\left(1-\frac{1}{q_0}\right)\sum_{n\leq x} a_n\right| > \left(\frac{1}{x}\sum_{n\leq x}|a_n|^{\alpha}\right)^{1/\alpha}$$

is satisfied, and the corresponding summation in the second of the norm terms over those q for which the opposite inequality holds.

Here $n \cong 0(\mathrm{mod}\, q)$ denotes that q divides n, but qq_0 does not. This is sometimes expressed as $q \| n$.

Let s be the number of prime powers q not exceeding a real $x \geq 2$. For each $\alpha \geq 1$, let $L^\alpha(\mathbb{C}^s)$ denote the complex vector space of s-tuples f, with one coordinate $f(q)$ for each prime power $q \leq x$, topologised with the norm

$$\|f\|_\alpha = \left(\sum_{q \leq x} \frac{1}{q} \left(1 - \frac{1}{q_0}\right) |f(q)|^\alpha \right)^{1/\alpha}$$

We may regard $L^\alpha(\mathbb{C}^s)$ as a space of functions, with an underlying measure $d\mu$ that assigns weight $q^{-1}(1 - q_0^{-1})$ to the q-coordinate point. The dual space $(L^\alpha)'$ of $L^\alpha(\mathbb{C}^s)$ can then be identified with $L^{\alpha'}(\mathbb{C}^s)$, where $\alpha^{-1} + (\alpha')^{-1} = 1$.

In a similar way, if t denotes the number of positive integers n up to x, $M^\alpha(\mathbb{C}^t)$ denotes the space of t-tuples with one coordinate for each integer n, topologised by

$$\|y\|_\alpha = \left(\frac{1}{[x]} \sum_{n \leq x} |y_n|^\alpha \right)^{1/\alpha}$$

This may be viewed as a measure space, with underlying probability measure.

The classical Turán–Kubilius inequality may be restated in the form that the operator $A : L^2(\mathbb{C}^s) \to M^2(\mathbb{C}^t)$ given by

$$(Af)(n) = f(n) - \sum_{q \leq x} \frac{1}{q} \left(1 - \frac{1}{q_0}\right) f(q)$$

with $f(n)$ the additive function determined by the $f(q)$, has norm bounded uniformly in $x \geq 2$. Thus the notion of an additive function is equivalent to that of a linear operator on $\mathbb{C}^s$ to $\mathbb{C}^t$.

When $\alpha = 2$, $L^2(\mathbb{C}^s)$ and $M^2(\mathbb{C}^t)$ are Hilbert spaces, with inner products

$$(f, k) = \sum_{q \leq x} \frac{1}{q} \left(1 - \frac{1}{q_0}\right) f(q)\overline{k(q)}, \quad [a, b] = [x]^{-1} \sum_{n \leq x} a_n \bar{b}_n$$

respectively.

For applications to number theory, especially to problems involving the characterisation of arithmetic functions, it is necessary to have lower bounds for $\|Af\|$. These seem to lie deeper than upper bounds. If A^* denotes the adjoint of A, then in terms of the inner product on L^2, $(f, A^*Af) = \|Af\|^2$, so that a lower bound for $\|Af\|$ naturally involves the consideration of the self-adjoint operator A^*A of $L^2(\mathbb{C}^s)$ into itself.

When $\alpha \geq 2$ the version of the Turán–Kubilius inequality which I gave in my paper, [42], Theorem 1, it is the first part of the present Theorem 2.1, shows that with norm $\max(\| \quad \|_2, \| \quad \|_\alpha)$ on $\mathbb{C}^s$, and in an obvious notation, the map $A_\alpha : L^2 \cap L^\alpha \to M^\alpha$ is (uniformly) bounded. In the situation $\alpha > 2$ there is no adjoint to A_α. After adopting the philosophy:

$$\text{operator} \to \text{sufficiency}, \qquad \text{dual operator} \to \text{necessity},$$

and identifying the spaces M^α and $(M^{\alpha'})'$, we consider the composition $A'_{\alpha'}A_\alpha$, where $'$ again denotes the dual. Suppose that for some normed space K on $\mathbb{C}^s$, $1 < \alpha < 2$, the operator $A_\alpha : K \to M^\alpha$ were bounded. Then we should have the chain of maps

$$K \xrightarrow{A_\alpha} M^\alpha \simeq (M^{\alpha'})' \xrightarrow{A'_{\alpha'}} (L^2 \cap L^{\alpha'})'.$$

This suggests that the spaces at the extreme ends should be the same. Moreover, if $\alpha > 2$, we should then have

$$L^2 \cap L^\alpha \xrightarrow{A_\alpha} M^\alpha \simeq (M^{\alpha'})' \xrightarrow{A'_{\alpha'}} (L^2 \cap L^\alpha)''.$$

This last space can be identified with (the isometric embedding) $L^2 \cap L^\alpha$, since all these spaces have finite dimension. Thus the picture would be complete. We are reduced to accurately estimating the norm on the space $(L^2 \cap L^{\alpha'})'$, $1 < \alpha < 2$. It is convenient to do this within the format of Lemma 2.7.

On $\mathbb{C}^k$ define a norm $\|a\|_\alpha = \left(\sum_{j=1}^{k} c_j |a_j|^\alpha\right)^{1/\alpha}$, where a has coordinates a_j, and each $c_j > 0$. Let S^α be the space $\mathbb{C}^k$ topologised with this norm, and $S^\alpha \cap S^\beta$ the same space topologised with the norm $\max(\| \quad \|_\alpha, \| \quad \|_\beta)$. Let $\| \quad \|$ denote the norm on $\mathbb{C}^k$ which renders it isometrically equivalent to $(S^2 \cap S^{\alpha'})'$.

Lemma 3.2. *Let* $1 < \alpha \leq 2$. *In the notation of Lemma 2.7,* $\|w\| \leq \inf_{y \geq 0} h(y) \leq h(\|w\|) \leq 2\|w\|$ *for all* w *in* $\mathbb{C}^k$.

Proof. Every linear map of $S^2 \cap S^{\alpha'}$ into $\mathbb{C}$ has a representation $x \mapsto (x, w)$, where $w \in \mathbb{C}^k$, and the inner product is given by

$$(a, b) = \sum_{j=1}^{k} c_j a_j \bar{b}_j.$$

Define

$$u_j = \begin{cases} w_j & \text{if } |w_j| \le y, \\ 0 & \text{otherwise,} \end{cases} \qquad v_j = \begin{cases} w_j & \text{if } |w_j| > y, \\ 0 & \text{otherwise,} \end{cases}$$

so that $w_j = u_j + v_j$. Then applications of Cauchy's and Hölder's inequalities, respectively, show that $|(x,u)| \le \|x\|_2\|u\|_2$, $|(x,v)| \le \|x\|_{\alpha'}\|v\|_\alpha$. By linearity $|(x,w)| \le \max(\|x\|_2, \|x\|_{\alpha'})h(y)$. Since this holds for all x in $\mathbb{C}^k$, $\|w\|$ does not exceed $h(y)$ for any y. This gives the lower bound for h in the lemma.

To obtain the upper bound we choose x to effect equality in (one of) these inequalities. Let $y = \|w\|$. Define $z_j = |w_j|^{\alpha-1}\exp(-i\arg w_j)$ if $|w_j| > y$, $z_j = 0$ otherwise. Then $\|z\|_{\alpha'} = \|v\|_\alpha^{\alpha-1}$, and $|(z,w)| = \|z\|_{\alpha'}\|v\|_\alpha = \|v\|_\alpha^\alpha$. Moreover $\|z\|_2 =$

$$\left(\sum_{|w_j|>y} c_j|w_j|^{2(\alpha-1)}\right)^{1/2} \le \left(y^{\alpha-2}\sum_{j=1}^{k} c_j|w_j|^\alpha\right)^{1/2} = y^{(\alpha/2)-1}\|v\|_\alpha^{\alpha/2}.$$

Hence

$$\|v\|_\alpha^\alpha \le y\max(\|z\|_2, \|z\|_{\alpha'}) \le y\max(y^{(\alpha/2)-1}\|v\|_\alpha^{\alpha/2}, \|v\|_\alpha^{\alpha-1})$$

from which $y \ge \|v\|_\alpha$. A similar argument gives $y \ge \|u\|_2$, and the proof of the lemma is complete.

Remarks. (i) Since u and z are supported on disjoint sets, the choice $x = \bar{u} + z$ effects equality in both inequalities. Then

$$(\bar{u} + z, w) = \|v\|_\alpha^\alpha + \|u\|_2^2.$$

(ii) For vectors a, b in $\mathbb{C}^k$, let ab denote the vector with j-coordinate a_jb_j. For any z in $\mathbb{C}^k$, $(bw, z) = (w, \bar{b}z)$. The first argument in the proof of Lemma 3.2 shows that

$$|(\bar{b}z, w)| \le \max(\|\bar{b}z\|_2, \|\bar{b}z\|_{\alpha'})h(y)$$

uniformly in $y \ge 0$. In terms of the norm on $(S^2 \cap S^{\alpha'})'$, we set $y = \|w\|$. Since $\|\bar{b}z\|_\beta \le \|z\|_\beta \max|b_j|$ for every $\beta \ge 1$, an appeal to Lemma 3.2 gives

$$\|bw\| \le 2\|w\| \max_{1\le j\le k} |b_j|.$$

This often allows the vectors w in $(S^2 \cap S^{\alpha'})'$ to be rescaled, coordinate by coordinate.

Let $1 < \alpha < 2$. Bearing in mind Lemma 2.7 and Lemma 3.2, we reappraise the second estimate in Theorem 2.1. Taking $\eta = \|f\|$ in the two sums there in terms of the norm on $(L^2 \cap L^{\alpha'})'$ gives $\|A_\alpha f\| \ll \|f\|$. We may indeed set $K = (L^2 \cap L^{\alpha'})'$ to close the loops in our map diagrams:

$$L^2 \cap L^\alpha \xrightarrow{A_\alpha} M^\alpha \simeq (M^{\alpha'})' \xrightarrow{A'_{\alpha'}} (L^2 \cap L^\alpha)'' \simeq L^2 \cap L^\alpha \quad \text{if } \alpha \geq 2,$$

$$(L^2 \cap L^{\alpha'})' \xrightarrow{A_\alpha} M^\alpha \simeq (M^{\alpha'})' \xrightarrow{A'_{\alpha'}} (L^2 \cap L^{\alpha'})' \qquad \text{if } 1 < \alpha \leq 2.$$

Proof of Theorem 3.1. The operator A' dual to A is given by

$$(A'(a))(q) = \frac{q}{[x]}\left(1 - \frac{1}{q_0}\right)^{-1}\left(\sum_{\substack{n \leq x \\ n \cong 0 (\mathrm{mod}\, q)}} a_n - \frac{1}{q}\left(1 - \frac{1}{q_0}\right)\sum_{n \leq x} a_n\right).$$

Since operators and their duals have the same norms, the assertion of Theorem 3.1 for $\alpha \geq 2$ is immediate. For $1 < \alpha < 2$ we obtain

$$h(\|A'(a)\|) \ll \left(x^{-1}\sum_{n \leq x} |a_n|^\alpha\right)^{1/\alpha},$$

where we identify the spaces L^α, S^α by taking $c_j = q_j^{-1}$, the prime powers q_j not exceeding x. With

$$c = \left(x^{-1}\sum_{n \leq x} |a_n|^\alpha\right)^{1/\alpha}$$

we have $h(y) \leq \gamma c$ for $y = \|A'(a)\|$, and some constant γ depending upon α only. Hence by Lemma 2.7, $h(\gamma c) \leq 8\gamma c$, and by the remark following that lemma, $h(c) \ll c$. Since $\|A'(a)\|$ does not exceed the infimum of $h(y)$ taken over all real y, this inequality is not substantially different from $\|A'(a)\| \ll c$, which is the best to be expected. This completes the proof of Theorem 3.1.

Let $\theta(n, q)$ be 1 if $q \parallel n$, and be zero otherwise. The trace of the (s-by-s matrix representation underlying the) operator A^*A is

$$\sum_{q \leq x}\sum_{n \leq x} \frac{q}{[x]}\left(\theta(n,q) - \frac{1}{q}\left(1 - \frac{1}{q_0}\right)\right)^2 = (1 + o(1))s, \quad x \to \infty.$$

Since this trace is the sum of the eigenvalues of A^*A, at least one eigenvalue exceeds $1+o(1)$. Let h be an eigenvector of A^*A with corresponding eigenvalue $\delta > 1/2$. The matrix representing A^* has real entries. The operator A' is represented by the same matrix, and h is also an eigenvector of $A'A$. From our loop diagrams

$$\|\delta h\| = \|A'_{\alpha'}A_\alpha h\| \le \|A'_{\alpha'}A_\alpha\|\|h\|,$$

the outermost norms taken in $L^2 \cap L^\alpha(\mathbb{C}^s)$, $(L^2 \cap L^{\alpha'}(\mathbb{C}^s))'$ as the case may be. In particular, $\|A'_{\alpha'}\|\|A_\alpha\| \ge \|A'_{\alpha'}A_\alpha\| \ge \delta$. Combined with Theorem 2.1 this shows that for each $\alpha > 1$, and all x sufficiently large, $1 \ll \|A_\alpha\| \ll 1$. To this extent Theorems 2.1 and 3.1 are best possible.

A detailed discussion of the operators A_α and their action upon the various spaces L^α I gave in my paper, [59]. For our immediate purposes it is enough to know that the norms $\|A_\alpha\|$ are bounded uniformly in $x \ge 2$. I consider these matters further in Chapters 20 and 21.

Once found, estimates like those in Theorem 3.1 can be viewed as merely a collection of inequalities, and established as such, eschewing all mention of functional analysis and background philosophy. This seductive approach can also be disingenuous, even obfuscating, with a sophistication perilously close to that of the student who when asked to prove that the sum of the interior angles in a euclidean triangle comes to two right angles protests, "Why should I? We know it is true."

However, in several proofs in this volume only the case $\alpha = 2$ of Theorems 2.1 and 3.1 is employed. A direct proof of the equivalence of the basic Turán–Kubilius inequality and its dual by means of the Cauchy–Schwarz inequality is so rapid that I give it.

We represent the operator A by a matrix $\mathbf{A}$ with t rows, one for each positive integer n not exceeding x, and s columns, one for each prime power q not exceeding x. A typical entry for $\mathbf{A}$ is $\theta(q,n) - q^{-1}(1-q_0^{-1})$. The map $A : \mathbb{C}^s \to M^t$ is given by $\mathbf{f} \mapsto \mathbf{Af}$, where (now) $\mathbf{f}$ denotes a typical point in $\mathbb{C}^s$. The equivalence of inequalities (1.1) and (1.2) follows from the following result.

Lemma 3.3. *An m-by-n matrix* $\mathbf{C}$ *satisfies* $|\mathbf{Ca}|^2 \le \lambda|\mathbf{a}|^2$ *for all* $\mathbf{a}$ *in* $\mathbb{C}^n$ *if and only if it satisfies* $|\mathbf{b}^T\mathbf{C}|^2 \le \lambda|\mathbf{b}^T|^2$ *for all* $\mathbf{b}$ *in* $\mathbb{C}^m$*. Here* ^{T} *denotes transposition.*

Proof. Assume the validity of the first hypothesis. Beginning with the Cauchy–Schwarz inequality

$$|\mathbf{b}^T\mathbf{C}|^2 = \mathbf{b}^T\mathbf{C}\bar{\mathbf{C}}^T\bar{\mathbf{b}} \le |\mathbf{b}^T||\mathbf{C}\bar{\mathbf{C}}^T\bar{\mathbf{b}}| \le |\mathbf{b}^T|\lambda^{1/2}|\bar{\mathbf{C}}^T\bar{\mathbf{b}}|.$$

Since $|\mathbf{b}^T \mathbf{C}| = |\bar{\mathbf{C}}^T \bar{\mathbf{b}}|$, the second hypothesis is clearly valid. A similar proof goes in the other direction.

When $\alpha = 2$, the inequalities in Theorem 3.1 are of Large Sieve type, save that the moduli q include powers of primes as well as primes, and run all the way up to x rather than to the traditional $x^{1/2}$. Other mean square inequalities involve more or different residue classes to prime moduli, or restrict the variable n to run over selected integers, such as the shifted primes. Examples may be found in Chapter 4 of my book, [41], and in Chapters 4, 18, 23, 29 and 34 of the present work.

Any reasonably-normed freely-generated commutative semigroup will give rise to inequalities of Turán–Kubilius type, which with their duals will allow the application of the whole of the method presented in this volume.

4

Exercises: Including the Large Sieve

I discovered the simple but useful argument of Lemma 2.6 when I was a graduate student at Cambridge, England. Although the occurrence of the sum $\sum q^{-1}t(q)$ taken over the prime powers is appropriate, equally important is the size of the constant k^k which occurs with it. This is comparable to the size of the constant appearing in the estimation of the moments for small values of $f(q)$, and leads to the bound $\|A_\alpha\| \ll \alpha$ for $\alpha \geq 2$. My idea at the time was to apply the bound of Lemma 2.6 in conjunction with Hölder's inequality, taking t to be 1 on the set of all prime powers and k to be a large function of x. In this way I estimated $\omega(n)$ on a thin sequence of integers. The following exercises illustrate the general procedure within the format of Theorems 2.1 and 3.1.

1. Let B be a sequence of integers not exceeding x, of number $B(x) > 0$. This is not the expression $B(x)$ considered in Chapter 1. Prove that

$$\left(B(x)^{-1} \sum_{n\leq x}{}' |f(n) - E(f)|^2 \right)^{1/2} \leq (x/B(x))^{1/\alpha} \left(x^{-1} \sum_{n\leq x} |f(n) - E(f)|^\alpha \right)^{1/\alpha}$$

for all $\alpha \geq 2$, where $'$ denotes that n belongs to the sequence B.

2. Prove that

$$\left(B(x)^{-1} \sum_{n\leq x}{}' |f(n) - E(f)|^2 \right)^{1/2}$$

$$\ll \alpha (x/B(x))^{1/\alpha} \left\{ \left(\sum_{q\leq x} |f(q)|^2 q^{-1} \right)^{1/2} + \left(\sum_{q\leq x} |f(q)|^\alpha q^{-1} \right)^{1/\alpha} \right\},$$

the implied constant absolute.

3. Prove that

$$\left(\frac{1}{B(x)^2}\sum_{q\le x}'' q\left|\sum_{\substack{n\le x\\ n\cong 0(\operatorname{mod} q)}}' a_n - \frac{1}{q}\left(1-\frac{1}{q_0}\right)\sum_{n\le x}' a_n\right|^2\right)^{1/2} \ll c,$$

where $''$ indicates that summation is confined to those prime powers q for which

$$\frac{q}{B(x)}\left|\sum_{\substack{n\le x\\ n\cong 0(\operatorname{mod} q)}}' a_n - \frac{1}{q}\left(1-\frac{1}{q_0}\right)\sum_{n\le x}' a_n\right| \le c,$$

and c is any number as large as

$$\beta(x) = \alpha\left(\frac{x}{B(x)}\right)^{1/\alpha}\left(\frac{1}{B(x)}\sum_{n\le x}' |a_n|^2\right)^{1/2}.$$

Hint: Regard the inequality of exercise 2 as a bound for the norm of an operator between $\mathbb{C}^s$ with norm $\max(\| \;\|_2, \| \;\|_\alpha)$, and $M^2(\mathbb{C}^t)$, but with t replaced by $B(x)$, and the norm of $M^2(\mathbb{C}^t)$ by

$$\left(B(x)^{-1}\sum_{n\le x}' |a_n|^2\right)^{1/2}.$$

4. Prove that

$$\left(\frac{1}{B(x)^2}\sum_{q\le x} q\left|\sum_{\substack{n\le x\\ n\cong 0(\operatorname{mod} q)}}' a_n - \frac{1}{q}\left(1-\frac{1}{q_0}\right)\sum_{n\le x}' a_n\right|^2\right)^{1/2}$$

$$\ll \beta(x) + \max_{q\le x}\frac{q}{B(x)}\left|\sum_{\substack{n\le x\\ n\cong 0(\operatorname{mod} q)}}' a_n - \frac{1}{q}\left(1-\frac{1}{q_0}\right)\sum_{n\le x}' a_n\right|$$

uniformly for all complex a_n, real $\alpha \ge 2$, nonempty sets B and real $x \ge 1$. By adjusting the definition of the additive function f in the beginning steps, we may restrict the prime powers q in any manner.

5. Suppose that $0 < \theta < 1$, and that $B(x,q)$, the number of members of B not exceeding x which satisfy $b_j \cong 0(\bmod q)$, is $\ll B(x)q^{-1}$ uniformly for $q \le x^{\theta}$. Prove that

$$\left(\frac{1}{B(x)^2} \sum_{q \le x^{\theta}} q \left| B(x,q) - \frac{1}{q}\left(1 - \frac{1}{q_0}\right) B(x) \right|^2 \right)^{1/2} \ll \alpha \left(\frac{x}{B(x)} \right)^{1/\alpha}$$

uniformly for $\alpha \ge 2$. We may choose α optimally. A direct application of Theorem 3.1 allows only the choice $\alpha = 2$. Translating B we may likewise treat those b_j which satisfy $b_j - r \cong 0(\bmod q)$ for a fixed integer r.

6. Let $B(x) \ge x(\log x)^{-c}$ for a positive c and all $x \ge 2$. Prove, without the assumption of exercise 5, that for each positive β

$$\sum_{b_j \le x} \omega(b_j)^{\beta} \ll B(x)(\log\log x)^{\beta}, \quad x \ge e^2,$$

the implied constant depending only upon c and β.

I employed variants of this argument a number of times, for example [25], [26]; I furnish some remarks in Chapter 15 of my book, [41].

The next eight exercises enable arithmetic interpretations of the Large Sieve.

Characters on finite abelian groups

The functions $f : G \to \mathbb{C}$, on a set G, form a vector space over $\mathbb{C}$ under the composition rules $(f_1 + f_2)(g) = f_1(g) + f_2(g)$, $(\rho f_1)(g) = \rho f_1(g)$ for g in G and ρ in $\mathbb{C}$. If G is an abelian group, then this space contains the characters on G.

7. Prove that any finite collection of characters on an abelian group is linearly independent over $\mathbb{C}$.

Hint: Adopt a proof method of Artin. Consider a possible dependence relation containing a minimal number of characters and use the fact that $\widehat{G}$ is a group.

G will continue to denote a finite abelian group and $\widehat{G}$ its dual until exercise 14.

8. Let $\lambda : H \to U$ be a homomorphism of a subgroup of G into the complex unit circle. Prove that some character on G coincides with λ on H.

Hint: For an element w of G, that does not belong to H, consider the subgroup Γ of elements $w^j h$, $0 \le j \le t$, $h \in H$, where t is the least positive integer for which $w^t \in H$. Define $\lambda_1 : \Gamma \to U$ by $\lambda_1(w^j h) =$

$\exp(2\pi ijt^{-1})\lambda(h)$. Allowing an axiom of choice, a modified version of this argument may be applied to any abelian group.

9. Prove that the only element of a group at which all characters have the value 1 is the identity.

10. Prove that for each g in G, $\chi \mapsto \chi(g)$ defines a character T_g on $\widehat{G}$. Prove that $g \mapsto T_g$ maps G isomorphically into $\widehat{\widehat{G}}$.

11. Prove that $|G| = |\widehat{G}|$.

Hint: Consider the row/column rank of the matrix $(\chi(g))$, $g \in G$, $\chi \in \widehat{G}$.

12. Prove that $G \cong \widehat{\widehat{G}}$ (exemplifying the Pontryagin duality theorem).

In this volume I do not employ the decomposition of a finite abelian group as a direct product of cyclic groups, nor an isomorphism between G and $\widehat{G}$.

13. Prove that

$$\frac{1}{|G|}\sum_{g\in G}\chi(g) = \begin{cases} 1 & \text{if } \chi \text{ is principal,} \\ 0 & \text{otherwise.} \end{cases}$$

$$\frac{1}{|G|}\sum_{\chi\in \widehat{G}}\chi(g) = \begin{cases} 1 & \text{if } g \text{ is the identity,} \\ 0 & \text{otherwise.} \end{cases}$$

14. Define the Fourier transform $\hat{f}$ of a function $f : G \to \mathbb{C}$, by

$$\hat{f}(\chi) = |G|^{-1/2}\sum_{g\in G} f(g)\overline{\chi(g)}.$$

Prove that

$$\sum_{\chi\in\widehat{G}} |\hat{f}(\chi)|^2 = \sum_{g\in G} |f(g)|^2.$$

This may be viewed an analogue of the theorems of Parseval and Plancherel.

I continue the general study of Fourier analysis on finite abelian groups in Chapters 23 and 24.

The results of the foregoing section touch upon the abstract construction of dual objects. Exercise 8 is an analogue of the Hahn–Banach theorem in functional analysis. A direct corollary of such extension theorems is that given two elements in the object, there is a character on the object that will distinguish between them. In particular, the natural map of a group or a Banach space into its second dual is one-to-one (a bijection). The proof requires that the characters assume values in a(nother) sufficiently well structured object. Regarding abelian groups as $\mathbb{Z}$-modules, the image group for characters should be $\mathbb{Z}$-divisible. To order subgroups of a group by inclusion more is needed; characters should map into a group that is extra divisible, cf. [56], Lemmas 15.1, 15.5 respectively. Thus for the bare algebraic study of products in Q^*, it suffices to consider homomorphisms of Q^* into the multiplicative group of roots of unity, a countable subgroup of U. Defining $\widehat{G}$ more generally in terms of U allows larger groups G to be embraced, and appeal to the methods of analysis.

A complex Banach space B may be viewed as an abelian group of vectors with a $\mathbb{C}$-module structure. The field $\mathbb{C}$ is itself the simplest $\mathbb{C}$-divisible additive group. The use of $\mathbb{C}$ as image object already allows the ordering of subspaces of B by means of characters; if some ($\mathbb{C}$-module) power of a vector belongs to a subspace of B, then so does the vector itself.

Related remarks of number theoretic interest may be found in [56], Chapter 15.

The Large Sieve via the duality principle and Poisson summation

15. Use Poisson summation to prove that

$$\sum_{n=-\infty}^{\infty} e^{-\pi n^2 y + 2\pi i n\alpha} = y^{-1/2} \sum_{n=-\infty}^{\infty} e^{-\pi(n+\alpha)^2/y}$$

is valid for real α and $y > 0$.

16. Prove that each eigenvalue λ of a matrix (a_{ij}), $1 \le i, j \le n$, over $\mathbb{C}$, lies in a Gershgorin disc

$$|\lambda - a_{kk}| \le \sum_{\substack{j=1 \\ j \ne k}}^{n} |a_{kj}|,$$

for some k.

Hint: Consider a coordinate, maximal in absolute value, of an eigenvector corresponding to λ.

17. Let x_j, $j = 1, \ldots, J$, be real numbers which satisfy $\|x_j - x_k\| \ge \delta > 0$ for all $j \ne k$. Here $\|y\|$ denotes the distance of the real y from a nearest integer. For complex numbers a_n, $M < n \le M + N$, $N \ge 1$, define

$$S(\alpha) = \sum_{M<n\le M+N} a_n e^{2\pi i n\alpha}.$$

Prove that

$$\sum_{j=1}^{J} |S(\alpha_j)|^2 \ll (N + \delta^{-1}) \sum_{M<n\le M+N} |a_n|^2,$$

the implied constant absolute.

Hint: Without loss of generality $M = 0$. Consider the Hermitian form

$$\sum_{j=1}^{J} \left| \sum_{n=-\infty}^{\infty} b_n e^{-\pi n^2/N^2 + 2\pi i n x_j} \right|^2 .$$

The mean square bound for $|S(x_j)|$ is the inequality of the Large Sieve in the form given it by Davenport and Halberstam, [20], with a different proof. The argument suggested here is carried out in [56], Chapter 6.

18. In the notation of exercise 17, prove that

$$\sum_{q \le Q} \sum_{\substack{r=1 \\ (r,q)=1}}^{q} \left| S\left(\frac{r}{q}\right) \right|^2 \ll (N + Q^2) \sum_{M < n \le M+N} |a_n|^2$$

uniformly in $N \ge 1, M, Q$, complex a_n. Here r, q denote positive integers.

For an additive function f, define

$$f_1(n) = \sum_{p|n, p \le \sqrt{x}} f(p), \quad E = \sum_{p \le \sqrt{x}} p^{-1} f(p).$$

19. Prove that

$$|f_1(n) - E|^2 = \sum_{p \le \sqrt{x}} \frac{\overline{f(p)}}{p} \sum_{r=1}^{p-1} (f_1(n) - E) e^{2\pi i r n/p}.$$

20. In the notation of exercise 17, prove that with a suitable choice for the a_n

$$\left(\sum_{n \le x} |f_1(n) - E|^2 \right)^2 \le \sum_{p \le \sqrt{x}} p^{-1} |f(p)|^2 \sum_{p \le \sqrt{x}} \sum_{r=1}^{p-1} \left| S\left(\frac{r}{p}\right) \right|^2,$$

and deduce that

$$\sum_{n \le x} |f_1(n) - E|^2 \ll x \sum_{p \le \sqrt{x}} p^{-1} |f(p)|^2.$$

With the approach of exercises 19 and 20 I joined the Turán–Kubilius inequality to that of the Large Sieve. Methods from probabilistic number theory allowed the asymptotic distribution of an appropriate renormalised additive function to be computed, to give a lower bound limitation for the Large Sieve. Because of his interest in such matters, I wrote to Rényi.

He expressed surprise at the connection between the two inequalities. It was 1968, before the inequality of the Large Sieve became an exercise in spectral approximation and began to go around with a twin; cf. [30]. And an elaboration of the argument was apparent:

An additive function f is *strongly additive* if it satisfies $f(p^m) = f(p)$, $m \geq 1$, p prime.

21. Let F be a polynomial in $\mathbb{Z}[x]$, positive on the integers exceeding b. Let $\rho(D)$ denote the number of incongruent solutions to $F(n) \equiv 0 \pmod{D}$. Let $A = \sum p^{-1} f(p)\rho(p)$ taken over the primes up to x. Prove that

$$\sum_{b<n\leq x} |f(F(n)) - A|^2 \ll x \sum_{p\leq x} p^{-1}|f(p)|^2 \rho(p)$$

for real strongly additive functions $f, x \geq 2$, the implied constant depending only upon F and b.

22. In the notation of exercise 17, for each k define δ_k to be the minimum of $\|x_j - x_k\|$ taken over the j distinct from k. Prove that

$$\|w - x_k\| \leq \frac{3}{2} \|x_j - x_k\|, \quad j \neq k,$$

uniformly for $x_j - \frac{1}{2}\delta_j \leq w \leq x_j + \frac{1}{2}\delta_j$. Deduce that

$$\sum_{j=1}^{J} \delta_j N e^{-c\|x_j - x_k\|^2 N^2} \ll 1, \quad j = 1, \ldots J,$$

the implied constant depending only upon the (positive) value of c.

Hint: Compare the sum with an integral over disjoint intervals.

23. In the notation of exercise 22, prove that

$$\sum_{j=1}^{J} \left(N + \frac{1}{\delta_j} \right)^{-1} |S(x_j)|^2 \ll \sum_{M<n\leq M+N} |a_n|^2$$

holds with the uniformities of exercise 17. This refinement of the inequality of Davenport and Halberstam is due to Montgomery and Vaughan, [126], with a different proof. For a further remark see [56], Chapter 6.

24. What is the analogue of exercise 18 when we apply exercise 23 in place of 17? The resulting inequality was obtained by Montgomery, [124], with an again different proof.

Why is the Large Sieve a sieve?

The square integers in the interval $(N^{1/2}, N]$ are $N^{1/2} + O(N^{1/4})$ in number. They do not represent any non-square reduced residue class to a prime modulus p not exceeding $N^{1/2}$. Let us estimate the density of a sequence of integers using only local information of this type. Let B be a sequence of integers, not exceeding N, which do not represent some (chosen) $\frac{1}{2}(p-1)$ residue classes to each odd prime p not exceeding $N^{1/2}$. We denote by $B(N)$ its cardinality.

25. In the notation of exercise 17, prove that

$$\sum_{r=1}^{p-1}\left|S\left(\frac{r}{p}\right)\right|^2 = \sum_{s=0}^{p-1} p \left| \sum_{\substack{M<n\le M+N \\ n\equiv s (\mathrm{mod}\, p)}} a_n - \frac{1}{p}\sum_{M<n\le M+N} a_n \right|^2$$

for prime moduli p.

26. Prove that

$$\sum_{p\le\sqrt{N}} p \sum_{s=0}^{p-1} \left| \sum_{\substack{M<n\le M+N \\ n\equiv s (\mathrm{mod}\, p)}} a_n - \frac{1}{p}\sum_{M<n\le M+N} a_n \right|^2 \ll N \sum_{M<n\le M+N} |a_n|^2$$

uniformly in $N \ge 1, M$, complex a_n. This inequality may be compared with those of Theorem 3.1.

27. Prove that $B(N) \ll N^{1/2}\log N$. This is almost best possible but we can do better.

Informally: To display the primes amongst the positive integers, the (sieve) method of Eratosthenes, third century B.C., begins with 2 and successively strikes all proper multiples of the last revealed prime. The method can be made precise but is often inefficient. By extension, an argument which estimates the number of integers remaining in an interval after segments of arithmetic progressions have been removed, is a sieve method. Additive translation is so introduced. Further extensions are possible. For a thoroughgoing introduction to sieve methods, particularly those of Brun and Selberg, see Halberstam and Richert, [93].

Around the Selberg sieve

Let P be a (squarefree) product of primes, each not exceeding z. By the Chinese Remainder theorem, to estimate the number of integers in an interval $(M, M+N]$ which avoid certain residue classes to the prime divisors of P, we may consider

$$\sum_{\substack{M\le n\le M+N \\ (F(n),P)=1}} 1,$$

where F is a polynomial in $\mathbb{Z}[x]$. F may have very high degree in terms of M, N, z.

Given any sequence of integers a_n, Selberg's sieve method begins with the step

$$\sum_{(a_n,P)=1} 1 \le \sum_n \left(\sum_{\substack{d|a_n \\ d|P}} \lambda_d \right)^2,$$

valid for all real λ_d which satisfy $\lambda_1 = 1$. The square is expanded, the summation inverted, and the λ_d chosen to minimise the quadratic form

$$\sum_{d_j|P} \lambda_{d_1}\lambda_{d_2} \sum_{a_n \equiv 0 (\mathrm{mod}[d_1,d_2])} 1.$$

To render this problem tractable, Selberg sets $\lambda_d = 0$ for $d > z$, and assumes the inner sum to have a form $Xf([d_1,d_2]) + R_{[d_1,d_2]}$ for some $X \ge 0$, function f multiplicative on the divisors of P, and 'small' term $R_{[d_1,d_2]}$. The λ_d are determined to minimise the quadratic form

$$\Delta = \sum_{\substack{d_1|P \\ d_2|P}} \lambda_{d_1}\lambda_{d_2} f([d_1,d_2]).$$

The function which is 1 when $F(n) \equiv 0 \pmod q$, zero otherwise, is defined on the additive group $\mathbb{Z}/q\mathbb{Z}$, and by exercise 7 is representable as a sum of exponentials $\exp(2\pi i n r/q)$. There is then a representation

$$\sum_{d|(F(n),P)=1} \lambda_d = \sum_{h \le z} \sum_{\substack{s=1 \\ (s,h)=1}}^{h} \omega_{s,h} e^{2\pi i s n/h},$$

valid for all integers n. We assume the λ_d to vanish for $d > z$, but rather than follow Selberg's procedure to diagonalise Δ, we estimate the Hermitian form

$$\sum_n \left| \sum_{h \le z} \sum_{\substack{s=1 \\ (s,h)=1}}^{h} \omega_{s,h} e^{2\pi i s n/h} \right|^2$$

for the $\omega_{s,h}$ to hand. We recognise this Hermitian form as the conjugate/dual of the form

$$\sum_{q \le z} \sum_{\substack{r=1 \\ (r,q)=1}}^{q} \left| S\left(\frac{r}{q}\right) \right|^2$$

appearing in exercise 18. However, we do not appeal to that result, nor to the duality principle. The Selberg sieve and the Large Sieve meet on the common ground that the form in $\omega_{s,h}$ is somewhat tractable.

28. Prove that

$$\sum_{n=-\infty}^{\infty} e^{-\pi n^2/N^2} \left| \sum_{d|(F(n),P)} \lambda_d \right|^2 = (N + O(z^2)) \sum_{h\le z} \sum_{\substack{s=1 \\ (s,h)=1}}^{h} |\omega_{s,h}|^2$$

holds uniformly for $N \geq 1$, z and all complex λ_d.

What are the $\omega_{s,h}$ arising from the Selberg sum involving the divisors of $(F(n), P)$?

29. Prove that

$$\frac{1}{d} \sum_{h|d} \sum_{\substack{s=1 \\ (s,h)=1}}^{h} e^{2\pi i s n/h} = \begin{cases} 1 & \text{if } \; n \equiv 0 (\operatorname{mod} d), \\ 0 & \text{otherwise.} \end{cases}$$

30. Prove that the arithmetic function which is 1 when $F(n) \equiv 0 (\operatorname{mod} d)$, and zero otherwise, has a representation

$$\sum_{n_t (\operatorname{mod} d)} \frac{1}{d} \sum_{h|d} \sum_{\substack{s=1 \\ (s,h)=1}}^{h} e^{2\pi i \frac{s}{h}(n-n_t)},$$

where n_t runs through a set of least positive solutions to $F(n) \equiv 0 (\operatorname{mod} d)$.

31. Prove that

$$\omega_{s,h} = \sum_{\substack{d|P \\ d\equiv 0 (\operatorname{mod} h)}} \frac{\lambda_d}{d} \sum_{n_t (\operatorname{mod} d)} e^{-\frac{2\pi i s n_t}{h}}.$$

32. Prove that for divisors h of a squarefree d,

$$\sum_{F(m)\equiv 0 (\operatorname{mod} d)} e^{-2\pi i s m/h} = \rho(d/h) \sum_{F(b)\equiv 0 (\operatorname{mod} h)} e^{-2\pi i s b/h},$$

where the sum runs through a set of incongruent solutions $(\operatorname{mod} d)$, $(\operatorname{mod} h)$ as the case may be, and $\rho(w)$ denotes the number of such solutions $(\operatorname{mod} w)$.

33. Prove that

$$\sum_{h\le z} \sum_{\substack{s=1 \\ (s,h)=1}}^{h} |\omega_{s,h}|^2 = \sum_{h\le z} \psi_h \left| \sum_{\substack{d|P \\ d\equiv 0 (\operatorname{mod} h)}} \frac{\lambda_d \rho(d)}{d} \right|^2,$$

where

$$\psi_h = \sum_{\substack{s=1 \\ (s,h)=1}}^{h} \rho(h)^{-2} \left| \sum_{F(b)\equiv 0 (\mathrm{mod}\, h)} e^{-2\pi i s b/h} \right|^2.$$

34. Prove that

$$\sum_{h|k} \psi_h = \frac{k}{\rho(k)}, \quad k \mid P.$$

Hint: Set $\lambda_k = 1$, $\lambda_d = 0$ for $d \neq k$ in exercise 28 and appeal to the Poisson sum relation of exercise 15 as $y^{-1/2} = N \to \infty$.

Deduce that

$$\psi_k^{-1} = \prod_{p|k} \left(\frac{\rho(p)}{p - \rho(p)} \right), \quad k \mid P.$$

The function ψ_k^{-1} is known in the literature as $g(k)$, cf. Halberstam and Richert, [93].

35. Prove that

$$\sum_{n=-\infty}^{\infty} e^{-\pi n^2/N^2} \left| \sum_{d|(F(n),P)} \lambda_d \right|^2 = (N + O(z^2)) \sum_{\substack{h\leq z \\ h|P}} \frac{1}{g(h)} \left| \sum_{\substack{d|P \\ d\equiv 0 (\mathrm{mod}\, h)}} \frac{\lambda_d \rho(d)}{d} \right|^2,$$

uniformly for $N \geq 1, z$ and all complex λ_d which vanish for $d > z$. For real λ_d, the quadratic form on the right-hand side is in the classical form of Selberg, arrived at in a manner quite different from that of Selberg. I employ the bare bones of Selberg's argument in Chapter 25, Lemma 25.7.

We coast home along one of many routes.

36. Prove that if

$$\Gamma = \left(\sum_{\substack{h|P \\ h\leq z}} g(h) \right)^{-1},$$

then

$$\sum_{h\leq z} \frac{1}{g(h)} \left| \sum_{\substack{d|P \\ d\equiv 0 (\mathrm{mod}\, h)}} \frac{\lambda_d \rho(d)}{d} \right|^2$$

$$= \sum_{h\leq z} g(h) \left| \frac{1}{g(h)} \sum_{d\equiv 0 (\mathrm{mod}\, h)} \frac{\lambda_d \rho(d)}{d} - \mu(h)\Gamma\lambda_1 \right|^2 + \lambda_1^2 \Gamma.$$

37. Choose the λ_d to effect equality with $\lambda_1^2\Gamma$ alone, and

$$\sum_{n=-\infty}^{\infty} e^{-\pi n^2/N^2} \left| \sum_{d|(F(n),P)} \lambda_d \right|^2 = (N + O(z^2))\lambda_1^2 \left(\sum_{\substack{h|P \\ h\le z}} g(h) \right)^{-1}$$

uniformly in $\lambda_1, N \ge 1, z$.

In these exercises about the Selberg sieve it was tacitly assumed that P contains only primes p for which $0 < \rho(p) < p$. The argument works equally well for an arbitrary polynomial F in $\mathbb{Z}[x]$. Moreover, we may uniformly translate the integers n.

38. Returning to the sequence B described preceding exercise 25, prove that in this case $g(h) \ge (h^{-1}\phi(h))^2$ on the divisors of P when P contains all odd primes not exceeding $N^{1/2}$. Deduce that $B(N) \ll N^{1/2}$, best possible with respect to size.

Hint: Express the function $h \mapsto (h^{-1}\phi(h))^2$ as a Dirichlet convolution $\mathbf{1} * u$, where $\mathbf{1}$ denotes the arithmetic function identically 1.

39. Suppose that the members of B omit only $\frac{1}{2}(p-3)$ of the residue classes to odd primes p not exceeding $N^{1/2}$. Prove that $B(N) \ll N^{1/2}$ still holds.

... and so on ...

In part, the Large Sieve estimates the maximum of an Hermitian form on a unit sphere. Selberg's sieve method minimises a quadratic form on a plane. Within the topology induced by the standard euclidean metric, the sphere is compact, the plane is not. We may move the plane to a punctured sphere.

40. Let $\rho > 0$. Prove that in a Hilbert space, the inversion $x \to \rho x|x|^{-2}$ of the space with the origin removed, transforms the sphere $|x-a| = |a|^2 > 0$ with the origin removed, into the plane $(x, a) = \rho/2$.

41. Let F, G be complex valued functions on a Hilbert space. Prove that if $c \ne 0$, then

$$\min_{(x,c)=1} F(x) = \min_{\substack{|z|=1 \\ z\ne -c|c|^{-1}}} F\left(\frac{2(c+z|c|)}{|c+z|c||^2} \right).$$

Moreover

$$\min_{|x|=1, x\ne -b} G(x) = \min_{(w,b)=1} G\left(\frac{2w}{|w|^2} - b \right)$$

for any b with $|b| = 1$.

In particular, the Selberg sieve minimises

$$4|e_1 + y|^{-4} \sum_n \left(1 + \sum_{d|(F(n),P)} y_d\right)^2$$

over the unit sphere $|y| = 1$ in a space $\mathbb{R}^m$ with the negative of the first standard basis vector e_1 in $\mathbb{R}^m$ removed; as may be verified directly.

The arithmetic applications of Selberg's sieve are explicit, and determine the course of its development. Arithmetic applications of the Large Sieve are less apparent. Arithmetic applications of the duality principle are implicit, not apparent. It is helpful to direct them by a background philosophy.

Rényi, [141], applied an additive Large Sieve to estimate sums involving Dirichlet characters. After exercise 7, an analogue of the inequality of exercise 18 for Dirichlet characters is to be expected.

Let $\tau(\chi)$ denote the Gauss sum

$$\sum_{a=1}^{q} \chi(a) e^{2\pi i a/q}$$

associated with the Dirichlet character $\chi(\text{mod}\, q)$.

42. Prove that

$$\tau(\bar{\chi})\chi(n) = \sum_{b=1}^{q} \bar{\chi}(b) e^{2\pi i b n/q}$$

for $(n, q) = 1$.

A Dirichlet character $(\text{mod}\, q)$ is said to be *induced* by a character $(\text{mod}\, d)$, d a divisor of $q, d < q$, if the values of the characters coincide on the integers prime to q. Otherwise the initial character is *primitive*. Each character $(\text{mod}\, q)$ is either primitive or induced by a unique primitive character to a modulus which is a proper divisor of q.

In Chapter 9 of his book, [19], Davenport proves that for a primitive character $\chi(\text{mod}\, q)$, and integers $(n, q) > 1$, the sum

$$\sum_{a=1}^{q} \chi(a) e^{2\pi i a n/q}$$

vanishes, and consequently $|\tau(\chi)| = q^{1/2}$; cf. Prachar, [134], Chapter VII, Lemma 1.1. Let us traverse in the opposite direction, and with different argument.

43. Prove that if χ is a primitive Dirichlet character $(\bmod q)$, and d a proper divisor of q, then there is a c, $(c,q)=1$, $c \equiv 1 (\bmod d)$, for which $\chi(c) \neq 1$. Deduce that

$$\sum_{\substack{u=1 \\ u\equiv 1 (\bmod d)}}^{q} \chi(u) = 0.$$

Hint: Consider the transformation $u \mapsto cu$.

Let $c_r(n)$ denote the Ramanujan sum

$$\sum_{\substack{b=1 \\ (b,r)=1}}^{r} e^{2\pi i n b/r}.$$

44. Prove that

$$c_r(n) = \sum_{d|(n,r)} \mu\left(\frac{r}{d}\right) d.$$

45. Prove that if χ is primitive, then $|\tau(\chi)| = q^{1/2}$.

Hint: First prove that

$$|\tau(\chi)|^2 = \sum_{m=1}^{q} \chi(m) e^{2\pi i m/q} \sum_{r=1}^{q} \overline{\chi(mr)} e^{-2\pi i m r/q}.$$

46. Prove that

$$\sum_{n=1}^{q} \chi(n) e^{-2\pi i n b/q} = 0$$

for primitive characters $\chi (\bmod q)$ and $(n,q) > 1$.

Hint: Regard the sum as a Fourier coefficient of χ viewed on the group $\mathbb{Z}/q\mathbb{Z}$, and apply Parseval.

47. Prove that for a primitive character χ, the identity of exercise 42 continues to hold when $(n,q) > 1$.

48. Apply exercise 47 to establish the Pólya–Vinogradov inequality:

$$\sum_{n \leq y} \chi(n) \ll q^{1/2} \log q$$

uniformly for non-principal characters $(\bmod q)$, $y > 0$.

Hint: Geometric progression.

49. Prove that if χ is primitive $(\operatorname{mod} q)$ and $(r,q)=1$, then

$$\sum_{\substack{m=1\\(m,qr)=1}}^{qr} \bar{\chi}(m) e^{2\pi i n m/(qr)} = \bar{\chi}(r)\chi(n)\tau(\bar{\chi})c_r(n).$$

Deduce that in the notation of exercise 17,

$$q\left|\sum_{M<n\le M+N} a_n \chi(n) c_r(n)\right|^2 = \left|\sum_{\substack{m=1\\(m,qr)=1}}^{qr} \bar{\chi}(m) S\left(\frac{m}{qr}\right)\right|^2.$$

50. With * denoting summation over primitive characters, prove that

$$\sum_{qr\le Q} \frac{1}{\phi(qr)} \sum_{\chi(\operatorname{mod} q)}^{*} \left|\sum_{\substack{m=1\\(m,qr)=1}}^{qr} \bar{\chi}(m) S\left(\frac{m}{qr}\right)\right|^2$$

$$= \sum_{t\le Q} \frac{1}{\phi(t)} \sum_{\chi(\operatorname{mod} t)} \left|\sum_{\substack{b=1\\(b,t)=1}}^{t} \bar{\chi}(b) S\left(\frac{b}{t}\right)\right|^2$$

$$= \sum_{t\le Q} \sum_{\substack{b=1\\(b,t)=1}}^{t} \left|S\left(\frac{b}{t}\right)\right|^2.$$

51. Prove that

$$\sum_{\substack{qr\le Q\\(q,r)=1}} \frac{q}{\phi(qr)} \sum_{\chi(\operatorname{mod} q)}^{*} \left|\sum_{M<n\le M+N} a_n \chi(n) c_r(n)\right|^2 \ll (N+Q^2) \sum_{M<n\le M+N} |a_n|^2$$

uniformly for $N \ge 1, M, Q$, complex a_n.

The explicit derivation from an additive Large Sieve of a version involving Dirichlet characters was effected with increasing simplicity, Bombieri, [6], Bombieri and Davenport, [7], Gallagher, [89], culminating in inequality 51 of Selberg. Successive results sought to squeeze more into the sum to be bounded, without degrading the upper bound. An advantage of introducing Ramanujan sums is illustrated in the following exercises.

52. Prove that for $(n,r)=1$, $c_r(n)=\mu(r)$, the Möbius function at r.

53. Prove that

$$\frac{q}{\phi(q)} \sum_{\substack{r \le y \\ (r,q)=1}} \frac{\mu^2(r)}{\phi(r)} > \log y$$

uniformly for $y \ge 1$, $q \ge 1$. An argument is given in Chapter 25, Lemma 25.7.

54. Let $\varepsilon > 0$. Prove that

$$\sum_{q \le Q} \sideset{}{^*}\sum_{\chi (\bmod q)} \left| \sum_{M < p \le M+N} a_p \chi(p) \right|^2 \ll \left(\frac{N}{\log N} + Q^{2+\varepsilon} \right) \sum_{M < p \le M+N} |a_p|^2,$$

uniformly for $M \ge 1$, $N \ge 2, Q$, complex a_p.

55. Let $\varepsilon > 0$. Prove that

$$\sum_{\substack{q \le N^{\frac{1}{2}-\varepsilon} \\ q \text{ prime}}} \phi(q) \sum_{r=1}^{q-1} \left| \sum_{\substack{M < p \le M+N \\ p \equiv r (\bmod q)}} a_p - \frac{1}{\phi(q)} \sum_{\substack{M < p \le M+N \\ (p,q)=1}} a_p \right|^2$$
$$\ll \frac{N}{\log N} \sum_{M < p \le M+N} |a_p|^2,$$

uniformly for $M \ge 1$, $N \ge 2$, complex a_p. This inequality may be compared with that of exercise 26, again with those of Theorem 3.1.

The simplicity, even elegance, of standard Large Sieve inequalities owes much to their involvement with squares. To treat arithmetic functions in general spaces $\mathcal{L}^\alpha$, particularly to pursue the characterisation of additive functions in terms of their discrete derivatives, a version of the Large Sieve corresponding to exponents between 1 and 2 is desirable. The difficulty of establishing such an inequality is bound to the want of an appropriate formulation. I turn to this nexus in Chapter 23.

5

The method of the stable dual: Deriving the approximate functional equations

I continue with the general discussion of the method of the stable dual, employing the inequality

$$(5.1) \qquad \sum_{p\le x} p \left| \sum_{\substack{n\le x \\ n\equiv 0(\mathrm{mod}\, p)}} d_n - \frac{1}{p}\sum_{n\le x} d_n \right|^2 \ll x \sum_{n\le x} |d_n|^2$$

for prime moduli p. This inequality may be derived without difficulty from the dual of the Turán–Kubilius inequality (1.2) by applying the Cauchy–Schwarz inequality to the sums in the decomposition

$$\sum_{\substack{n\le x \\ n\equiv 0(\mathrm{mod}\, p)}} d_n - \sum_{\substack{p^j\le x \\ j\ge 1}} \frac{1}{p^j}\left(1-\frac{1}{p}\right)\sum_{n\le x} d_n$$

$$= \sum_{\substack{p^j\le x \\ j\ge 1}} \frac{1}{p^{(j-1)/2}} \cdot p^{(j-1)/2} \left(\sum_{\substack{n\le x \\ p^j \| n}} d_n - \frac{1}{p^j}\left(1-\frac{1}{p}\right)\sum_{n\le x} d_n \right),$$

noting that the sums $\sum p^{-j+1}$ are uniformly bounded. Alternatively we may establish directly a form of the Turán–Kubilius inequality for strongly additive functions f, and obtain the inequality (5.1) by duality.

Coarsely, in the method of the stable dual the d_n are chosen to be the values of an arithmetic function, depending upon the problem at hand, so that the average of $|d_n|^2$ is controlled by hypothesis. The dependence of the mean values

$$x^{-1} \sum_{n\le x,\ n\equiv 0(\mathrm{mod}\, p)} d_n$$

upon the modulus p is then factored out as far as possible, using the arithmetic nature of the d_n, reducing the sum to a mean value over the range

$1 \le n \le x/p$. The resulting inequality is regarded as an approximate functional equation for the mean value $x^{-1} \sum d_n$, $1 \le n \le x$, and its variants. This functional equation is to be largely solved, say by varying x, and the information so obtained fed back into the initial inequality to gain the desired control over the original arithmetic function on the primes.

Example. *Let f be a complex completely additive arithmetic function,*

$$\|f\| = \left(x^{-1} \sum_{n \le x} |f(n)|^2 \right)^{1/2} \ge 0.$$

Then

$$\sum_{p \le x} \frac{1}{p} \left| f(p) + m\left(\frac{x}{p}\right) - m(x) \right|^2 \ll \|f\|^2. \tag{5.2}$$

with

$$m(y) = [y]^{-1} \sum_{n \le y} f(n). \tag{5.3}$$

Proof. We set $d_n = f(n)$ in (5.1) and note that from the functional property of f

$$\sum_{\substack{n \le x \\ n \equiv 0 \,(\mathrm{mod}\, p)}} d_n = f(p) \left[\frac{x}{p} \right] + \sum_{m \le x/p} f(m).$$

The proof is now completed using the estimate

$$\sum_{p \le x} |f(p)|^2 \ll x \|f\|^2.$$

The general method presents a number of features. It begins with information on the integers and derives information on the primes. This is the direction taken in the proof of the prime number theorem. That complications should arise in its implementation is to be expected, for arguments from integers to primes are usually more difficult and lie deeper than arguments from primes to integers, cf. [65]. To compensate for this, the method is extremely flexible, and can usually be modified to take into account new information gained along the way; it lends itself to iteration.

There is no initial requirement that the a_n be attached to a particular structural form, say an Euler product. This is particularly valuable when dealing with certain problems in analytic and probabilistic number theory.

For example, Kac' simulation of additive functions by independent random variables may be replaced by the application of the notion of self-adjointness of an operator on a Hilbert space, [59].

The celebrated Erdős-Kac theorem in probabilistic number theory asserts that if an additive function f satisfies $|f(p)| \leq 1$ and $f(p^m) = f(p)$ on the primes, and $B(N) \to \infty$, as $N \to \infty$, then

$$\nu_N(n; f(n) - A(N) \leq zB(N)) \Rightarrow \frac{1}{\sqrt{2\pi}} \int_{-\infty}^{z} e^{-u^2/2} du, \quad N \to \infty.$$

Accounts may be found in their original papers, [84], [85], and my book, [41], Vol. 2, Chapter 12. Whilst their method was clarified and given an axiomatic probabilistic basis by Kubilius, and their result generalised to admit some non-normal limit laws, the functions f were still restricted on the primes at the outset; Kubilius, [109], an account may be found in my book, [41], Vol. 1, Chapter 3, Vol. 2, Chapter 12. Broadly speaking, Kac' idea to simulate additive functions by sums of independent random variables has the direction *primes to integers.* Experience in the theory of probability proper guides what you wish to prove concerning the distribution of f on the integers, and if you assume enough concerning f on the primes, then the implementation of Kac' idea gives a method that will likely succeed. Of course, the technical details can be more or less formidable.

I was for a long time interested in going the other way. Assume (in an obvious notation) that for some $\alpha(x), \beta(x) > 0$ the frequencies $\nu_x(n; f(n) - \alpha(x) \leq z\beta(x))$ converge weakly as $x \to \infty$, and *deduce* the behaviour of f on the primes. I well remember driving Erdős down Nineteenth Street in Boulder, on the way to the university, it was in late 1971 or early 1972, and saying "In such a frequency $\alpha(x)$ and $\beta(x)$ should be completely free."

"With $\beta(x)$ smaller than $\alpha(x)$" was his immediate comment.

"Why not completely free?"

"Oh well, then."

It was clear to me before Spring 1972 that any method that would avoid an appeal to Kac' idea must necessarily take the same direction as the proof of the prime number theorem. Moreover, with an eye to longer range considerations, such as the difference of additive functions $f(n+1) - f(n)$, I sought a method that did not require the existence of an associated Euler product. By early 1972 I had begun to consciously devise such a method, cf. [32], St. Louis 1972. It was two and a half years before I had developed the method sufficiently that it would deliver substantial progress on the first of these problems. It was two and a half further years before I made a substantial inroad upon the second problem. With each development further ideas were needed.

Although I originally developed the method of the stable dual using L^2 norms, the method extends quite naturally to include other norms. For a number theorist all that is required is the existence of suitable duals, and the possibility of interpreting arithmetically at least one of the various general inequalities arising. Indeed, one might seek to apply this mixture of structural and dynamical ideas in other mathematical disciplines.

The background philosophy and the use of duality, I am sure, I derived ultimately from the intensive study of projective geometry that I made at grammar (in high) school.

6

The method of the stable dual: Solving the approximate functional equations

Consider the functional inequality

$$\sum_{z^c < p \le z^d} \frac{1}{p}\left| f(p) + \alpha\left(\frac{z}{p}\right) - \alpha(z)\right| \le B,$$

where $0 < c < d$, and the functions f, α are, for the moment, arbitrary complex valued. In accordance with the remarks made concerning the general method, I solve this inequality by varying z; I assume that it holds uniformly for $x^{2-\tau} \le z \le x^\tau$, $\tau > 1$. If now $(c+d)\tau < 2d$, then $(x^{c\tau}, x^{(2-\tau)d}]$ is the intersection of the intervals $(z^c, z^d]$, $x^{2-\tau} \le z \le x^\tau$. In particular, with $\gamma = c\tau$, $\delta = d(2-\tau)$,

$$\sum_{x^\gamma < p \le x^\delta} \frac{1}{p}\left| f(p) + \alpha\left(\frac{z}{p}\right) - \alpha(z)\right| \le B, \tag{6.1}$$

uniformly for $x^{2-\tau} \le z \le x^\tau$. This small step is important; I have introduced the free variable z. In what follows I concentrate on this free variable.

Theorem 6.1. *Let $K \ge 6$, $P \ge 2$. Let the $\mathbb{R} \to \mathbb{C}$ functions f, ψ, ϕ satisfy*

$$\sum_{P < p \le P^K} \frac{1}{p} \left| f(p) + \psi(pw) - \phi(w)\right| \le \varepsilon \log K \tag{6.2}$$

uniformly for $P^{-\tau} \le w \le P^\tau$, where $0 < \tau \le K^{1/2}$ and the sum is not empty. Let ψ satisfy the asymptotic continuity condition

$$|\psi(v) - \psi(u)| \le \varepsilon \tag{6.3}$$

uniformly for $P^{1-\tau} \le u \le v \le P^{K+\tau}$, $v \le u(1 + P^{-\frac{1}{3}})$.

Then there is a representation

$$\phi(w) = D \log w + \phi(1) + O(\varepsilon)$$

with the same uniformity in w, and $D\tau \log P = \phi(P^\tau) - \phi(1)$.

Remarks. The functions ψ, ϕ need only be defined on the intervals $[P^{1-\tau}, P^{K+\tau}]$, $[P^{-\tau}, P^\tau]$ respectively, and f on the primes p in (P, P^K). They may otherwise be set zero.

It follows from the asymptotic continuity condition on ψ that there is a continuous function ψ_1 satisfying $|\psi_1(u) - \psi(u)| \le \varepsilon$. At the expense of multiplying ε by an absolute constant we may replace ψ in the inequalities of the hypothesis by ψ_1. Without loss of generality ψ may be assumed continuous.

The asymptotic condition on ψ is needlessly strong. It would suffice that ψ be measurable and satisfy $|\psi(vw) - \psi(uw)| \le \varepsilon$ uniformly for $P^{-\tau} \le w \le P^\tau$, $P \le u \le v \le P^K$, $v \le u + u^{2/3}$. In practice even stronger continuity conditions are often available.

It is possible to restrict the primes p to some reduced residue class (mod a). The error term $O(\varepsilon)$ then depends upon a, and for a up to a small positive power of P this dependence can be made largely explicit. The only property of primes that is needed is their occurrence within a (constant) multiple of the expected density in intervals large enough to be covered by the asymptotic continuity condition on ψ. In the present circumstances a sufficient result is provided by Satz (3.2) of Prachar, [134] p. 323, bearing in mind the remarks which he makes following Satz (3.3) on the same page.

Lemma 6.2. *Let $c > 0$, $5/8 < \alpha \le 1$. Then*

$$\sum_{\substack{x<p\le x+y \\ p\equiv r (\text{mod}\, a)}} \log p = \frac{y}{\phi(a)}\left(1 + O\left(\frac{1}{\log x}\right)\right), \qquad x \to \infty,$$

uniformly for $x^\alpha \le y \le x$, $(r, a) = 1$ and $1 \le a \le (\log x)^c$.

As I indicate later, much weaker results suffice in actual practice.

Similar remarks may be made concerning the following two theorems.

For use in the proof of Theorem 6.1 I need the following result, due essentially to Ruzsa, [149].

Lemma 6.3. *If the measurable complex $u(x)$ satisfies $|u(x+y) - u(x) - u(y)| \le \varepsilon$ for $|x| \le 1$, $|y| \le 1$, $|x+y| \le 1$, then $|u(x) - xu(1)| \le 3\varepsilon$ for $|x| \le 1$.*

It is clear that

$$|u(1)| \le \inf_{0<|x|\le 1} |x|^{-1}(|u(x)| + 3\varepsilon),$$

and in applications this can be a useful bound.

Ruzsa deduces his result from a theorem of Hyers [106], that is almost fifty years old. Hyers answered affirmatively for Banach spaces the following question raised by Ulam: *If a function approximately satisfies Cauchy's functional equation, then must it be near to a genuine solution of that equation?* I reproduce their arguments here.

Proof of Lemma 6.3. As Ruzsa remarks, the proof of Hyers works equally well if the function u is defined on an abelian group G, with values satisfying $\|u(x+y)-u(x)-u(y)\| \le \varepsilon$ in a Banach space B. There is then a homomorphism $v : G \to B$ so that $\|u(x)-v(x)\| \le \varepsilon$ on the whole of the group.

Indeed, repeated application of the fundamental inequality shows that $\|2^{-n}u(2^n x) - 2^{-m}u(2^m x)\| \le 2^{-m}\varepsilon$ for integers $n \ge m \ge 0$, and all x in G. The function

$$v(x) = \lim_{n\to\infty} 2^{-n}u(2^n x)$$

is well defined by Cauchy's criterion, and satisfies

$$\begin{aligned}\|v(x+y)-v(x)-v(y)\| &= \lim_{n\to\infty} \|2^{-n}\{u(2^n(x+y)) - u(2^n x) - u(2^n y)\}\| \\ &\le \lim_{n\to\infty} 2^{-n}\varepsilon = 0.\end{aligned}$$

As is well known, measurable solutions to Cauchy's functional equation on the additive group of reals are necessarily linear. A proof may be found in Chapter 1 of my book, [41]. An extensive account of exact functional equations may be found in Aczél, [1].

We return to the present circumstances, with u defined on the interval $[-1,1]$. Define $w(2k+t) = 2ku(1) + u(t)$ for $-1 \le t < 1$ and all integers k. This extends u to the whole of $\mathbb{R}$, except that $w(1) = 2u(1) + u(-1)$. We shall show that $d = w(x+y) - w(x) - w(y)$ satisfies $|d| \le 3\varepsilon$ for all x, y in $\mathbb{R}$, and may then appeal to the above result of Hyers.

Since $w(x) - xu(1)$ is periodic, with period 2, we may assume x, y to be in $[-1,1)$. If $x+y$ belongs to $[-1,1)$, then $|d| \le \varepsilon$. If $x+y \ge 1$, then $d = u(x+y-2) + 2u(1) - u(x) - u(y)$ which is $\{u(x+y-2) - u(x-1) - u(y-1)\} + \{u(x-1) + u(1) - u(x)\} + \{u(y-1) + u(1) - u(y)\}$, giving $|d| \le 3\varepsilon$. The case $x+y < -1$ may be dealt with similarly.

According to the result of Hyers $|w(x) - v(x)| \le 3\varepsilon$ with $v(x) = \lim 2^{-n}w(2^n x) = u(1)x$. This gives $|u(x) - u(1)x| \le 3\varepsilon$ for $-1 \le x < 1$, and for $x = 1$ it is trivial.

During the proof of Theorem 6.1 and the following Theorems 6.4 and 6.6, Lebesgue measurable subsets E of the half-line $[2, \infty)$ will be assigned measure

$$|E| = \int_E \frac{du}{u \log u}.$$

Proof of Theorem 6.1. Elimination between the functional inequality (6.2) with w, and with w replaced by z, gives

$$\sum_{P<p\le P^K} \frac{1}{p} |\phi(w) - \phi(z) - \psi(wp) + \psi(zp)| \le 2\varepsilon \log K.$$

The continuity condition (6.3) on ψ, and the distribution of primes in short intervals guaranteed by Lemma 6.2 allows a reformulation in terms of an integral

$$\int_P^{P^K} |\phi(w) - \phi(z) - \psi(wu) + \psi(zu)| \frac{du}{u \log u} \ll \varepsilon \log K.$$

In this part of the argument, we do not use the continuity condition again.

Suppose that θ and $w\theta^{-1}$ both belong to the interval $[P^{-\tau}, P^{\tau}]$. We set $z = 1$ and make the change of variables $w \to w\theta^{-1}$, $u \to y\theta$. Then, shortening the range,

$$\int_{A(\theta)}^{B(\theta)} |\phi(w\theta^{-1}) - \phi(1) - \psi(wy) + \psi(y\theta)| \frac{dy}{y \log y\theta} \ll \varepsilon \log K,$$

where $A(\theta) = \max(P, P\theta^{-1})$, if $6 \le K \le 10$, $\max(\theta, P, P\theta^{-1})$ if $K > 10$, $B(\theta) = \min(P^K, P^K\theta^{-1})$. If $y \ge A$, then $y\theta \le yP^{\tau} \le y^5$ when $6 \le K \le 10$, $y\theta \le y^2 \le y^5$ when $K > 10$. With a permissible error, the factor θ may be removed from the logarithm in this integral.

Elimination between the resulting integral with θ, and θ replaced by 1, yields

$$\int_{A(\theta)}^{B(\theta)} |\phi(w) - \phi(w\theta^{-1}) + \psi(y) - \psi(y\theta)| \frac{dy}{y \log y} \ll \varepsilon \log K.$$

We employ this with $w = \theta_1\theta_2$, $\theta = \theta_1$, and then $w = \theta_2$, $\theta = \theta_2$, where the θ_j and $\theta_1\theta_2$ all belong to $[P^{-\tau}, P^{\tau}]$. In the second application we change the variable y to $v\theta_1$, and again shorten:

$$\int_C^D |\phi(\theta_2) - \phi(1) + \psi(v\theta_1) - \psi(v\theta_1\theta_2)| \frac{dv}{v \log v\theta_1} \ll \varepsilon \log K,$$

where $C = \max(\theta_1, A(\theta_2), A(\theta_2)\theta_1^{-1})$, $D = \min(B(\theta_2), B(\theta_2)\theta_1^{-1})$. Once again the factor θ_1 in the logarithm in the integrand may be removed. Restricting to the intersection J of the ranges $[A(\theta_1), B(\theta_1)]$ and $[C, D]$, and employing the identity

$$\psi(y) - \psi(y\theta_1\theta_2) = \psi(y) - \psi(y\theta_1) + \{\psi(y\theta_1) - \psi(y\theta_1\theta_2)\}$$

we see that

$$\int_J |F(\theta_1\theta_2) - F(\theta_1) - F(\theta_2)| \frac{dy}{y \log y} \ll \varepsilon \log K$$

where $F(\theta) = \phi(\theta) - \phi(1)$. Here the integrand does not depend upon y. Moreover, J contains $[P^{1+\tau}, P^{K-\tau}]$ if $6 \le K \le 10$, and the intersection of this interval with $[P^{2\tau}, P^{K-\tau}]$ if $K > 10$. In either case it has measure $\gg \log K$. Hence

$$|F(\theta_1\theta_2) - F(\theta_1) - F(\theta_2)| \ll \varepsilon.$$

For all sufficiently large $P^{K^{1/3}}$

$$\sum_{P^{K^{1/3}} < p \le P^K} \frac{1}{p} = \frac{2}{3} \log K + O(K^{-\frac{1}{3}}(\log P)^{-1}) > \frac{1}{2} \log K.$$

From the initial inequality of the proof

$$|\phi(w) - \phi(z)| \sum_{P < p \le P^K} \frac{1}{p} \le 2\varepsilon \log K + \sum_{P < p \le P^K} \frac{1}{p} |\psi(wp) - \psi(zp)|,$$

so that for $w \le z \le w(1 + P^{-\frac{1}{3}})$, $P^{-\tau} \le w \le z \le P^{\tau}$, we have $|\phi(w) - \phi(z)| < 5\varepsilon$. The function ϕ is also asymptotically continuous, and may be assumed continuous.

The asserted result of Theorem 6.1 follows from an application of Lemma 6.3 with $u(t) = F(P^{t\tau})$.

Remarks on Theorem 6.1. In order for the proof to proceed it suffices if $\tau \le K^{1-\delta}$ and K is sufficiently large in terms of δ. Likewise, the lower end P of the range of summation in the fundamental inequality (6.2) may be replaced by $P^{K^{1-\delta}}$. This is valuable in certain applications. For $1 < K_0 \le K < 6$ the argument still works provided $0 < \tau \le \tau_0 < (K_0 - 1)/2$. The implied constants then depend upon K_0, τ_0.

The argument establishing Theorem 6.1 is not particularly bound to functions with complex values. For real y, let $\|y\|$ denote the distance of y from a nearest integer. Then $\|\ \|$ induces a translation invariant metric on the group $\mathbb{R}/\mathbb{Z}$, and satisfies $\|-y\| = \|y\|$.

Theorem 6.4. *Let $K \geq 6$, $P \geq 2$. Let the $\mathbb{R} \to \mathbb{R}/\mathbb{Z}$ functions f, ψ, ϕ satisfy*

$$\sum_{P<p\leq P^K} \frac{1}{p} \|f(p) + \psi(pw) - \phi(w)\| \leq \varepsilon \log K$$

uniformly for $P^{-\tau} \leq w \leq P^{\tau}$, where $0 < \tau \leq K^{1/2}$, and the sum is not empty. Let ψ satisfy the asymptotic continuity condition

$$\|\psi(v) - \psi(u)\| \leq \varepsilon$$

uniformly for $P^{1-\tau} \leq u \leq v \leq P^{K+\tau}$, $v \leq u(1 + P^{-\frac{1}{3}})$.

Then there is a representation

$$\|\phi(w) - D \log w - \phi(1)\| \ll \varepsilon$$

with the same uniformity in w. If λ is a continuous real function whose image $\bar{\lambda}$ under the canonical map $\mathbb{R} \to \mathbb{R}/\mathbb{Z}$ satisfies $\|\bar{\lambda}(w) - \phi(w)\| \leq 5\varepsilon$ uniformly for $P^{-\tau} \leq w \leq P^{\tau}$, then we may set $D\tau \log P = \lambda(P^{\tau}) - \lambda(1)$.

Proof. The proof of Theorem 6.1 only uses the triangle property of the absolute value on $\mathbb{R}$. It leads at once to a bound

$$\|F(\theta_1\theta_2) - F(\theta_1) - F(\theta_2)\| \ll \varepsilon$$

with $F(\theta) = \phi(\theta) - \phi(1)$.

In view of the penultimate remark in the proof of Theorem 6.1, we may assume ϕ to satisfy the continuity condition $\|\phi(w) - \phi(z)\| < 5\varepsilon$. There is a continuous function ϕ_1 for which $\|\phi_1(w) - \phi(w)\| \leq 5\varepsilon$, uniformly for $P^{-\tau} \leq w \leq P^{\tau}$. The real function $\lambda : \mathbb{R} \to \mathbb{R}$ for which $\bar{\lambda}(w) = \phi_1(w)$ is unique up to translation by a fixed integer, and $H(w) = \lambda(w) - \lambda(1)$ satisfies $\|H(\theta_1\theta_2) - H(\theta_1) - H(\theta_2)\| \ll \varepsilon$. The proof of Theorem 6.4 is now completed by an appeal to the following lemma which reduces us to the case of real-valued functions, to which Lemma 6.3 applies.

Lemma 6.5. *Let $f : S \to \mathbb{R}$ be a continuous function on a connected topological space, and satisfy $\|f(u)\| \leq c < 1/2$ on S. Then there is an integer N so that $|f(u) - N| \leq \|f(u)\|$ on S.*

Proof. Cover each point of S with an open set on which the variation of f is less than $d < \frac{1}{2}(1 - 2c)$. For each u there is an integer $N(u)$ so that $|f(u) - N(u)| \leq \|f(u)\|$. The variation of $N(u)$ on any member of the cover is $\leq \operatorname{var} f + 2c < 1$, so N is constant there. Similarly N is constant on any two overlapping open sets.

Let $z \in S$. Let X denote the subset of S which can be reached using an overlapping chain of finitely many members of the cover, the first containing z. Since the last member of such a chain is open, so is X. Let Y denote the subset of S which cannot be so reached. If w belongs to Y, and w belongs to the set O of the cover, then O is contained in Y. Y is open. Any non-empty Y leads to a separation of S as $X \cup Y$. Thus $X = S$ and the lemma is established.

Theorem 6.6. *Let $K \geq 6$, $P \geq 2$. Let the $\mathbb{R} \to \mathbb{C}$ functions g, ψ, ϕ satisfy*

$$\sum_{P<p\leq P^K} \frac{1}{p}|g(p)\psi(pw) - \phi(w)| \leq \varepsilon \log K$$

uniformly for $P^{-\tau} \leq w \leq P^{\tau}$ with $0 < \tau \leq K^{1/2}$, the sum not empty, and $|g(p)| = 1$. Let

$$|\psi(v) - \psi(w)| \leq \varepsilon$$

uniformly for $P^{1-\tau} \leq u \leq v \leq P^{K+\tau}$, $v \leq u(1 + P^{-\frac{1}{3}})$.

Then there is a representation

$$\phi(w) = w^{i\alpha}\phi(1) + O(\varepsilon)$$

with the same uniformity in w. For real ϕ, $\alpha = 0$ may be taken. In general either $|\alpha| \leq 1$, or

$$|\phi(1)| \leq 2\left|\phi\left(t\left(1 + \frac{1}{2|\alpha|}\right)\right) - \phi(t)\right|$$

uniformly for $2P^{-\tau} \leq t \leq 2P^{\tau}/3$. In every case $|\alpha| < P^{1/3}$.

Proof. For real β let $e(\beta) = \exp(2\pi i\beta)$. Then $e(\beta/2) - e(-\beta/2) = 2\pi \sin \pi\beta$ $= \pm 2i \sin \pi\|\beta\|$, so that

$$4\|\beta\| \leq |e(\beta) - 1| \leq 2\pi\|\beta\|.$$

For values of the argument at which they don't vanish let $\psi(u) = |\psi(u)|e(s(u))$, $\phi(w) = |\phi(w)|e(h(w))$, and let $g(p) = e(f(p))$. If ψ and ϕ were unimodular, then we could reduce our fundamental inequality to

$$\sum_{P<p\leq P^K} \frac{1}{p}\|f(p) + s(pw) - h(w)\| \leq \frac{\varepsilon}{4}\log K,$$

and apply Theorem 6.4 at once.

We shall show that if ϕ is not uniformly large, then it is uniformly small, a situation which can actually occur. In connection with the characterisation of the exponential function by an approximate functional equation, a similar dichotomy, together with a method for its actualisation, are exemplified in my paper, [49]. When ϕ is large, the fundamental inequality ensures that ψ is usually large, also. In particular ϕ and ψ are not zero, and the above reduction argument proceeds.

Suppose first that ψ and ϕ are real and non-negative, $g(p)$ identically 1.

This is a special case of Theorem 6.1, and therefore $\phi(w) - \phi(1) = D \log w + O(\varepsilon)$ over the whole range of w. The fundamental equation becomes

$$\sum_{P<p\le P^K} \frac{1}{p}\Big|\psi(pw) - D\log w - \phi(1)\Big| \ll \varepsilon \log K,$$

and we are reduced to showing that the constant D satisfies $D\tau \log P \ll \varepsilon$. Instead of this appeal to Theorem 6.1 a simple direct argument can be given. In integral terms our temporary hypothesis asserts that

$$\int_P^{P^K} |\psi(wu) - \phi(w)| \frac{du}{u\log u} \ll \varepsilon \log K.$$

If the sets E and wE belong to $[P, P^K]$, then

$$|E||\phi(w) - \phi(1)| \le \int_E |\phi(w) - \psi(wu)| \frac{du}{u\log u} + \int_E |\psi(wu) - \phi(1)| \frac{du}{u\log u}.$$

The change of variable $u \to yw^{-1}$ shows the second integral to be

$$\int_{wE} |\psi(y) - \phi(1)| \frac{dy}{y \log yw^{-1}}.$$

For $w \le 1$ we may remove the factor w^{-1} in the logarithm and, from our initial inequality, bound the resultant integral $\ll \varepsilon \log K$. If $w > 1$ and every member of E is as large as w (say), then $yw^{-1} \ge y^{1/2}$ for y in wE and the integral is again $\ll \varepsilon \log K$.

For $P^{-\tau} \le w \le P^{\tau}$ such a set E is provided by the interval $[P^{1+\tau}, P^{K-\tau}]$. Moreover, $|E| \gg \log K$, so that $\phi(w) = \phi(1) + O(\varepsilon)$.

Since $||a| - |b|| \le |a - b|$,

$$\sum_{P<p\le P^K} \frac{1}{p}||\psi(pw)| - |\phi(w)|| \le \varepsilon \log K \tag{6.4}$$

uniformly for $P^{-\tau} \le w \le P^{\tau}$. Moreover, $|\psi(u)|$ satisfies an appropriate asymptotic continuity condition. Hence $|\phi(w)| = |\phi(1)| + O(\varepsilon)$ with the same uniformity in w.

If $\phi(1) \ll \varepsilon$, then $\phi(w) \ll \varepsilon$ uniformly in w and, since ψ may be identically zero, no more is to be expected from the hypothesis of the theorem. In this case the assertion of the theorem is valid with $\alpha = 0$.

Otherwise, we may assume that $|\phi(1)| \ge c\varepsilon$ for a positive constant c. This constant will be assumed sufficiently (absolutely) large at a finite number of places in the following argument. In particular, we may assume that $|\phi(1)|/2 \le |\phi(w)| \le 2|\phi(1)|$ over the whole w range. A real valued ϕ with $c > 1$ will therefore have constant sign. The theorem is established for ϕ real.

Cover the interval $[P, P^K]$ with adjoining smaller intervals I of the form $[y, y(1+P^{-1/3}]$, coalescing the last complete interval with any part interval. For each w, $P^{-\tau} \le w \le P^{\tau}$, let $G(w)$ be the union of those intervals which contain a point u for which $||\psi(uw)| - |\phi(w)|| > |\phi(w)|/4$. If p is a prime in such an interval, then, by the continuity condition, $||\psi(pw) - |\phi(w)|| > |\phi(w)|/8$ provided the constant c is sufficiently large. Hence

$$|G(w)| \ll \sum_{I} \sum_{p \in I} \frac{1}{p} \ll \varepsilon_1 \log K, \qquad \varepsilon_1 |\phi(1)| = \varepsilon,$$

the last step from an application of (6.4).

For any prime q, $P < q \le P^K$, not in $G(w)$, $2|\phi(1)| \|f(q)+s(qw)-h(w)\| \le 4|\phi(w)||e(f(q)+s(qw))-e(h(w))| \le ||\phi(w)|-|\psi(pw)||+|g(q)\psi(qw)-\phi(w)|$, so that

$$\sum_{P<q\le P^K} \frac{1}{q} \|f(q) + s(qw) - h(w)\| \le \varepsilon_1 \log K.$$

Moreover, on any interval I not contained in $G(w)$, $|\psi(u)|4\|s(v) - s(u)\| \le |\psi(u)||e(s(v))-e(s(u))| \le |\psi(v)-\psi(u)|+||\psi(u)|-|\psi(v)|| \le 4\varepsilon$, and $\psi(u)| \ge 3|\phi(w)|/4 \ge 3|\phi(1)|/8$, from which $\|s(v) - s(u)\| < 3\varepsilon_1$. We can now adapt the proof of Theorem 6.4, the fundamental inequality being

$$\int_P^{P^K} \|h(w) - h(z) - s(wu) + s(uz)\| \frac{du}{u \log u} \ll \varepsilon_1 \log K \tag{6.5}$$

with the understanding that $G(w)$ and $G(z)$ are to be removed from the range of integration. As a consequence

$$\int_{J-L} \|F(\theta_1\theta_2) - F(\theta_1) - F(\theta_2)\| \frac{dy}{y \log y} \ll \varepsilon \log K$$

where $F(\theta)$ is now $h(\theta) - h(1)$. Here L is the union of the sets $\theta_1^{-1}G(\theta_1\theta_2)$, $G(\theta_1\theta_2)$, $(\theta_1\theta_2)^{-1}G(\theta_2)$, $\theta_1^{-1}G(\theta_2)$ and four similar sets involving $G(1)$. Since J contains the interval $E \cap [P^{2\tau}, P^{K-\tau}]$, of measure $\gg \log K$, the proof may be continued as before provided we show that the measure of a typical set $S = E \cap \theta G(w)$ with $P^{-\tau} \le \theta \le P^{\tau}$ is relatively small. In fact

$$|S| = \int_{\theta^{-1}S} \frac{dv}{v \log v\theta}.$$

For $\theta \ge 1$ we may omit θ from the logarithm without decreasing the value of the integral, so that $|S| \le |G(w)| \ll \varepsilon_1 \log K$. If $\theta < 1$, then

$$|S| - \int_{\theta^{-1}S} \frac{dv}{v \log v} = \int_{\theta^{-1}S} \frac{1}{v \log v\theta}\left(-\frac{\log\theta}{\log v}\right) dv.$$

For v in $\theta^{-1}S$, $v \ge \theta^{-1}P^{1+\tau} > \theta^{-2}$ so that $\log v > -2\log\theta$ and the integral involving $\log\theta$ does not exceed $|S|/2$. In this case $|S| \le 2|G(w)| \ll \varepsilon_1 \log K$. As a consequence

$$\|F(\theta_1\theta_2) - F(\theta_1) - F(\theta_2)\| \ll \varepsilon_1$$

obtains, and hence a representation

$$\|h(\theta) - \alpha \log w - h(1)\| \ll \varepsilon_1,$$

the necessary continuity condition on h following from (6.5). Otherwise expressed

$$|\phi(w) - w^{i\alpha}\phi(1)| \le |\phi(1)||e(h(w)) - e(\alpha \log w + h(1))| + ||\phi(w)| - |\phi(1)|| \ll \varepsilon.$$

In particular,

$$|\phi(1)\{(1+\theta)^{i\alpha} - 1\}| \le |\phi(t(1+\theta)) - \phi(t)| + O(\varepsilon)$$

uniformly for $0 < |\theta| \le 1/2$, $2P^{-\tau} \le t \le 2P^{\tau}/3$. If $|\alpha| > 1$ and $\theta = (2|\alpha|)^{-1}$, then

$$|(1+\theta)^{i\alpha} - 1| = 2\left|\sin\frac{\alpha}{2}\log(1+\theta)\right| > \frac{2}{3}.$$

We may assume the constant c so large that the error term $O(\varepsilon)$ does not exceed $|\phi(1)|/6$.

Suppose that $|\alpha| \ge P^{1/3}$, so that $(2|\alpha|)^{-1} < P^{-\frac{1}{3}}$. Again as in the penultimate remark during the proof of Theorem 6.1, we may assume ϕ to satisfy the continuity condition $|\phi(w) - \phi(z)| < 5\varepsilon$ uniformly for $P^{-\tau} \le w \le z \le P^{\tau}$, $z \le w(1 + P^{-\frac{1}{3}})$. Except for the particular constant 5 as much was already obtained earlier in the present proof. In particular, $\phi(1+\theta)-\phi(1) \ll \varepsilon$. For a sufficiently large value of c this leads to a contradiction with $|\phi(1)| \geq c\varepsilon$.

Theorem 6.6 is proved.

Remark. It is the continuity property of ϕ that controls the size of α.

A paradigm for the approximate functional equations which arise in the application of the method described in Chapter 5 occurs as Lemma 2 of my 1972 St. Louis paper, [27]. It asserts that

$$\sum_{p \leq x}{}' \frac{1}{p} \to 0, \qquad x \to \infty \tag{6.6}$$

where the sum is taken over those primes p for which the inequality $|f(p) - \alpha(x) + \alpha(x/p)| > \varepsilon\beta(x)$ is satisfied. Here ε is an arbitrary positive number and, apart from the requirement that $\beta(x)$ be non-decreasing and unbounded, $\beta(x), \alpha(x)$ and f are arbitrary.

From its context in the study of the law of large numbers for additive arithmetic functions I hoped that this functional inequality would ultimately lead to a representation for $\alpha(x)$ of the form $\alpha_1(x)+\alpha_2(x)$, where α_1 behaved like a logarithm, and α_2 was relatively small. If the obstacle α_1 were not present, then the additive function under consideration should behave like a sum of independent random variables, and α_2 like its truncated mean. And so it proved.

It took me a couple of years to fully understand the equation (6.6). By 1974 I had derived from it that *there exist* $\omega(x), \varepsilon(x)$ *and* $\delta = \delta(x)$, *satisfying* $\varepsilon(x) \to 0$, $\delta(x) \to 0$ *as* $x \to \infty$, *such that the approximate relation*

$$\alpha(ab) = \alpha(a) + \alpha(b) + \omega(x) + o(\beta(x)) \tag{6.7}$$

holds uniformly for all pairs of real numbers (a, b) *in the box* $x^{\delta} \leq a, b \leq x^{1/\delta}$, *with the possible exception of a set of pairs* E *for which* $\mu_2 E \leq \varepsilon(x)$. To this end I subjected $\alpha(x)$ and $\beta(x)$ to the weak constraints that for each $C > 1$

$$\sup_{x/C \leq w \leq x} |\alpha(w) - \alpha(x)| \Big/ \beta(x) \to 0, \qquad x \to \infty, \tag{6.8}$$

and $\beta(x^y)/\beta(x)$ is uniformly bounded for each positive y. In fact this last condition allows the non-decreasing property of β to be omitted. That β is unbounded I did not use. This may be found as Lemma (3.2) of my 1975 paper on the law of large numbers, [34], or as Lemma (14.4), Chapter 14, in the second volume of my book, [41].

The map $u \to \log u/\log x$ gives an isomorphism between $\mathbb{R}^*$, the multiplicative group of positive reals, and $\mathbb{R}$, the additive group of reals. By using Lebesgue measure on its image under this map we induce a (one-dimensional) Haar measure on $\mathbb{R}^*$. In the above assertion μ_2 denotes the corresponding product measure on $\mathbb{R}^* \times \mathbb{R}^*$. Thus if $r \le s$, then the set of points (a, b) which lie in the box $x^r \le a$, $b \le x^s$ has μ_2 measure $(s - r)^2$; independent of x so-to-speak. Of course there is essentially only one Haar measure on $\mathbb{R}^*$, but it can be rescaled, and the precise rescaling was one of the difficulties. Need each marginal measure have the same rescaling? It was not obvious from the hypothesis (6.6). In the event the Fourier analytic context in which I was considering the law of large numbers again helped. The treatment of additive arithmetic functions given in my 1974 paper, [33], is here relevant.

It is clear from (6.7) that, as a function of z, $\alpha(z) + \omega(x)$ satisfies a (continuous) approximate functional equation of Cauchy type on $\mathbb{R}^*$. Since $\mu_2 E \to 0$ as $x \to \infty$, we can find a common pair (a, b) so that the relation (6.7) holds for both x, and x replaced by x^y. As a consequence $\omega(x^y) = \omega(x) + o(\beta(x))$ as $x \to \infty$, which is acceptable as a notion of small relative to $\beta(x)$.

In my 1975 paper I obtained for the $\alpha(x)$ under consideration a decomposition $\alpha(x) = \alpha_1(x) - \omega(x)$ where, for each fixed $y > 0$, $\alpha_1(x^y) = y\alpha_1(x) + o(\beta(x))$ is satisfied as $x \to \infty$. In particular

$$\alpha(x^{1+\varepsilon}) = \alpha(x) + \varepsilon(\alpha(x) - \omega(x)) + o(\beta(x)), \qquad x \to \infty,$$

for each $\varepsilon > -1$. This is an estimate of the type that arises in the study of stability in mechanics, or the perturbation of operators. It is not difficult to show that such a decomposition has the equivalent form: there is a $D(x)$ so that

$$\alpha(xt) = \alpha(x) + D(x)\log t + o(\beta(x)), \qquad x \to \infty,$$

uniformly on every interval $1 \le t \le x^k$, where $k > 0$. This may be viewed as a local asymptotic estimate.

The argument in my St. Louis paper is quite general, and may be used to treat functions defined on a wide class of semigroups, and with values in groups other than $\mathbb{R}$, such as $\mathbb{R}/\mathbb{Z}$. The method developed to derive the equation (6.7) also serves to treat analogues of the discrete equation (6.6), and is not tied to (the groups) $\mathbb{R}^*$ and $\mathbb{R}$. Consider the following example from $\mathbb{R}^*$ to $\mathbb{R}/\mathbb{Z}$. As earlier in this section, let $\|\beta\|$ denote the distance of the real β to a nearest integer. We change the summation condition for (6.6) to $\|f(p) - \alpha(x) + \alpha(x/p)\| > \varepsilon$, and replace the asymptotic continuity condition (6.8) by its analogue

$$\sup_{x/C \le w \le x} \|\alpha(w) - \alpha(x)\| \to 0, \qquad x \to \infty. \tag{6.9}$$

Then with only trivial changes the proof of (6.7) yields

$$\|\alpha(ab) - \alpha(a) - \alpha(b) - \omega(x)\| \to 0, \qquad x \to \infty, \tag{6.10}$$

with the same uniformities as before.

Important features of this method for treating these approximate functional equations are

(i) *The elimination of $f(p)$ by keeping a range of primes fixed and varying x.*

(ii) *The derivation of continuous analogues of the various discrete equations obtained, so as to permit manipulation of the variables using calculus.*

(iii) *Application of the transformations $u \mapsto zu$, $u \mapsto z/u$, $z > 0$, with respect to which the underlying measure (on $\mathbb{R}^*$) is invariant.*

In particular, the aim of applying the inversion $u \mapsto z/u$, otherwise expressed as 'turning a variable upside down', I held in the forefront of my considerations from the outset. It makes manifest that the approximate functional equation (6.6) is defined on $\mathbb{R}^*$ rather than on $\mathbb{R}$. The replacement of the primes p by a continuous variable in my 1975 paper, [34], I effected using only the elementary bound

$$\sum_{p \le x} \frac{1}{p} = \log\log x + c + O((\log x)^{-1}), \qquad x \ge 2.$$

If we allow sharper forms of the prime number theorem, then the condition (6.8) can be correspondingly weakened. Thus an error term $O((\log x)^{-B-1})$ here would allow the supremum at (6.8) to be taken over the shorter interval $x - Cx(\log x)^{-B} \le w \le x$ for a certain positive constant C.

These devices have proved effective in all subsequent considerations of discrete approximate functional equations obtained by the arithmetic function analytic method of Chapter 5.

This then, is what became of my will to find another approach to the study of additive arithmetic functions. In particular, I was able to solve a fifty-nine year old problem of Hardy and Ramanujan concerning the normal order of additive functions, [37], and settle in the affirmative a related conjecture of Narkiewicz, [36].

And what happened to my further aim, the study of the discrete derivative $f(n+1) - f(n)$? Omitting several stages, let us step ahead.

I gave another elaborate application of the general method of the stable dual in my 1985 book on arithmetic functions and integer products, [56]. It involves the equation

$$\sum_{x^\gamma < p \le x^\delta} \frac{1}{p}\left|f(p) - \alpha(x) + \alpha(x/p)\right|^2 = o(\beta(x)^2), \qquad x \to \infty, \tag{6.11}$$

where $0 < \gamma < \delta < 1$. I have changed the notation slightly to conform with the usage of the present section. Equation (6.11) is an L^2 version of the in-measure equation (6.6), save that it is restricted to the interval $(x^\gamma, x^\delta]$. It is more convenient (and was so in the book) to consider the L^1 form

$$\sum_{x^\gamma < p \le x^\delta} \frac{1}{p}\left| f(p) - \alpha(x) + \alpha(x/p)\right| = o(\beta(x)), \qquad x \to \infty, \tag{6.12}$$

obtained by applying the Cauchy–Schwarz inequality and an estimate from elementary number theory.

As with the earlier treatment in-measure, I eliminated between (6.12), and (6.12) with x replaced by z, $x \le z \le x^k$. There will be a common range of primes in the two summations provided $k\gamma < \delta$, and one obtains an estimate of the form

$$\int_{x^\gamma}^{x^\delta} \left|\alpha(z) - \alpha\Big(\frac{z}{u}\Big) - \Big\{\alpha(w) - \alpha\Big(\frac{w}{u}\Big)\Big\}\right| \frac{du}{u} = o(\beta(x)\log x) \tag{6.13}$$

uniformly for $x \le z$, $w \le x^k$. This is Lemma (9.4), Chapter 9 of [56]. By adapting the earlier argument to the shorter interval, but following similar steps, I deduced from (6.13) that for a suitably chosen constant η, $0 < \eta < 1$, and function $\lambda(x)$,

$$\int_{x^{\delta(1-\eta)}}^{x^\delta} \left|\alpha(x) - \alpha\Big(\frac{x}{u}\Big) - \alpha(u) - \lambda(x)\right| \frac{du}{u} = o(\beta(x)\log x), \; x \to \infty. \tag{6.14}$$

This is Lemma (9.7) of the same reference. In the more striking form

$$\int_{x^{\delta(1-\eta)}}^{x^\delta} \left|\alpha(z) - \alpha\Big(\frac{z}{u}\Big) - \alpha(u) - \lambda(x)\right| \frac{du}{u\log x} = o(\beta(x)) \tag{6.15}$$

which may be obtained immediately by combining Lemmas (9.4) and (9.7) of that reference (i.e., (6.13) with $w = x$, and (6.14) here), we see an L^1 analogue of the approximate functional equation (6.7). It is valid uniformly for $x \le z \le x^\tau$ with a certain $\tau > 1$. This last functional equation involves one integration but two variables: u over $x^{\delta(1-\eta)} \le u \le x^\delta$, and z over $x \le z \le x^\tau$.

By analogy with the treatment of $\omega(x)$ in the in-measure case, elimination between the case of (6.15) with x replaced by x^y, and (6.15) itself, shows that $\lambda(x^y) = \lambda(x) + o(\beta(x))$ as $x \to \infty$, at first for $1 \le y \le \tau$, and then, by iteration, for all positive y.

In fact this last together with the estimate (6.14) contains essentially all the available information. Combining the discrete form of (6.14) with the hypothesis (6.12) we see that

$$\sum_{x^{\delta(1-\eta)}<p\le x^{\delta}} \frac{1}{p}\Big|f(p)-\alpha(p)-\lambda(x)\Big| = o(\beta(x)), \qquad x\to\infty.$$

This shows that f may be viewed as the restriction of an asymptotically continuous function. Moreover, since the choice $f(p)=\alpha(p)+\lambda(x)$ is possible, indeed it arises naturally when $f(p)=\log p$, $\alpha(x)=\log x$, little further information can be gleaned from the hypothesis.

Exactly similar results may be obtained when, in both the hypothesis and the argument, the absolute value $|\ \ |$ on $\mathbb{R}$ is replaced by $\|\ \ \|$ on $\mathbb{R}/\mathbb{Z}$, and $\beta(x)$ by 1. This includes the analogue of (6.15):

$$\int_{x^{\delta(1-\eta)}}^{x^{\delta}} \left\|\alpha(z)-\alpha\left(\frac{z}{u}\right)-\alpha(u)-\lambda(x)\right\|\frac{du}{u\log x} = o(1), \qquad x\to\infty.$$

The essential point, carried out between Lemma (9.6) and Lemma (9.7) of my book, [56], is to detach the outer integral in the triple integral appearing there by rendering the inner double integral independent of the outer variable u. This I do by shrinking the region of integration. $\lambda(x)$ may be defined by a minimum rather than an average. The argument, from my 1975 paper, which establishes the existence of $\omega(x)$ $(=\lambda(x))$ also avoids any appeal to division.

In the application of the equation (6.12) that I consider in Chapter 9 of [56], I have the representation $\alpha(x)=\sum f(p)p^{-1}, p\le x$, to hand. It enables me to cut the range of summation in (6.12) down to the variable interval $(x^{\delta(1-\eta)}, t]$, and then use (6.14) to construct an approximate differential equation. This I solve to give a representation

$$\alpha(t) = G(x)\log t - \lambda(x) + o(\beta(x)) \tag{6.16}$$

uniformly for $x^{\delta(1-\eta)}\le t\le x^{\delta}$. Returned to the original inequality (6.12) we have

$$\sum_{x^{\gamma}<p\le x^{\delta}} \frac{1}{p}\Big|f(p)-G(x)\log p\Big| = o(\beta(x)),$$

and a similar version for (6.11). With this we may consider the original functional equations (6.11) and (6.12) to be solved.

This derivation of a local asymptotic estimate by means of a differential equation reminds me of the study of the perturbation of planetary orbits in celestial mechanics using differentiation with respect to the boundary coordinates, within the circle of ideas of the Hamilton–Jacobi equation, which I studied as part of my undergraduate curriculum.

I solved the equation (6.12), in the continuous form that I have given here, in 1980, whilst visiting Imperial College, London, on a Guggenheim Fellowship. I gave accounts of it and of its application, within the general method to the study of differences of additive arithmetic functions, that same Spring, at a number of places including Oxford, England; Ulm, Germany; Budapest, Hungary and Orsay, France. Highlights of my talk in the Journée de Théorie Analytique et Élémentaire des Nombres held in Orsay, June 2–3, 1980 are given in my paper in the proceedings, [45].

I rendered the whole procedure into inequalities, and had obtained the basic inequality

$$\sum_{\substack{q\le x\\(q,aA)=1}} \frac{1}{q}|f(q)-F(x)\log q|^2 \ll \sup_{x\le y\le x^c} y^{-1}|f(an+b)-f(An+B)|^2$$

for additive functions certainly by 1982. Here q runs through prime powers and $F(x)$ is an explicitly defined function of f and x. The four integers $a>0, b$, $A>0, B$, upon which the implied constant depends, satisfy $aB\ne bA$. An account of this inequality, together with the necessary abstract norm generalisations of the Bombieri–Vinogradov theorem to additive functions, the approximate functional equation and the discussion of a number of applications I gave at Orsay, France in June 1982, during the two day Colloque in honor of the official retirement of Professor H. Delange. A summary of this lecture appears in the proceedings, [50]. I give a complete detailed account of this application of the method and related matters, including the representation of rationals by products of given rationals, in my book, [56]. The results which I obtain in that volume concerning problems of Kátai, [108], and Wirsing, [172], as well as the torsion properties of certain groups can at the moment only be obtained by the method of the stable dual.

In the present volume I do not pursue the arithmetic particularities attached to the study of additive functions on sequences $(an+b)/(An+B)$. In part they may be viewed in my AMS memoir on the correlations of multiplicative and the sums of additive functions, [77]. Rather, beginning in Chapter 23, I set out to formulate and make precise the method for studying differences $f(n+1)-f(n)$ when norms other than that of the mean square are considered.

7

Exercises: Almost linear, almost exponential

When I began actively pursuing the application of approximate functional equations to number theory, in the early seventies, results of the Ulam–Hyers type were sparse. Moreover, they did not lend themselves to the problems which I had to hand.

It should be emphasised that the method of the stable dual is not concerned with the approximate functional equations that arise, for example, in the theory of the Riemann zeta function. In that theory approximate functional equations are established for certain given functions, mainly sums of exponentials. In a sense an analytic reciprocity law is derived. In the method of the stable dual an unknown function is assumed to satisfy a weak global constraint, and as far as possible the local nature of the function is then determined.

As applied to number theory the method of the stable dual typically gives rise to a complicated approximate functional equation involving several functions and many variables. The first step is to tease out an approximate equation of a more manageable type. This step depends upon the number theoretic and distributional properties of the objects under consideration. The appropriate notion of stability is then determined by the number theoretic application in view. My aim was usually towards an equation with continuous rather than discrete variables. Although by 1980 I had developed a tolerable technique for treating approximate functional equations arising in the study of arithmetic functions, I felt the need to better understand some of the arguments. The following results, from the Spring of 1980, illuminated some of the arguments that I had devised in/for [34], [54] and [56].

Theorem 7.1. *Let p be real, $p \geq 1$. Let the complex valued function $w(t)$ belong to the Lebesgue class $L^p(0,z)$ for each real $z > 0$. Then*

$$\lim_{z\to\infty} z^{-2} \int_0^z \int_0^z |w(x+y) - w(x) - w(y)|^p dx dy = 0,$$

if and only if for some A

$$\lim_{z\to\infty} z^{-1}\int_0^z |w(x) - Ax|^p dx = 0.$$

I indicate a proof that employs an approximate differential equation. Define

$$J(z) = \int_0^z w(t)dt.$$

1. Prove that

$$zJ(z) - 3\int_0^z J(u)du = o(z^2), \quad z \to \infty.$$

Hint: Consider the integral of $w(x+y) - w(x) - w(y)$ over the region $x + y \leq z$, $x \geq 0$, $y \geq 0$, and change to appropriate polar coordinates.

In terms of

$$J_1(z) = \int_0^z J(u)du,$$

exercise 1 shows that

$$z\frac{dJ_1}{dz} - 3J_1 = o(z^2), \quad z \to \infty.$$

2. Prove that for some constant C, $J_1(z) = Cz^3 + o(z^2)$ hence that $J(z) = 3Cz^2 + o(z)$, as $z \to \infty$.
3. Establish Theorem 7.1.
4. Under the initial hypotheses of Theorem 7.1, prove that

$$\lim_{z\to\infty} z^{-1}\int_0^z\int_0^z |w(x+y) - w(x) - w(y)|^p dxdy = 0$$

if and only if for some A, $f(x) = Ax$ almost surely on $x \geq 0$.

In a sense Theorem 7.1 is best possible. A proof by way of a simple differential equation, is given in [57]. Important features of the result are that the hypothesis is weak and of a global nature, and the argument may be given a quantitative aspect that is significant as soon as z is sufficiently large. Another presentation, with remarks concerning its relevance to a number theoretic problem, may be found in the Third Motive of [56].

A like characterisation of the complex exponential function apparently lies deeper.

Theorem 7.2. *For a real $p > 1$, let the complex-valued function $f(x)$ belong to the Lebesgue class $L^p(0, z)$ for each real $z > 0$, and satisfy*

$$\lim_{z\to\infty} e^{-\varepsilon z} \int_0^z \int_0^z |f(x+y) - f(x)f(y)|^p dx dy = 0 \tag{7.1}$$

for each fixed $\varepsilon > 0$.

Then either there is a (possibly complex) constant β so that

$$f(x) = e^{\beta x} \tag{7.2}$$

almost surely for $x \geq 0$, or

$$\lim_{z\to\infty} e^{-\varepsilon z} \int_0^z |f(x)|^p dx = 0 \tag{7.3}$$

for each fixed $\varepsilon > 0$.

Each of the conditions (7.2), (7.3) *guarantees the validity of* (7.1).

I indicate a proof of this theorem that depends upon the possibility of analytically continuing the solution of certain Riccati differential equations in the complex plane.

Since $||a| - |b|| \leq |a - b|$, condition (7.1) is also satisfied by $|f(x)|$. For the time being I shall assume that $f(x)$ is real and non-negative. It is convenient to define

$$J(z) = \int_0^z f(x)dx, \quad z \geq 0.$$

5. Prove that $J(z) \leq e^{Bz}$ holds for some non-negative B and all $z \geq 3$.
Hint: Show that

$$J(2z - 1) \leq (J(z))^2 + c_1 e^z$$

for some c_1 and all $z \geq 1$.

6. Prove that

$$J(z + h) - J(z) \ll h^\theta e^{Bz}$$

with $\theta = \min(1 - p^{-1}, 1/2)$, uniformly for $z \geq 0$, $0 \leq h \leq 1$.

For complex $s = \sigma + i\tau$, $\sigma = \mathrm{Re}\,(s)$, define the Laplace transform

$$w = w(s) = \int_0^\infty e^{-sx} f(x)dx. \tag{7.4}$$

Exercise 4 shows w to be well defined in the half-plane $\sigma > B$. If the integral converges for every positive real value of s, then we shall already have the result

$$\lim_{z\to\infty} e^{-\varepsilon z} \int_0^z f(x)dx = 0$$

for each $\varepsilon > 0$.

7. Prove that under this last circumstance, proposition (7.3) of Theorem 7.2 is valid.

Hint: Consider

$$\int_0^z |J(z)f(x) - \{J(z+x) - J(x)\}|^p dx.$$

We may therefore assume the integral at (7.4) to converge absolutely for $\mathrm{Re}(s) > \alpha > 0$, and not for $\mathrm{Re}(s) < \alpha$. In particular, the convergence is uniform in each half-plane $\sigma \geq \alpha + \delta$, $\delta > 0$, and so defines an analytic function $w(s)$ in the half-plane $\sigma > \alpha$.

Define

$$g(s) = \int_0^\infty \int_0^\infty e^{-s(x+y)}\{f(x+y) - f(x)f(y)\}dxdy.$$

8. Prove that $g(s)$ is analytic in $\sigma > 0$, bounded above uniformly in every half-plane $\sigma \geq \delta > 0$. Moreover,

$$\lim_{|\tau|\to\infty} g(\sigma + i\tau) = 0$$

uniformly in every strip $(0 <)\sigma_1 \leq \sigma \leq \sigma_2(< \infty)$.

Hint: Besides the standard Riemann–Lebesgue argument, consider that of Phragmén–Lindelöf in the theory of complex variables, e.g., Titchmarsh [163] p. 180.

9. Prove that w satisfies the Riccati differential equation

$$\frac{dw}{ds} + w^2 = -g(s)$$

in the half-plane $\sigma > \alpha$.

Hint: $x = r(\cos\theta)^2$, $y = r(\sin\theta)^2$.

The line is drawn: Analytically continue a solution w of this differential equation into the strip $0 < \mathrm{Re}(s) < \alpha$. This is to adopt Riemann's procedure for treating the function $\zeta(s) = \sum_{n=1}^{\infty} n^{-s}$, and its logarithmic derivative $\zeta'(s)/\zeta(s)$, associated with the primes. Riemann had the structural properties of the integers at his disposal; we have a non-linear differential equation. And a question suggests itself.

10. Define b_n to be $-\log n + \tau(n) - 2\gamma$ if $n > 1$, and b_1 to be $(1-\gamma)^2$, where γ is Euler's constant. Prove by elementary means that

$$\sum_{n\leq x} b_n \ll x^{1/2}, \quad x \geq 1.$$

Hence show that $\zeta(s)-\gamma$ satisfies a Riccati differential equation of the same type as that in exercise 9, save we know only that $g(s)$ is analytic in the half-plane $\sigma > 1/2$, and $\ll |s|(\sigma - 1/2)^{-1}$ there.

In pursuit of f we shall formally continue the Riemann zeta function across the line $\sigma = 1$. Some trouble is to be expected!

Let $s_0 = \sigma_0 + i\tau_0$ be a point in the half-plane $\sigma > \alpha$. Let c_1, c_2, t_1 be positive numbers and define a rectangle

$$\Delta : \alpha - c_1 \leq \sigma \leq \alpha + c_2, \quad |\tau - \tau_0| \leq t_1.$$

We shall assume that $\sigma_0 < \alpha + c_2$, so that s_0 lies inside Δ. It is convenient to use L to denote $c_1 + c_2 + 2t$, so that $2L$ is the length of the perimeter of Δ.

Let M be a positive number. The space S of all functions $h(x, y)$, considered as functions of the pair (x, y), that are continuous and bounded by M on Δ, is complete with respect to the sup norm

$$\|h\| = \sup |h(x, y)|, \quad (x, y) \in \Delta.$$

11. Prove that if $3LM \leq 1$, $3L\|g\| \leq M$ and w_0 is a complex number not exceeding $M/3$ in absolute value, then the map T given by

$$(Th)(s) = w_0 + \int_{s_0}^{s} \{-h(s)^2 - g(s)\} ds,$$

the integration taken along the half-rectangle T : s_0 to $\sigma_0 + i\tau$ to s, is a contraction map on S into itself.

There is a unique fixed point w for which

$$w(s) = w_0 - \int_{s_0}^{s} \{w(s)^2 + g(s)\} ds.$$

If, instead, we integrate along the (other) half-rectangle $\widetilde{\Gamma}$, s_0 to $\sigma + i\tau_0$ to s, then we obtain a fixed point $\tilde{w}$ of the corresponding map $\widetilde{T}$.

12. Prove that $w = \tilde{w}$.

Hint: Prove $T^n(M)$, $\widetilde{T}^n(M)$, $n = 1, 2, \ldots$, analytic inside Δ.

As a consequence of exercise 12,

$$\int_{\Gamma} \{w(s)^2 + g(s)\} ds = \int_{\widetilde{\Gamma}} (w(s)^2 + g(s)\} ds.$$

13. Prove that w is analytic in the interior of Δ.

Hint: Begin by proving that $\partial w/\partial \sigma$, $-i\partial w/\partial \tau$ exist.

14. Prove that for each $\delta > 0$ there is a real τ_1 so that $w(s)$ may be analytically continued into the semi-infinite strip(s), $\delta < \sigma < \alpha+1$, $|\tau| > \tau_1$. Moreover, $w(s)$ is uniformly bounded in these strips.

Hint: Fix σ_0 and slide τ_0.

The next three results concerning differential equations over the complex plane go much as their counterparts over the real line.

Let D be a connected open set in the complex plane. A non-trivial function analytic on D can vanish at most finitely many times on each compact subset of D.

15. Let p, q be analytic on D. Prove that any three solutions y_j, $1 \le j \le 3$, to $y' = py+q$, analytic on D, satisfy a relation $(y_3-y_1)(y_2-y_1)^{-1} =$ constant there. This is interpreted as meaningful when $y_1 \neq y_2$. Given solutions y_1, y_2 to the equation, we may essentially parametrise further solutions by the value of this constant.

16. Let r be a further function analytic on D. Let y_j, $1 \le j \le 3$, be solutions to the Riccati equation $y' = py^2 + qy + r$, analytic on D. Prove that for any further solution y_4, analytic on D, the cross-ratio $(y_4 - y_2)(y_3 - y_1)(y_4 - y_1)^{-1}(y_3 - y_2)^{-1}$ is constant. Again, this statement is to be suitably interpreted.

Hint: Set $y = y_1 + v^{-1}$.

17. Prove that the above results still hold if the solutions y_j are allowed to be meromorphic on D. In particular, meromorphic solutions to the complex Riccati equation in exercise 16 are essentially parametrised by the value of a cross–ratio.

To analytically continue $w(s)$ into the box $\delta < \sigma < \alpha + 1$, $|\tau| \le \tau_1$, precisely in the style of the exercises leading to 14 is not possible, since we expect a singularity at the point $s = \alpha$. Even if we didn't, the earlier argument need not apply directly since larger values of $\|g\|$ and $|w(s_0)|$ are to be expected. These will be compatible with the conditions of exercise 11 only if L is small; not allowing us to reach as far left as the line $\sigma = \delta$.

We apply our above space S with a box Δ_1 defined by

$$\sigma_0 - \frac{1}{12\sqrt{c_0}} \le \sigma \le \sigma_0 + \frac{1}{12\sqrt{c_0}}, \quad |\tau - \tau_0| \le \frac{1}{12\sqrt{c_0}},$$

in our old notation $c_1 = -\sigma_0 + \alpha + (12c_0^{1/2})^{-1}$, $c_2 = \sigma_0 - \alpha + (12c_0^{1/2})^{-1}$, $t_1 = (12c_0^{1/2})^{-1}$. Assume that $\sigma_0 - \alpha \le (24c_0^{1/2})^{-1}$. For the upper bound M in the definition of the space S we take $c_0^{1/2}$. Then $3L = c_0^{-1/2}$, so that $3LM = 1$. Moreover, $3L\|g\| \le c_0^{-1/2}c_0 = M$.

18. Prove that for $3|\lambda| \le c_0^{1/2}$ there is a unique solution to the Riccati equation $w' + w^2 = -g$, valid in the box Δ_1, analytic in the interior of Δ_1, and satisfying the boundary condition $w(s_0) = \lambda$.

19. In the notation of exercise 14, prove that $w(s)$ may be meromorphically continued into the strip $\delta < \sigma < \alpha + 1, |\tau| < \tau_1 + 1$. Prove that any pole of w in this region must be simple and have residue 1 (surprise!). Moreover, for each $\theta > 0$ there is a line segment $\mathrm{Re}\,(s) = \sigma_1$, $|\tau| < \tau_1 + 1$, $\delta < \sigma_1 < \delta + \theta$, on which $w(s)$ is uniformly bounded.

Let $\varepsilon > 0$ be given, and choose δ so that $\varepsilon/2 \le \delta \le 2\varepsilon/3$ and $w(\delta + i\tau)$ is bounded uniformly for all real τ. Let $\rho_1, \ldots, \rho_n$ denote the poles of w in the half-plane $\sigma > \delta$. Define

$$J_1(z) = \int_0^z J(u)du, \quad z \ge 0.$$

20. Prove that $J_1(z)$ has Laplace transform $w(s)s^{-2}$, and an approximate representation

$$J_1(z) = \sum_{j=1}^{n} \rho_j^{-2} e^{\rho_j z} + O(e^{\varepsilon z}), \quad z \ge 0.$$

The formal Fourier/Laplace inversion of $s^{-1}w(s)$,

$$J(z) = \frac{1}{2\pi i} \int_{c-i\infty}^{c+i\infty} \frac{e^{sz}w(s)}{s}\, ds, \quad c > \alpha,$$

gives rise to an integral not necessarily absolutely convergent.

21. Prove that

$$J(z) = \sum_{j=1}^{n} \rho_j^{-1} e^{\rho_j z} + O(e^{\varepsilon z}), \quad z \ge 0.$$

Hint: Deform the contour $c - i\infty \to c + i\infty$ into $c - i\infty \to c - it \to \delta - it \to \delta + it \to c + it \to c + i\infty$ and take the line segments $\mathrm{Re}\,(s) = c$, $|\tau| \ge t$ together.

22. Let

$$\left| \sum_{j=1}^{k} c_j e^{d_j z} \right| \le A e^{wz}$$

hold for some real A, w, and all real $z \ge 0$. Prove that for each j with $c_j \ne 0$, $\mathrm{Re}\,(d_j) \le w$.

23. Prove that without loss of generality the ρ_j in exercise 21 may be assumed to have a common real part.

24. In our present circumstances $w(s)$ is the Laplace transform of a non-negative function. Prove that its defining integral diverges at $\sigma = \alpha$. This

is the analogue for Dirichlet series of a classical result in the theory of power series.

25. Prove that

$$J(z) = \alpha^{-1}e^{\alpha z} + O(e^{\varepsilon z}), \quad z \geq 0,$$

is valid for each fixed $\varepsilon > 0$.

26. Prove that $f(x) = e^{\alpha x}$ almost surely on $x \geq 0$.

Towards the general case we have shown that unless condition (7.3) is satisfied, $|f(x)|$ almost surely has the value $e^{\alpha x}$.

27. Prove that either $f(x)$ has an almost sure representation $e^{\beta x}$ on $x \geq 0$, or for each positive ε

$$\int_0^z f(x)dx \ll e^{\varepsilon z}, \quad z \geq 0.$$

28. Prove that the second possibility in exercise 27 leads to the bound $J(z) \ll e^{(\varepsilon - 2\alpha)z}$, $z \geq 0$, valid for each positive ε. Moreover, $w(s)$ may be analytically continued into the half-plane $\sigma > -2\alpha$ and has a zero at the origin.

29. Prove that if both (7.2) and (7.3) fail, then w vanishes on the imaginary axis and f is almost surely zero on $x \geq 0$.

Hint: Consider $e^{i\tau x}f(x)$, τ real.

This contradiction completes a proof for Theorem 7.2, taken from the proof in [49].

30. Let $h(s)$ be analytic in $\sigma = \text{Re}\,(s) > 1$. Prove that if $\frac{dh}{ds} + h^2$ is analytically continuable into $\sigma > 1 - \delta$ for some $\delta > 0$, then h is meromorphically continuable into the same region. In particular, approximate functional relations between the coefficients of a Dirichlet series can sometimes serve to analytically continue the series. To pursue this topic would lead us too far afield.

Before we leave the study of approximate functional equations, let us look back to the results of Chapter 6. In view of my remarks concerning the estimate (6.14) it should be possible to derive the asymptotic representation (6.16) using the asymptotic constancy of λ, but without the intervention of any particular representation for the function $\alpha(x)$ in terms of the $f(p)$. Indeed, there are several ways to deduce such an estimate from (6.15). To set out from (6.14) is more challenging. The following method is almost formal. We begin with (6.14):

$$\int_{x^{\delta(1-\eta)}}^{x^\delta} \left|\alpha(x) - \alpha\left(\frac{x}{u}\right) - \alpha(u) - \lambda(x)\right| \frac{du}{u} = o(\beta(x)\log x), \quad x \to \infty,$$

where α, β, λ are measurable complex valued functions defined for $x \geq 1$; $\beta(x) \geq 1$ and for each positive y, $\beta(x^y) \ll \beta(x)$. Besides this, λ is asymptotically constant: for each positive y, $\lambda(x^y) - \lambda(x) = o(\beta(x))$ as $x \to \infty$.

31. Prove that the function

$$Y(t) = \int_{x^{\delta(1-\eta)}}^{t} (\alpha(u) + \lambda(x)) \frac{du}{u}$$

satisfies

$$Y(t) - Y\left(\frac{x}{t}\right) = (\alpha(x) + \lambda(x)) \log t + H(x) + o(\beta(x) \log x), \quad x \to \infty,$$

uniformly for $x^{\delta(1-\eta)} \leq t \leq x^{\delta}$. Here $H(x)$ is a function of x alone.

32. Prove that

$$Y(t\theta) - Y(t) = A \log t + B + o(\beta(x) \log x), \quad x \to \infty,$$

where $A = A(\theta) = \alpha(x\theta) - \alpha(x)$, B is a function of x, θ alone, and the relation holds uniformly for $x^{-\delta\eta/4} \leq \theta \leq x^{\delta\eta/4}$, $x^{\delta(1-3\eta/4)} \leq t \leq x^{\delta(1-\eta/4)}$. Only here do we apply the asymptotic constancy of λ.

33. Prove that as $x \to \infty$,

$$\{A(\theta_1\theta_2) - A(\theta_1) - A(\theta_2)\} \log t$$

differs from a function of x, θ_1, θ_2 alone by $o(\beta(x) \log x)$, uniformly for $x^{-\delta\eta/8} \leq \theta_j \leq x^{\delta\eta/8}$, $j = 1, 2$, $x^{\delta(1-5\eta/8)} \leq t \leq x^{\delta(1-3\eta/8)}$.

34. Prove that

$$A(\theta_1\theta_2) = A(\theta_1) + A(\theta_2) + o(\beta(x)), \quad x \to \infty,$$

uniformly for $x^{-\delta\eta/8} \leq \theta_j \leq x^{\delta\eta/8}$, $j = 1, 2$.

35. Prove that

$$\alpha(xt) = F(x) \log t + \alpha(x) + o(\beta(x)), \quad x \to \infty,$$

for some function F of x alone, uniformly for $x^{-\delta\eta/8} \leq t \leq x^{\delta\eta/8}$.

For a treatment of the equation (6.12) beginning along the lines of that given in my book, [56], but replacing the derivation of the approximate differential equation there by the use of a local L^∞ approximate Cauchy functional equation on $\mathbb{R}^*$, see Hildebrand, [102].

The account in the previous chapter of the approximate functional equation (6.12) assumes $k > 1$ given (in (6.13), say,) and γ fixed at a value sufficiently small that the subsequent argument applies. This is the case that arises in every application of the general method of which I know. In fact the treatment of approximate functional equations presented in Chapter 9 of my book, [56], is applicable for any fixed pair of values (γ, δ) which satisfy $0 < \gamma < \delta < 1$. A better choice of the parameters appearing in Lemma (9.6) of that account is then $h = ab/\delta$, $b = \delta^2(1 - \delta + \delta^2)^{-1} - \tau^{1/2}$. Lemma (9.5) there holds uniformly for $x^{h-\tau} \le u \le x^{h+\tau}$, (ignoring ℓ), and a may be given any positive value. Then Lemma (9.6) holds uniformly for z in the range $x^\theta \le z \le x^{\theta+\tau}$, $\theta = a(b+\delta(1-b))$, and the choice $a\delta = 1-\delta+\delta^2$ allows the treatment to continue with the differential equation in Lemma (9.8) and Lemma (9.9). As before τ is chosen small; z lies in a shorter range, but a range that still has μ_1 measure independent of x.

The whole process for treating approximate functional equations can be somewhat refined. To gain more from a weaker hypothesis the bottom end x^γ of the range of integration is lowered. As γ is lowered, and the variable u allowed to move away from x, the measure $du/(u \log x)$ becomes increasingly coarse. In fact from (6.12) the method of proof, involving only integration by parts, actually yields

$$\int_{x^\gamma}^{x^\delta} \left| \alpha(z) - \alpha\left(\frac{z}{u}\right) - \left\{ \alpha(w) - \alpha\left(\frac{w}{u}\right) \right\} \right| \frac{du}{u \log u} = o(\beta(x)).$$

For fixed γ, $du/(u \log u)$ may be replaced by $du/(u \log x)$ at little loss, and at the great gain that the second measure is a Haar measure, and so invariant under the group operations on $\mathbb{R}^*$. When I began to write Chapter 9 of my book, [56], the chapter dealing with approximate functional equations of the type (6.12), I in fact used the measure $du/(u \log u)$. The complications of dealing with expressions of the form $du/(u \log(x/u))$ soon put me off, so I introduced a wasteful factor of $\log p$ into the hypothesis, and reduced myself to using the measure $du/(u \log x)$. The treatment of approximate functional equations with the more sensitive measure can still be carried out provided that after each application of a group operation, the appropriate factors $\log z/u$, $\log x/u$, say, are used to replace one another. This is done in the proofs of Theorems 6.1, 6.4 and 6.6 in the present text.

I sketch an alternative procedure. The solution of the approximate functional equation (6.12) employs a (small) bound for

$$\Delta = \int\int_B \int_{x^\gamma}^{x^\delta} \psi(z, w, u) \frac{du}{u \log x} d\mu(z, w)$$

where $\psi(z, w, u) = |\alpha(z) - \alpha(z/u) - \alpha(w) + \alpha(w/u)|$, the set B does not depend upon the variable u, the measure $d\mu(z, w)$ is the sum of two further measures, one purely atomic with a(n absolutely) bounded number of atoms, the other a multiple of $z^{-1}w^{-1}dzdw$, a product of Haar measures.

The hypothesis implicit in the treatment views the corresponding integral, with $(u \log u)^{-1} du$ in place of $(u \log x)^{-1} du$, as small. We cover the range $(x^\gamma, x^\delta]$ by intervals $(x_{\nu+1}, x_\nu]$, $x_\nu = \exp(\delta\theta^\nu \log x)$, where $0 < \theta < 1$, $j = 0, 1, \ldots, r-1$, $\delta\theta^r = \gamma$; and choose a value of ν so that

$$\int\int_B \int_{x_{\nu+1}}^{x_\nu} \psi(z, w, u) \frac{du}{u \log u} d\mu(z, w)$$
$$\leq \frac{1}{r} \int\int_B \int_{x^\gamma}^{x^\delta} \psi(z, w, u) \frac{du}{u \log u} d\mu(z, w).$$

This allows a treatment replacing x^γ, x^δ in Δ by $x_{\nu+1}, x_\nu$ respectively, the measure $(u \log x)^{-1} du$ by $(u \log x_\nu)^{-1} du$, and with the upper bound on Δ 'improved' by a factor $\log \frac{1}{\theta} \left(\log \frac{\gamma}{\delta}\right)^{-1}$.

Nonetheless, underlying all these treatments is the invariance of the Haar measure on $\mathbb{R}^*$ under translation and inversion.

A form of Theorem 6.6 with $\psi = \phi$ (and so no problems with the vanishing of ψ), the smaller uniformity $1 \leq w \leq P$ in place of $P^{-\tau} \leq w \leq P^\tau$, and without the localisation of α in terms of ϕ, was given by Hildebrand [103]. He gives a direct proof, without reducing to theorems of the type 6.1, 6.4, and in place of appeal to results of the type Lemma 6.3, Lemma 6.5 here, devises a variant of the original argument of Hyers, [106]. I have adopted a presentation in part similar to his because it simplifies the exposition of the proof. The inversion of the studied variable then comes in the application: generally the study of $\alpha(z)$ with z near to x is replaced by that of $\phi(u) = \alpha(x/u)$ with u near to 1. There is a loss here, however. In the application to functions on intervals $(x - y, x]$, which is the case that Hildebrand has to hand, the simplified presentation does not allow the independent variation of end points $x, x - y$.

The complete method for treating approximate functional equations, as exemplified in Theorems 6.1, 6.4, and 6.6, may be applied whenever the function f takes values in an abelian group G which possesses a map, from G into the non-negative reals, and satisfying the triangle inequality. It is sufficient that G have a translation invariant metric. One is then reduced to the study of approximate versions of the Cauchy functional equation from $\mathbb{R}^*$ to G.

In the following chapters I begin to illustrate the method of the stable dual by applying it to the study of additive and multiplicative functions on the natural numbers.

8

Additive functions of class $\mathcal{L}^\alpha$. A first application of the method

A complex valued arithmetic function h belongs to the class $\mathcal{L}^\alpha(\mathbb{Z})$, $\alpha > 0$, or more shortly, $\mathcal{L}^\alpha$, if

$$\limsup_{x\to\infty} x^{-1} \sum_{n\le x} |h(n)|^\alpha$$

is finite. These functions form a vector space over the complex numbers.

Theorem 8.1. *A complex valued additive function f belongs to the class $\mathcal{L}^\alpha(\mathbb{Z})$ with $\alpha \ge 2$ if and only if the series*

$$\sum_q q^{-1}|f(q)|^2, \qquad \sum_q q^{-1}|f(q)|^\alpha,$$

taken over the prime powers, converge, and the partial sums

$$\sum_{q\le x} \frac{1}{q}\left(1 - \frac{1}{q_0}\right) f(q)$$

are bounded uniformly in x. When $1 < \alpha < 2$, there is a similar criterion, but with summation conditions $|f(q)| \le 1$, $|f(q)| > 1$ attached to the two infinite series, respectively.

Proof. I treat the more complicated case, $1 < \alpha < 2$, and illustrate both the method of the stable dual and its background philosophy.

I begin in the opposite way to the usual, i.e., not by demonstrating the sufficiency of the conditions, but by deriving their necessity. Thus I begin with the one defined condition. Assume that f belongs to $\mathcal{L}^\alpha(\mathbb{Z})$, $1 < \alpha < 2$.

We set $a_n = f(n)$ in the appropriate dual of the generalised Turán–Kubilius inequality given in Theorem 2.1, in the notation of Chapter 3, the dual of the operator $A_{\alpha'}$ between $L^2 \cap L^{\alpha'}$ and $M^{\alpha'}$. Why?

The long answer first. If we view f in the functional analytic model of Chapter 3, then the hypothesis asserts that $\|A_\alpha f\| \ll 1$ uniformly in $x \ge 1$.

Since the operators $A'_{\alpha'}$ have uniformly bounded norms, $\|A'_{\alpha'}A_\alpha f\| \ll 1$. The operator $A'_{\alpha'}A_\alpha$ from $L^2 \cap L^\alpha$ into itself has f as an approximate eigenvector, with eigenvalue zero. Note that we may view $\|f\|$ with the norm in $(L^2 \cap L^{\alpha'})'$ as large, otherwise we are done. Since the spectrum of $I - A'_{\alpha'}A_\alpha$ and so $A'_{\alpha'}A_\alpha$ can be determined with some accuracy, we may estimate $\|f\|$. I show how this may be carried out in my paper, [59], and in Chapter 21 of the present volume. Here I note that the use of stability is then avoided since the approximate functional equations are susceptible to a direct functional analytic approach. To the extent that it adopts the philosophy of Chapter 1, the proof structure still follows that of the method of the stable dual. In this view the individual additive function is considered in the light of the study of all additive functions.

The short answer next. The choice $a_n = f(n)$ in the second inequality of Theorem 3.1 allows us to employ our hypothesis immediately. The resulting restriction upon the trapped sums of the a_n necessarily embodies some information concerning the function f, if only we can untangle it. This is the philosophy of the great chess player Alekhine: 'When all else fails, put yourself in jeopardy'.

And now the proof. We know that each of the sums

$$\sum_{q\le x}{}^{*} q^{\alpha-1}\left|\frac{1}{x}\sum_{\substack{n\le x\\ n\cong 0\,(\mathrm{mod}\,q)}} f(n) - \frac{1}{q}\left(1-\frac{1}{q_0}\right)\frac{1}{x}\sum_{n\le x} f(n)\right|^\alpha$$

$$\sum_{q\le x}{}^{**} q\left|\frac{1}{x}\sum_{\substack{n\le x\\ n\cong 0\,(\mathrm{mod}\,q)}} f(n) - \frac{1}{q}\left(1-\frac{1}{q_0}\right)\frac{1}{x}\sum_{n\le x} f(n)\right|^2$$

is bounded uniformly in x, where * indicates that summation is confined to those prime powers q for which the inequality

$$\frac{q}{x}\left|\sum_{\substack{n\le x\\ n\cong 0\,(\mathrm{mod}\,q)}} f(n) - \frac{1}{q}\left(1-\frac{1}{q_0}\right)\sum_{n\le x} f(n)\right| > \left(\frac{1}{x}\sum_{n\le x}|f(n)|^\alpha\right)^{1/\alpha} \tag{8.1}$$

is satisfied, and ** that it is confined to those for which the opposite condition holds. The additive nature of f allows a representation

$$\sum_{\substack{n\le x\\ q\| n}} f(n) = f(q)\left\{\left[\frac{x}{q}\right] - \left[\frac{x}{qq_0}\right]\right\} + \sum_{r\le x/q} f(r) - \sum_{\substack{r\le x/q\\ (r,q)>1}} f(r). \tag{8.2}$$

By Hölder's inequality and the initial hypothesis this last sum is

$$\ll \left(\sum_{\substack{r \le x/q \\ (r,q)>1}} 1 \right)^{1/\alpha'} \left(\sum_{\substack{r \le x/q \\ (r,q)>1}} |f(r)|^\alpha \right)^{1/\alpha} \ll xq^{-1}q_0^{-1/\alpha'}.$$

Altogether, if

$$m(y) = y^{-1} \sum_{r \le y} f(r),$$

then

$$\sum_{q \le x}{}^{*} \frac{1}{q} \left| \frac{q}{x} f(q) \left\{ \left[\frac{x}{q}\right] - \left[\frac{x}{qq_0}\right] \right\} + m\left(\frac{x}{q}\right) - m(x) \right|^\alpha, \tag{8.3}$$

$$\sum_{q \le x}{}^{**} \frac{1}{q} \left| \frac{q}{x} f(q) \left\{ \left[\frac{x}{q}\right] - \left[\frac{x}{qq_0}\right] \right\} + m\left(\frac{x}{q}\right) - m(x) \right|^2 \tag{8.4}$$

are bounded uniformly in $x \ge 1$.

The coefficient of $f(q)$ in these expressions is awkward to deal with, and depends upon x. In its first form the method of the stable dual requires the coefficient of $f(q)$ to be independent of x. We cannot replace the coefficient by the (valid) estimate $(1 - q_0^{-1}) + O(qx^{-1})$ without knowledge of the size of f on the prime powers, the very matter of our intent.

One way over this difficulty is to modify the definition of the operator $A'_{\alpha'}$ so that the dual inequality attached to it has weight $x^{-2}([x/q] - [x/qq_0])$ where at present it has $x^{-1}q^{-1}(1 - q_0^{-1})$. However, this is again the long answer.

Instead, I note that by (8.2) a typical inner sum

$$\sum_{\substack{n \le x \\ n \cong 0 \, (\mathrm{mod}\, p)}} f(n) - \frac{1}{q}\left(1 - \frac{1}{q_0}\right) \sum_{n \le x} f(n)$$

has the estimate

$$\left(\left[\frac{x}{q}\right] - \left[\frac{x}{qq_0}\right] \right) f(q) + O\left(\frac{x}{q}\right).$$

For a suitable positive number γ, the restriction $|f(q)| > \gamma$ not only amply guarantees the condition (8.1) of *, but also guarantees that a typical inner sum of the expression (8.3) will exceed $x|f(q)|/(4q)$ in absolute value. We obtain at once the convergence of the series $\sum q^{-1}|f(q)|^\alpha$ taken over the prime powers for which $|f(q)| > \gamma$. The ability to gain global control over the large values of the function under consideration is a characteristic and important feature of the method of the stable dual.

By further confining the summation to those prime powers for which $|f(q)| \le \gamma$ we may replace the coefficients of $f(q)$ in the expressions (8.3) and (8.4) by 1.

If $K \ge 6$, $P \ge 2$, $P^K \le x$, then applications of Hölder's inequality show that in preparation for an application of Theorem 6.1

$$\sum_{P<q\le P^K} \frac{1}{q}\left| f(q) + m\left(\frac{x}{q}\right) - m(x)\right| \ll (\log K)^{\frac{1}{2}}.$$

At first this approximate functional equation can be derived only for those q for which $|f(q)| \le \gamma$. However, in view of the information already obtained concerning the values of $|f(q)| > \gamma$, this restriction may be dropped.

I set $P = (\log\log x)^3$, $P^K = x^{1/3}$, $\tau = K^{1/2}$, $\varepsilon = (\log\log x)^{-1}$, $\psi(w) = \phi(w) = m(x/w)$ in Theorem 6.1. In view of the representation

$$m(y) = y^{-1}\sum_{q\le y} f(q)\Big\{[yq^{-1}] - [yq^{-1}q_0^{-1}]\Big\},$$

and the membership of $\mathcal{L}^\alpha$ by f, an application of Hölder's inequality gives

$$m(y) = \sum_{q\le y} f(q)q^{-1}(1 - q_0^{-1}) + O((\log y)^{-1/\alpha'}).$$

From this we readily obtain the desired continuity condition upon ϕ. Hence there is a representation

$$m(x) - m(x/q) = -D\log q + O(\varepsilon),$$

uniformly for $2 \le q \le P^\tau$. The function D is given by $D\tau\log P = \phi(P^\tau) - \phi(1)$, which from the hypothesis is bounded uniformly in x. Thus over the range $2 \le q \le P$, $D\log q \ll \tau^{-1} \ll \varepsilon$. From (8.3) and (8.4) the sums

$$\sum_{\substack{q\le P\\ |f(q)|\le\gamma}} q^{-1}|f(q)|^2, \qquad x \ge 1,$$

are uniformly bounded.

Arrived at the convergence of the two series involving f on the prime powers we look back at the appropriate generalisation of the Turán–Kubilius inequality to see what control this gives us over the values of f on the natural numbers. The moment-sums

$$x^{-1}\sum_{n\le x}\left|f(n)-\sum_{q\le x}\frac{f(q)}{q}\left(1-\frac{1}{q_0}\right)\right|^{\alpha}$$

are bounded uniformly for $x \ge 1$.

The completion of the proof is now rapid.

This is a proof that I think would have pleased that last great exemplar of the ancient tradition of Greek geometry, Pappus of Alexandria, fourth century A.D.

It is interesting to compare my present treatment of Theorem 8.1 with the treatment that I gave in my original paper, [42], which also employed the duals of generalised Turán–Kubilius inequalities.

An independent proof of Theorem 8.1 along different lines may be found in Hildebrand and Spilker, [105].

9

Multiplicative functions of the class $\mathcal{L}^\alpha$: First Approach

Let g be a multiplicative function with values in the complex unit disc. In this century, three results concerning the asymptotic behaviour of such functions have become classical.

(i) Delange proved that the (asymptotic) mean value

$$\lim_{x\to\infty} x^{-1} \sum_{n\le x} g(n)$$

exists and is non-zero if and only if the series $\sum p^{-1}(1 - g(p))$, taken over the primes, converges, and for some positive r, $g(2^r) \neq -1$. If only the convergence of the series over the primes is assumed, then the mean value exists and has the value

$$\prod_p \left(1 - \frac{1}{p}\right)\left(1 + \frac{g(p)}{p} + \frac{g(p^2)}{p^2} + \cdots\right),$$

possibly zero.

(ii) Assuming that g is real valued and that the series $\sum p^{-1}(1 - g(p))$ diverges, Wirsing proved that g has mean value zero. In particular, a real valued multiplicative function with values in the interval $[-1, 1]$ always possesses a mean value. This settled an old question raised by Erdős and Wintner.

(iii) Halász proved that the divergence of the series $\sum p^{-1}(1 - \operatorname{Re} g(p)p^{i\tau})$ for each real τ guarantees that a complex valued g has mean value zero.

Each of these authors established rather more. Thus Halász showed that there are always constants A, α and a slowly-oscillating function L of absolute value 1, so that

$$\sum_{n\le x} g(n) = Ax^{1+i\alpha} L(\log x) + o(x), \qquad x \to \infty,$$

as had been conjectured by Wirsing.

An account of these results and their historical background may be found in [41] Chapters 6 and 19. There they are established using Fourier analysis in the complex plane in the style of Halász. Alternative proofs will be furnished in Chapters 9, 10, 11, 14 and 15 of the present work. The celebrated prime number theorem is formally a consequence of Wirsing's theorem.

The condition $|g(n)| \le 1$ has an important consequence: either the series $\sum p^{-1}(1 - \operatorname{Re} g(p))$ converges, or it doesn't. If we abandon the restriction on the size of g, then this dichotomy is lost. We can no longer classify the asymptotic behaviour of the means of g in a straightforward manner.

In the present chapter I consider multiplicative functions which belong to the class $\mathcal{L}^\alpha$, $\alpha > 1$. I apply the method of the stable dual to both formulate and derive appropriate analogues of the classical results. The functions considered by Delange, Wirsing and Halász belong to every class $\mathcal{L}^\alpha$.

Theorem 9.1. *In order for a multiplicative function g to belong to $\mathcal{L}^\alpha$, $\alpha > 1$, and possess a non-zero mean value*

$$\lim_{x\to\infty} x^{-1} \sum_{n\le x} g(n),$$

it is both necessary and sufficient that the series

$$\sum p^{-1}(g(p) - 1) \tag{9.1}$$

$$\sum_{|g(p)|\le 3/2} p^{-1}|g(p) - 1|^2, \quad \sum_{|g(p)|>3/2} p^{-1}|g(p)|^\alpha, \quad \sum_{p,k\ge 2}\sum p^{-k}|g(p^k)|^\alpha \tag{9.2}$$

converge, and that for each prime p

$$\sum_{k=1}^{\infty} p^{-k} g(p^k) \ne -1. \tag{9.3}$$

When these conditions are satisfied the mean value is

$$\prod\left(1 - \frac{1}{p}\right)\left(1 + \frac{g(p)}{p} + \frac{g(p^2)}{p^2} + \cdots\right)$$

Note that membership of $\mathcal{L}^\alpha$ and the existence of a non-zero mean value are together equivalent to a set of explicit conditions on the prime powers. There are no auxiliary enabling conditions upon the multiplicative function under consideration beyond that it takes its values in the complex plane.

Proof of Theorem 9.1. First Approach. I treat the cases $1 < \alpha \le 2$. When $\alpha > 2$, the details are similar but simpler. As in the treatment of additive functions given in Chapter 8, I begin with the defined condition: g belongs to $\mathcal{L}^\alpha$.

From Theorem 3.1 each of the sums

$$\sum_{q\le x}{}^{*} q^{\alpha-1}\left|\frac{1}{x}\sum_{\substack{n\le x\\ n\equiv 0 \pmod q}} g(n) - \frac{1}{q}\left(1-\frac{1}{q_0}\right)\frac{1}{x}\sum_{n\le x} g(n)\right|^{\alpha}$$

$$\sum_{q\le x}{}^{**} q\left|\frac{1}{x}\sum_{\substack{n\le x\\ n\equiv 0 \pmod q}} g(n) - \frac{1}{q}\left(1-\frac{1}{q_0}\right)\frac{1}{x}\sum_{n\le x} g(n)\right|^{2}$$

is bounded uniformly in x, where * indicates that summation is confined to those prime powers q for which the inequality

$$(9.4)\qquad \frac{q}{x}\left|\sum_{\substack{n\le x\\ n\equiv 0 \pmod q}} g(n) - \frac{1}{q}\left(1-\frac{1}{q_0}\right)\sum_{n\le x} g(n)\right| > \left(\frac{1}{x}\sum_{n\le x}|g(n)|^{\alpha}\right)^{1/\alpha}$$

is satisfied, and ** that it is confined to those for which the opposite condition holds. In the present situation g is multiplicative, and there is a representation

$$\sum_{\substack{n\le x\\ n\equiv 0 \pmod q}} g(n) = g(q)\sum_{r\le x/q} g(r) - g(q)\sum_{\substack{r\le x/q\\ (r,q)>1}} g(r).$$

By Hölder's inequality and the initial hypothesis, the last of these sums is

$$\ll |g(q)|\left(\sum_{\substack{r\le x/q\\ (r,q)>1}} 1\right)^{1/\alpha'}\left(\sum_{r\le x/q}|g(r)|^{\alpha}\right)^{1/\alpha} \ll x|g(q)|q^{-1}q_0^{-1/\alpha'}.$$

If now

$$m(y) = y^{-1}\sum_{n\le y} g(n),$$

the local mean of g, then

$$\sum_{q\le x}^{*} \frac{1}{q}\left|g(q)m\left(\frac{x}{q}\right)-m(x)\right|^{\alpha} \ll 1+\sum_{q\le x}^{*}\frac{|g(q)|^{\alpha}}{qq_0^{\alpha-1}} \tag{9.5}$$

and

$$\sum_{q\le x}^{**} \frac{1}{q}\left|g(q)m\left(\frac{x}{q}\right)-m(x)\right|^{2} \ll 1+\sum_{q\le x}^{**}\frac{|g(q)|^{2}}{qq_0^{2/\alpha'}} \tag{9.6}$$

uniformly in x. Note that the range of the variable q in these inequalities may be cut down as required.

By hypothesis $m(y)$ approaches a non-zero limit as $y\to\infty$. If $|g(q)|>\gamma$, $q_0>\gamma$, $x\ge\gamma$ for a suitably chosen positive number γ, then the condition (9.4) is amply satisfied, uniformly for $q\le x^{1/2}$. It follows from (9.5), again for γ large enough, that the series

$$\sum_{\substack{|g(q)|>\gamma\\ q_0>\gamma}} q^{-1}|g(q)|^{\alpha}$$

is convergent.

Likewise from (9.6), for a suitably small positive δ, the series

$$\sum_{|g(q)-1|\le\delta} q^{-1}|g(q)-1|^{2}$$

is convergent. Here there is no auxiliary condition $q_0>\gamma$ since the $g(q)$ are uniformly bounded, and so therefore is the sum involving $q_0^{-2/\alpha'}$.

Since any prime power q must satisfy the condition (9.4) or its complement, in order to establish the convergence of the three series at (9.2) in the statement of the theorem, it will suffice to establish that of $\sum p^{-k}|g(p^k)|^{\alpha}$ taken over the powers of a fixed prime p. There are several ways to do this, somewhat related to each other. The following method employs Euler products.

Let g_1 be the multiplicative function which coincides with g except on the powers of p, when it is zero. By examining their corresponding Euler products, we see that the Dirichlet series for $|g|$ and $|g_1|$ satisfy

$$\left(1+\sum_{k=1}^{\infty}p^{-k\sigma}|g(p^k)|^{\alpha}\right)\sum_{n=1}^{\infty}n^{-\sigma}|g_1(n)|^{\alpha}=\sum_{n=1}^{\infty}n^{-\sigma}|g(n)|^{\alpha},\quad \sigma>1. \tag{9.7}$$

The series are absolutely convergent since, integrating by parts,

$$\sum_{n=1}^{\infty} n^{-\sigma}|g(n)|^{\alpha} = \sigma \int_1^{\infty} y^{-\sigma-1} \sum_{n \le y} |g(n)|^{\alpha} dy. \tag{9.8}$$

Moreover, from the hypothesis that g belongs to $\mathcal{L}^{\alpha}$, the integral and so the sum at (9.8) is $\ll (\sigma - 1)^{-1}$, $1 < \sigma \le 2$.

The relations (9.7) and (9.8) continue to hold with α replaced by 1. Since g has a non-zero mean value,

$$x^{-1} \sum_{n \le x} |g(n)| \ge c > 0$$

for some c and for all sufficiently large values of x. From (9.8) with 1 in place of α,

$$\sum_{n=1}^{\infty} n^{-\sigma}|g(n)| \gg (\sigma - 1)^{-1}, \qquad 1 < \sigma \le 2.$$

Moreover, again from the membership of g in $\mathcal{L}^{\alpha}$, $g(q) \ll q^{1/\alpha}$. The sum $\sum p^{-k}|g(p^k)|$ taken over the powers of p converges, is indeed bounded uniformly in p. From (9.7) with α replaced by 1

$$\sum_{n=1}^{\infty} n^{-\sigma}|g_1(n)| \gg (\sigma - 1)^{-1}, \qquad 1 < \sigma \le 2.$$

An application of Hölder's inequality, together with the well known estimate $\zeta(\sigma) = \sum n^{-\sigma} \ll (\sigma - 1)^{-1}$, $1 < \sigma \le 2$, cf. (13.3), allows us to raise the power of $|g_1|$:

$$\sum_{n=1}^{\infty} n^{-\sigma}|g_1(n)|^{\alpha} \gg (\sigma - 1)^{-1}, \qquad 1 < \sigma \le 2.$$

Combining the lower and upper bounds for the second and third series in (9.7) we see that $\sum p^{-k\sigma}|g(p^k)|^{\alpha}$, $k \ge 1$, is bounded above uniformly for $1 < \sigma \le 2$. Let $\sigma \to 1+$.

I shall continue with the application of Euler products to derive the remaining conditions (9.1) and (9.3).

If in the relation (9.8) we replace $|g(n)|^{\alpha}$ by $g(n)$, and appeal to the estimate $\zeta(\sigma) \sim (\sigma - 1)^{-1}$ as $\sigma \to 1+$ (again cf. (13.3)), then the hypothesis that g has a non-zero mean value ensures that as $\sigma \to 1+$ the product

$$\zeta(\sigma)^{-1} \sum_{n=1}^{\infty} n^{-\sigma} g(n) = \prod_p (1 - p^{-\sigma}) \left(1 + \sum_{m=1}^{\infty} p^{-m\sigma} g(p^m)\right) \tag{9.9}$$

has a non-zero limit.

Suppose that some sum factor over the powers of a single prime ℓ vanishes. By appealing to the theory of analytic functions, or applying the mean value theorem of calculus separately to the real and imaginary parts of the function $\sum \ell^{-m\sigma} g(\ell^m)$, noting that the series $\sum \ell^{-m}|g(\ell^m)| \log \ell^m$ converges, we see that $1 + \sum \ell^{-m\sigma} g(\ell^m) \ll (\sigma - 1)$, uniformly for $1 < \sigma \le 2$. Moreover, the terms in (9.9) over the remaining primes are

$$\ll \exp\left(\sum_p \left(-p^{-\sigma} + \sum_{m=1}^{\infty} p^{-m\sigma}|g(p^m)|\right)\right)$$
$$\ll \exp\left(\sum_p p^{-\sigma}(|g(p)| - 1)\right),$$

since the convergence of the three series at (9.2) guarantees the convergence of $\sum p^{-k}|g(p^k)|$, $p, k \ge 2$.

By the Cauchy–Schwarz inequality

$$\sum_{|g(p)| \le 3/2} p^{-\sigma}||g(p)| - 1| \le \left(\sum_{|g(p)| \le 3/2} p^{-\sigma}||g(p)| - 1|^2 \sum p^{-\sigma}\right)^{1/2}.$$

Here $||g(p)| - 1| \le |g(p) - 1|$, and $\sum p^{-\sigma} \ll -\log(\sigma - 1)$, so that the upper bound is $\ll (-\log(\sigma - 1))^{1/2}$. Similarly, by Hölder's inequality

$$\sum_{|g(p)| > 3/2} p^{-\sigma}||g(p)| - 1| \le \left(\sum_{|g(p)| > 3/2} p^{-\sigma}||g(p)| - 1|^{\alpha}\right)^{1/\alpha} \left(\sum p^{-\sigma}\right)^{1-1/\alpha}$$
$$\ll \left(-\log(\sigma - 1)\right)^{1-1/\alpha}.$$

The product at (9.9) is $\ll (\sigma - 1)\exp(c_0(-\log(\sigma - 1))^{1/2})$ for some positive constant c_0, uniformly for $1 < \sigma \le 2$. For σ sufficiently near to 1 this gives a contradiction. No factor of the product (9.9) vanishes.

A similar argument now shows that

$$\lim_{\sigma \to 1+} \sum_{|g(p)| \le 3/2} p^{-\sigma}(g(p) - 1) \tag{9.10}$$

exists. We need only note that with the principal value of the logarithm, $\log(1 + z) = z + O(|z|^2)$ in the disc $|z| \le 1/2$. For sufficiently large primes p

$$\log\left(1 + \sum_{m=1}^{\infty} p^{-\sigma} g(p^m)\right) = \sum_{m=1}^{\infty} p^{-m\sigma} g(p^m) + O\left(\left(\sum_{m=1}^{\infty} p^{-m\sigma}|g(p^m)|\right)^2\right),$$

since for a certain constant c_1

$$\sum_{m=1}^{\infty} p^{-m}|g(p^m)| \le c_1 \sum_{m=1}^{\infty} p^{-m/\alpha'} \le 1/2.$$

This is applied when $|g(p)| \le 3/2$. From the second of the conditions at (9.2), the series $\sum p^{-1}(|g(p)|+1)$, taken over the primes for which $|g(p)| > 3/2$, converges. The terms in the infinite product at (9.9) which correspond to such primes may be ignored.

From the penultimate remark of the previous paragraph, to obtain the convergence of the series (9.1), and so satisfy all the conditions of Theorem 9.1 which involve the primes, it will suffice to derive from the existence of the limit (9.10) the convergence of the series $\sum p^{-1}(g(p)-1)$, $|g(p)| \le 3/2$. This is a result of tauberian type, and we may apply the later Lemma 10.1. Alternatively, the following argument suffices.

For notational simplicity set $a_p = g(p) - 1$ when $|g(p)| \le 3/2$, $a_p = 0$ on other primes. Let $\beta = \exp((\sigma-1)^{-1})$. Thus $\beta \to \infty$ as $\sigma \to 1+$. Let $\varepsilon > 0$. We employ the representation

$$\sum_p a_p p^{-\sigma} - \sum_{p \le \beta} a_p p^{-1}$$
$$= \sum_{p \le 1/\varepsilon} a_p(p^{-\sigma} - p^{-1}) + \sum_{1/\varepsilon < p \le \beta} a_p(p^{-\sigma} - p^{-1}) + \sum_{p > \beta} a_p p^{-\sigma},$$

valid if $\sigma \le 1 + (-\log \varepsilon)^{-1}$.

For those primes not exceeding $1/\varepsilon$, $p^{-\sigma} - p^{-1} \to 0$ as $\sigma \to 1+$.

An application of the Cauchy–Schwarz inequality shows that the sum over the primes $p > \beta$ is

$$\ll \left(\sum_{p > \beta} p^{-1} |a_p|^2 \sum_{p > \beta} p^{-\sigma} \right)^{1/2}.$$

From the well known Chebyshev estimate $\pi(y) \ll y(\log y)^{-1}$ and an integration by parts, $\sum p^{-\sigma}$, $p > \beta$ is bounded uniformly in $1 < \sigma \le 2$. From the first condition at (9.2), the series involving $|a_p|^2$ approaches zero as $\beta \to \infty$.

Over the range $p \le \beta$, $p^{-\sigma} - p^{-1} \ll (\sigma-1)p^{-1}\log p$. Another application of the Cauchy–Schwarz inequality shows that the sum over the primes in the interval $(1/\varepsilon, \beta]$ is

$$\ll \left(\sum_{p > 1/\varepsilon} p^{-1} |a_p|^2 \right)^{1/2} \left((\sigma-1)^2 \sum_{p \le \beta} p^{-1} (\log p)^2 \right)^{1/2}.$$

The second of these square roots is, again by the Chebyshev estimate, bounded uniformly in σ, $1 < \sigma \le 2$. Since ε may be chosen arbitrarily small,

$$\sum_{\substack{p \le \beta \\ |g(p)| \le 3/2}} \frac{g(p)-1}{p} = \sum_{|g(p)| \le 3/2} \frac{g(p)-1}{p^\sigma} + o(1), \quad \sigma \to 1+,$$

and the convergence of the series (9.1) is obtained.

Starting from the existence of a non-zero mean value for g we have arrived at the conditions (9.1) to (9.3) on the prime powers. How do we get back? The application of Euler products was elementary, but continuity of thought now suggests for a return route only the study of the Dirichlet series $\sum g(n)n^{-\sigma}$.

From the convergence of the series (9.1) and the second series at (9.2), we regain the existence of the limit (9.10). This is the analogue for Dirichlet series of a well known theorem of Abel in power series, and only needs an integration by parts. Combined with the rest of (9.2) we gain the asymptotic estimate

$$\text{(9.11)} \quad \lim_{\sigma \to 1+} \zeta(\sigma)^{-1} \sum_{n=1}^{\infty} n^{-\sigma} g(n) = \prod_p (1-p^{-1}) \left(1 + \sum_{m=1}^{\infty} p^{-m} g(p^m)\right),$$

and by (9.3) the limit is non-zero.

The development of good asymptotic estimates for the averages of arithmetic functions by applying Fourier analysis in the *complex* plane to the sum function of their corresponding Dirichlet series has been pursued intensively since the late nineteenth century. For multiplicative functions the associated Euler product has played a central rôle. The usual prerequisite for success was the existence of one or more important identifiable singularities of the sum function, and a way to analytically continue the sum function past these singularities into a region where it was still manageable. For many particular arithmetic functions this procedure has proved viable. However, serious difficulties arise when applying it to arithmetic functions of a general nature, concerning which only very weak local information is available.

The multiplicative function defined by $g(p^m) = (-1)^{m+1} p^{m-1}$ gives rise to a Dirichlet series $\sum g(n)n^{-s}$ absolutely convergent in $\operatorname{Re}(s) > 1$. The sum function $G(s)$ is analytic in $\operatorname{Re}(s) > 1$. It has zeros at the points $(2k\pi i + \log(p+1))/\log p$ for integers k and primes p, and every point of the lines $\operatorname{Re}(s) = 1$ is a limit point of these. $G(s)$ cannot be analytically continued.

Only in the late nineteen sixties did Halász develop a method for satisfactorily estimating the mean values of reasonably general multiplicative functions using a Mellin transform in the plane. His method was subtle as much as powerful. The existence of an Euler product representation for G was vital.

In order that the Euler product attached to G be manageable it is necessary to have some global control over the size of g on the prime powers. Sufficient control was posited by Halász at the outset. For this reason the method of Halász has the direction *primes to integers*. Since the method of the stable dual lends itself very well to gaining such control, it would be natural to combine the two methods. In the present circumstances it would be appropriate to show that as $\sigma \to 1+$

$$\sum_{n=1}^{\infty} n^{-s} g(n) = (s-1)^{-1} A + o(|s|(\sigma - 1)^{-1})$$

uniformly on $\operatorname{Re}(s) = \sigma$, where A is the infinite product appearing in (9.11). A treatment along these lines may be found in [41] Chapter 10, (see also [35], [43]). I shall instead fall back to an earlier point in the present argument and consider the implications of the method of the stable dual more deeply. This will give a better sense of direction.

10

Multiplicative functions of the class $\mathcal{L}^{\alpha}$: Second Approach

In this chapter it will be convenient to employ the following tauberian theorem of Hardy and Littlewood. A proof may found in [41] Chapter 2, Lemma 2.18, pp. 102–108.

Lemma 10.1. *Let $w(t)$ be a real valued function defined and of bounded variation on each finite interval $[0, t]$ with $t > 0$. Assume further that it satisfies $w(0) = 0$ and $\liminf(w(y) - w(x)) \geq 0$ as $x \to \infty$, $y > x$, $y/x \to 1$.*

If

$$\int_0^\infty e^{-yt} dw(t) \to \ell, \qquad y \to 0+,$$

where the integral exists for $y > 0$, then

$$w(t) \to \ell, \qquad t \to \infty.$$

Proof of Theorem 9.1. Second Approach I return to the point in the First Approach (keeping the same notation) where the convergence of the series

$$\sum_{|g(p)| \leq 3/2} p^{-1}|g(p) - 1|^2, \qquad \sum_{|g(p)|>3/2} p^{-1}|g(p)|^{\alpha},$$

is assured, before the introduction of the Euler products.

Let $K \geq 6$, $P \geq 2$, $P^K \leq x$. Applications of Hölder's inequality and Cauchy's inequality in conjunction with bounds (9.5), (9.6) show that

$$\sum_{P<p\leq P^K} \frac{1}{p}\left|g(p)m\left(\frac{x}{p}\right) - m(x)\right| \ll (\log K)^{1/2}. \tag{10.1}$$

Since $\sum p^{-1}|g(p) - 1|^2$ is convergent when taken over the primes for which $|g(p)| \leq 3/2$, on those primes we may replace $g(p)$ in (10.1) by 1. The analogous condition for the primes with $|g(p)| \geq 3/2$ enables a similar replacement to be made. In order then to apply Theorem 6.1, $m(x)$ must satisfy an approximate continuity condition.

For any arithmetic function a_n, and $\theta \ge 1$, $y \ge 1$

$$\left| \frac{1}{y\theta} \sum_{n \le y\theta} a_n - \frac{1}{y} \sum_{n \le y} a_n \right| \le \frac{1}{y\theta} \left| \sum_{y < n \le y\theta} a_n \right| + \left(1 - \frac{1}{\theta}\right) \frac{1}{y} \left| \sum_{n \le y} a_n \right|.$$

In the present circumstances, with $a_n = g(n)$, an application of Hölder's inequality shows that

$$\sum_{y<n\le y\theta} a_n \ll (y\theta - y + 1)^{1/\alpha'} (y\theta)^{1/\alpha} \ll y\theta \left(1 - \frac{1}{\theta} + \frac{1}{y\theta}\right)^{1/\alpha'}.$$

We apply Theorem 6.1 with $\psi(w) = \phi(w) = m(x/w)$, $\tau = K^{1/2}$, $P^{2K} = x$. If $P^{1-\tau} \le u \le v \le P^{K+\tau}$, $v \le u(1 + P^{-1/3})$, then $\psi(v) - \psi(u) \ll P^{-1/3}$. There is a representation

$$m(x/w) - m(x) = D \log w + O\Big(\max(P^{-1/3}, (\log K)^{-1/2}) \Big)$$

uniformly for $P^{-\tau} \le w \le P^{\tau}$. Moreover, $D\tau \log P = m(xP^{-\tau}) - m(x)$; and since g belongs to $\mathcal{L}^\alpha$, $DK^{1/2} \log P \ll 1$. Over the shorter range $1 \le w \le x^{1/(2K^{2/3})}$

$$m(x/w) - m(x) \ll x^{-1/(6K)} + (\log K)^{-1/2}. \tag{10.2}$$

A better approximation is had over a shorter range.

Set $K = (\log x)^{1/2}$. Then

$$m(x/w) - m(x) \ll (\log\log x)^{-1/2} \tag{10.3}$$

uniformly for $1 \le w \le \exp(\sqrt{\log x})$, for all sufficiently large x.

We may employ this slow oscillation to derive the convergence of the series $\sum p^{-m}|g(p^m)|^\alpha$, $m \ge 2$, for a fixed prime p. We obtain upper and lower bounds for the sum

$$S = \sum_{p^m \le \log x} |g(p^m)|^\alpha \sum_{n \le x/p^m} |g_1(n)|^\alpha.$$

Clearly $|g_1(n)|$ belongs to $\mathcal{L}^\alpha$. We may obtain analogues of (10.1), and so (10.3), with $g(n)$ everywhere replaced by $|g_1(n)|$. In particular we gain an approximate representation

$$\sum_{n\le x} |g(n)| = \sum_{m=1}^{\infty} p^{-m} |g(p^m)| \sum_{n \le x} |g_1(n)| + O(x(\log\log x)^{-\frac{1}{2}})$$

and a *uniform* lower bound

$$wx^{-1} \sum_{n \le x/w} |g_1(n)| \ge c_2 > 0, \qquad 1 \le w \le \log x.$$

Note that for x sufficiently large, $P \ge (\log x)^2 > p$; but the loss of a single prime from the range $(P, P^K]$ in (10.1) would not be significant.

An application of Hölder's inequality gives

$$wx^{-1} \sum_{n \le x/w} |g_1(n)|^\alpha \ge c_2^\alpha$$

with the same uniformity in w, from which

$$S \ge c_2^\alpha x \sum_{p^m \le \log x} p^{-m} |g(p^m)|^\alpha.$$

However,

$$S \le \sum_{n \le x} |g(n)|^\alpha \ll x,$$

since g belongs to the class $\mathcal{L}^\alpha$. The partial sums $\sum p^{-m}|g(p^m)|^\alpha$, $p^m \le \log x$, are bounded uniformly in $x > 0$, as asserted.

An argument in this style may be found in [35], and [41] Chapter 10, §4. It has the advantage over the appeal to Euler products made in the previous chapter that it is applicable whether g has a(n asymptotic) mean value or not. I shall employ this fact in Chapter 17.

Define the multiplicative function h by Dirichlet convolution: $g = \mathbf{1} * h$. I next derive from the slow oscillation (10.3) a convenient representation for the local mean of g.

Lemma 10.2. *If the multiplicative function g belongs to $\mathcal{L}^\alpha$, and the three series at (9.2) converge, then*

$$x^{-1} \sum_{n \le x} g(n) = \sum_{d \le x} h(d) d^{-1} + O((\log\log x)^{-\frac{1}{2}}) \tag{10.4}$$

uniformly in $x \ge 4$.

Proof of Lemma 10.2. I apply the estimate (10.3) for $1 \le w \le T \le \exp(\sqrt{\log x})$, multiply by w^{-1} and sum over the w:

$$\sum_{w \le T} x^{-1} \sum_{n \le x/w} g(n) = \sum_{w \le T} (xw)^{-1} \sum_{n \le x} g(n) + O\left(\frac{\log T}{\sqrt{\log\log x}}\right).$$

Here

$$x^{-1}\sum_{n\le x} g(n) \sum_{w\le \min(T,x/n)} 1 = \sum_{x/T<n\le x} g(n)n^{-1} + x^{-1}T\sum_{n\le x/T} g(n),$$

and since $\sum w^{-1}(w \le T)$ is $\log T + \gamma + O(T^{-1})$ with Euler's constant γ,

$$x^{-1}\sum_{n\le x} g(n) = (\log T+\gamma)^{-1}\sum_{x/T<n\le x} g(n)n^{-1} +$$

$$+ O((\log\log x)^{-1/2} + (\log T)^{-1}) + O((T(\log T)^2)^{-1}\sum_{n\le x}|g(n)|n^{-1}).$$

The function $g(n)$ has been replaced by $g(n)n^{-1}$, generally felt to be more manageable.

From the definition of h

$$\sum_{n\le y} g(n)n^{-1} = \sum_{n\le y} n^{-1}\sum_{d|n} h(d) = \sum_{d\le y} h(d)d^{-1}\sum_{m\le y/d} m^{-1}$$

$$= \sum_{d\le y} h(d)d^{-1}((\log(y/d)) + \gamma) + O\left(y^{-1}\sum_{d\le y}|h(d)|\right).$$

Hence

$$(\log T)^{-1}\sum_{x/T<n\le x} g(n)n^{-1} = \sum_{d\le x} h(d)d^{-1} + \sum_{x/T<d\le x} h(d)d^{-1}\log(x/Td)$$

$$+\gamma\sum_{x/T<d\le x} h(d)d^{-1} + \text{error terms},$$

with the error terms within the order of the four appearing in the preceding estimates. Altogether

$$x^{-1}\sum_{n\le x} g(n) = \sum_{d\le x} h(d)d^{-1} + E$$

where

$$E \ll (\log T)^{-1}\left|\sum_{d\le x} h(d)d^{-1}\right| + \sum_{x/T<d\le x}|h(d)|d^{-1}\log(Td/x)$$

$$+ \sum_{x/T<d\le x}|h(d)|d^{-1} + x^{-1}\sum_{d\le x}|h(d)| + x^{-1}T\sum_{d\le x/T}|h(d)|$$

$$+ (T(\log T)^2)^{-1}\sum_{n\le x}|g(n)|n^{-1} + (\log\log x)^{-1/2} + (\log T)^{-1}.$$

If we choose $T = \exp((\log\log x)^2)$ say, then each of these eight error terms is $\ll (\log\log x)^{-1/2}$. I treat the second as an example.

The average value of $|h(d)|$ over an interval $[2, y]$ may be estimated by Lemma 2.2. To this end I note that from the first part of (9.2) and an application of the Cauchy–Schwarz inequality

$$\sum_{\substack{q\le y\\ |g(q)|\le 3/2}} |g(q)-1|\log q \le \left(\sum_{\substack{q\le y\\ |g(q)|\le 3/2}} q^{-1}|g(q)-1|^2 \sum_{q\le y} q(\log q)^2 \right)^{1/2}$$
$$\ll y(\log y)^{1/2}.$$

From the second and third parts of (9.2) and an application of Hölder's inequality

$$\sum_{\substack{q\le y\\ |g(q)|>3/2}} |g(q)-1|\log q \ll y(\log y)^{1/\alpha}.$$

It is easy to see that

$$y^{-1}\sum_{q\le y} |h(q)|\log q \ll (\log y)^{1/\alpha}.$$

From Lemma 2.2

$$\sum_{d\le y} |h(d)| \ll y(\log y)^{1/\alpha} \exp\left(\sum_{q\le y} \frac{|h(q)|-1}{q} \right)$$
$$\ll y(\log y)^{(1/\alpha)-1} \exp\left(\sum_{p\le y} \frac{|g(p)-1|}{p} \right)$$
$$\ll y(\log y)^{1/\alpha-1}\exp(c_3(\log\log y)^{1/2}) \ll y(\log y)^{-c}$$

for certain positive constants c_3, c, where I have employed the convergence of the first two series at (9.2) in the penultimate step.

Integrating by parts

$$\sum_{x/T<d\le x} |h(d)d^{-1} \ll (\log x)^{-c} + \int_{x/T}^{x} y^{-2}\sum_{d\le y} |h(d)|dy$$
$$\ll \log T(\log x)^{-c}.$$

Hence

$$\sum_{x/T<d\le x} |h(d)|d^{-1}\log(dT/x) \ll (\log T)^2(\log x)^{-c} \ll (\log x)^{-c/2},$$

a much stronger bound than asserted.

Note that since

$$\sum_{d\le x} h(d)d^{-1} = x^{-1}\sum_{n\le x} g(n) + O\left((\log T)^{-1}\left|\sum_{d\le x} h(d)d^{-1}\right| + (\log\log x)^{-1/2}\right),$$

and g belongs to $\mathcal{L}^\alpha$, the partial sums involving $h(d)d^{-1}$ are bounded uniformly in x.

This completes the proof of Lemma 10.2.

It follows from the representation (10.4) and the existence of a mean value for g that the series $\sum h(d)d^{-1}$ converges. Again by the analogue of Abel's theorem, $\sum h(d)d^{-\sigma}$ approaches a limit as $\sigma \to 1+$. We are at (9.9), and may continue as before. This may seem an unnecessary diversion, but we have gained something; a proof of the converse proposition suggests itself.

As a first step I show that if the series at (9.1) and (9.2) converge, then g belongs to $\mathcal{L}^\alpha$. The following result is helpful.

Lemma 10.3. *Let $\beta > 0$. Then*

$$y^\beta - 1 = \beta(y-1) + O((y-1)^2)$$

and

$$|y-1| \ll |y^\beta - 1| \ll |y-1|$$

uniformly for $0 \le y \le 3/2$.

Proof of Lemma 10.3. If u is in the interval $[-\frac{1}{2}, \frac{1}{2}]$, then Taylor's theorem gives

$$(1+u)^\beta = 1 + \beta u + u^2 \cdot \frac{\beta(\beta-1)}{2!}(1+u_0)^{\beta-2}$$

for some u_0 also in $[-\frac{1}{2}, \frac{1}{2}]$. With $y-1$ in place of u the proof of the lemma is readily completed.

Define multiplicative functions g_j, $j = 2, 3$, by

$$g_2(q) = \begin{cases} |g(q)| & \text{if } |g(q)| \le 3/2, \\ 0 & \text{if } |g(q)| > 3/2, \end{cases} \qquad g_3(q) = \begin{cases} 0 & \text{if } |g(q)| \le 3/2, \\ |g(q)| & \text{if } |g(q)| > 3/2. \end{cases}$$

Then $|g|^\alpha$ is dominated by $g_2^\alpha * g_3^\alpha$ at each prime power, and therefore at each positive integer.

The function g_2 belongs to $\mathcal{L}^\beta$ for every $\beta \ge 0$. For clearly

$$\sum_{q\le y} g_2(q)^\beta \log q \ll y,$$

and by Lemma 2.2

$$x^{-1}\sum_{n\le x} g_2(n)^\beta \ll \exp\left(\sum_{p\le x} p^{-1}(g_2(p)^\beta - 1)\right).$$

Since $|1-|z|| \le |1-z|$ it follows from (9.2) that the series $\sum p^{-1}(1-g_2(p))^2$ converges. Moreover $|1-z|^2 - |1-|z||^2 - 2(1-\operatorname{Re} z) = 2(|z|-1)$, so that by (9.1) and (9.2), the series $\sum p^{-1}(g_2(p)-1)$ converges. In particular, its partial sums are uniformly bounded. Applying Lemma 10.3 our assertion is justified.

Then

$$\begin{aligned} x^{-1}\sum_{n\le x}|g(n)|^\alpha &\le x^{-1}\sum_{r\le x} g_3(r)^\alpha \sum_{m\le x/r} g_2(m)^\alpha \\ &\ll \sum_r g_3(r)^\alpha r^{-1} \ll \prod_p (1 + g_3(p)^\alpha p^{-1} + \cdots) \\ &\ll \exp\left(\sum_{|g(q)|\ge 3/2} |g(q)|^\alpha q^{-1}\right) \end{aligned}$$

which by (9.2) is well defined.

Since g belongs to $\mathcal{L}^\alpha$ and condition (9.2) is satisfied, the representation (10.4) becomes available in this direction also. For g to possess a mean value, the series $\sum h(d)d^{-1}$ must converge.

Using (9.1) and (9.2) alone, (9.11) gives the existence of $\lim \sum h(d)d^{-\sigma}$, $\sigma \to 1+$. We are reduced to a tauberian theorem: show that the (non-zero) existence of this limit, together with the convergence of the series at (9.2), guarantees the convergence of the series $\sum h(d)d^{-1}$.

I apply Lemma 10.1 with

$$w(t) = \operatorname{Re} \sum_{2\le d\le e^t} h(d)d^{-1},$$

$$\ell + 1 = \operatorname{Re} \prod_p \left(1 - \frac{1}{p}\right)\left(1 + \frac{g(p)}{p} + \frac{g(p^2)}{p^2} + \cdots\right),$$

and with a similar pair using imaginary parts. A sufficient condition of slow oscillation is obtained by taking K large enough in (10.2). The function g has a mean value given by the infinite product at (9.11). The proof of Theorem 9.1 is complete.

It is worthwhile reviewing the steps of the argument employed here to deduce conditions (9.1)–(9.3) from the existence of a non-zero mean value of the multiplicative function g of $\mathcal{L}^\alpha$.

(i) From the membership of g in $\mathcal{L}^\alpha$, and the existence of a non-zero mean value, I deduce the validity of condition (9.2) except for the convergence of $\sum p^{-m}|g(p^m)|^\alpha$ for a fixed prime p. This is *integers to primes*, and employs the method of the stable dual.

(ii) From the part of (9.2) obtained so far, stable duality gives the slow oscillation of the local mean of g. This is *primes to integers*.

(iii) The slow oscillation of local means gives the remainder of condition (9.2). This is *integers to primes*.

(iv) The validity of condition (9.2) together with the slow oscillation of the local means gives a representation of the local mean of g by the partial sum of the series $\sum h(d)d^{-1}$, where $g = \mathbf{1} * h$. This is *primes to integers*.

(v) Introducing Euler products, the convergence of $\sum h(d)d^{-1}$ to a non-zero value leads to the validity of conditions (9.1), (9.3). This is *integers to primes*.

These steps exhibit a characteristic of the method of the stable dual: argument in loops. The conditions (9.1)–(9.3) are not initially known. They have to be discovered and raised to clarity as the method proceeds. This is a procedure of perhaps geometric nature. I may add that I first discovered and established Theorem 9.1 in the direction *integers to primes*, for the case $\alpha = 2$, using arguments much along the lines given here. The main difference was that the slow oscillation of the local mean of an appropriate relative of g was obtained by the Fourier analytic method of Halász.

Although the application of the Hardy–Littlewood tauberian theorem effectively employs the slow oscillation of local means, there remains the feeling that an outside element has been introduced into the proceedings, that the existence of a mean value for g should be obtained from the validity of (9.1) and (9.2) by an argument more within the canons of stable duality and its underlying philosophy. I now give such an argument.

11

Multiplicative functions of the class $\mathcal{L}^\alpha$: Third Approach

Proof of Theorem 9.1, Third Approach. I consider only the direction: *primes to integers.* Let g be a multiplicative function for which the series (9.1) and (9.2) converge. To ease notational difficulties suppose for the moment that $|g(q) - 1| \le 1/2$ on all prime powers q.

We attach to g the additive function f defined by $f(q) = \log g(q)$, the argument of the logarithm being taken in $(-\pi, \pi]$. For each real $\tau > 4$ define the truncated additive function f_τ by $f_\tau(q) = f(q)$ if $q_0 \le \tau$, $f_\tau(q) = 0$ otherwise. Let g_τ denote the multiplicative function given by $g_\tau(n) = \exp(f_\tau(n))$. It is not difficult to see that g_τ possesses the mean value

$$D(\tau) = \prod_{p \le \tau} \left(1 - \frac{1}{p}\right)\left(1 + \frac{g(p)}{p} + \frac{g(p^2)}{p^2} + \cdots\right).$$

Indeed, if we define h by convolution, $g = \mathbf{1} * h$, then $h(q) = 0$ whenever $q_0 > \tau$, and the series $\sum h(n)n^{-1}$ converges absolutely. Instead of appealing to Lemma 10.2 we may argue directly:

$$\begin{aligned} x^{-1}\sum_{n \le x} g_\tau(n) &= x^{-1}\sum_{d \le x} h(d)\left[\frac{x}{d}\right] = \sum_{d \le x/\log x} h(d)d^{-1} + \\ &\quad + o\Bigg(\sum_{d \le x/\log x} |h(d)|d^{-1}\Bigg) + O\Bigg(\sum_{x/\log x < d \le x} |h(d)|d^{-1}\Bigg) \\ &= \sum_{d=1}^{\infty} h(d)d^{-1} + o(1), \quad x \to \infty. \end{aligned}$$

Furthermore, as $\tau \to \infty$, $D(\tau)$ approaches the expected mean value of g. I shall show that g has a mean value by proving that in terms of the norm on the space $M^1(\mathbb{C}^t)$ of Chapter 3,

$$\sup_{x \ge \tau} \|g - g_\tau\|_1 \to 0, \quad \tau \to \infty.$$

To this end I note that from the integral representation

$$e^z - 1 = \int_0^z e^w dw,$$

taken over the straight line from the origin to z in the complex plane, $|e^z - 1| \le |z| \max(e^{\mathrm{Re} z}, 1)$. For a typical integer $n > 1$

$$\begin{aligned} |g(n) - g_\tau(n)| &= |g_\tau(n)||\exp(f(n) - f_\tau(n)) - 1| \\ &\le |f(n) - f_\tau(n)| \max(|g(n)|, |g_\tau(n)|). \end{aligned}$$

Applications of the Cauchy–Schwarz and Minkowski inequalities give

$$\|g - g_\tau\|_1 \le \|f - f_\tau\|_2(\|g\|_2 + \|g_\tau\|_2),$$

provided that $\|f - f_\tau\|_2$ exists, the new norms on $M^2(\mathbb{C}^t)$.

The estimate $\log z = z - 1 + O(|z-1|^2)$ is valid in the disc $|z - 1| \le 1/2$. It therefore follows from (9.1) and (9.2) that the series $\sum f(q)q^{-1}$, $\sum |f(q)|^2 q^{-1}$ converge. Define

$$A(f, x) = \sum_{q \le x} f(q)q^{-1}.$$

Then

$$A(f, x) - A(f_\tau, x) = \sum_{\tau < q \le x} f(q)q^{-1} + O(\tau^{-1})$$

and

$$\lim_{\tau \to \infty} \sup_{x \ge \tau} |A(f, x) - A(f_\tau, x)| = 0.$$

From the Turán–Kubilius inequality (for complex additive functions)

$$x^{-1} \sum_{n \le x} |f(n) - f_\tau(n) - \{A(f, x) - A(f_\tau, x)\}|^2 \ll \left(\sum_{\substack{q \le x \\ q_0 > \tau}} q^{-1}|f(q)|^2 \right)^{1/2}.$$

We obtain both the existence of $\sup \|f - f_\tau\|_2$, $x \ge \tau$, and that it approaches zero as $\tau \to \infty$.

It is straightforward to apply Lemma 2.2 to show that $\|g_\tau\|_2$ and $\|g\|_2$ are bounded uniformly in all real $x \ge 2$, $\tau > 4$. Indeed, the necessary argument may be found in the Second Approach, in the elementary preparation for the application of the Hardy–Littlewood tauberian theorem. Thus $\|g - g_\tau\|_1 \to 0$ as $\tau \to \infty$, uniformly for $x \ge \tau$, and Theorem 9.1 is established in this case.

The removal of the condition $|g(q)-1| \le 1/2$ is achieved by manipulating Euler products in a manner more awkward to notate than to devise.

Let g satisfy condition (9.2) of Theorem 9.1. Define the truncated multiplicative function g_4 by

$$g_4(q) = \begin{cases} g(q) & \text{if } \ |g(q)-1| \le 1/2, \ q_0 \ge 3, \\ 1 & \text{on other prime powers.} \end{cases}$$

Define g_5 by Dirichlet convolution: $g = g_4 * g_5$. The argument hinges on the absolute convergence of the series $\sum g_5(n)n^{-1}$.

By definition $g_5(2^r) = g(2^r) - g(2^{r-1})$, and formally

$$\sum_{w=0}^{\infty} g_4(p^w)p^{-w\sigma} \sum_{t=0}^{\infty} g_5(p^t)p^{-t\sigma} = \sum_{m=0}^{\infty} g(p^m)p^{-m\sigma}, \quad p \ge 3.$$

For $\sigma \ge 1$, the sum

$$\eta = \sum_{w=1}^{\infty} g_4(p^w)p^{-w\sigma}$$

does not exceed $(3/2)(p-1)^{-1} \le 3/4$ in absolute value. The Dirichlet convolution inverse k of g_4 is given by

$$\sum_{r=0}^{\infty} k(p^r)p^{-r\sigma} = (1+\eta)^{-1} = 1 - \eta + \eta^2 - \cdots,$$

so that

$$\sum_{r=0}^{\infty} |k(p^r)|p^{-r\sigma} \le 1 + \eta_0 + \eta_0^2 + \cdots,$$

where

$$\eta_0 = \sum_{w=1}^{\infty} |g_4(p^w)|p^{-w\sigma}.$$

Here $\eta_0 \le 3/4$ uniformly for $\sigma \ge 1$. Hence

$$\sum_{r=0}^{\infty} |k(p^r)|p^{-r} \le 4.$$

If we remove the term corresponding to $r = 0$, then we may replace the bound 4 by $6p^{-1}$. If we remove the first two terms of the series, then we may replace 4 by $19p^{-2}$.

Since $g_5 = k * g$,

$$|g_5(p^t)| \le \sum_{u+v=t} |k(p^u)||g(p^v)|$$

and

$$\begin{aligned}\sum_{t=2}^{\infty} p^{-t}|g_5(p^t)| &\le \sum_{u+v\ge 2} |k(p^u)|p^{-u}|g(p^v)|p^{-v}\\ &= \sum_{v=0}^{\infty} |g(p^v)|p^{-v} \sum_{u\ge \max(2-v,0)} |k(p^u)|p^{-u}\\ &\le 4\sum_{v=2}^{\infty} |g(p^v)|p^{-v} + 6|g(p)|p^{-2} + 19p^{-2}.\end{aligned}$$

From (9.2) the series $\sum\sum p^{-t}|g_5(p^t)|$, taken over all prime powers p^t with $t \ge 2$, converges. Moreover, $g_5(p) = g(p) - g_4(p)$, so that again by (9.2)

$$\begin{aligned}\sum_{p\ge 3} |g_5(p)|p^{-1} &\le \sum_{|g(p)-1|>1/2} |g(p) - 1|p^{-1}\\ &\le 2\sum_{|g(p)|\le 3/2} |g(p) - 1|^2 p^{-1} + 2\sum_{|g(p)|>3/2} |g(p)|^{\alpha} p^{-1} < \infty.\end{aligned}$$

The series $\sum g_5(n)n^{-1}$ indeed converges absolutely.

If (9.1) and (9.2) are satisfied by a given multiplicative function g, then they are satisfied by g_4. From what we have already established g_4 has a mean value B, say. There is a representation

$$\sum_{n\le x} g(n) = \sum_{r\le x} g_5(r) \sum_{s\le x/r} g_4(s).$$

The inner sum of the double sum is $(B + o(1))x/r$ uniformly in $1 \le r \le x/\log x$, as $x \to \infty$, and $\ll x/r$ uniformly in $x/\log x < r \le x$. Hence

$$x^{-1}\sum_{n\le x} g(n) \to B\sum_{r=1}^{\infty} g_5(r)r^{-1}, \qquad x \to \infty.$$

Moreover, B has the expected product representation over the primes, so likewise the mean value of g.

This must be accounted the shortest and most elementary of the three arguments so far suggested for proceeding from (9.1) and (9.2) to the existence of a mean value; whence comes it?

Remarks.

(i) In the Second and Third approaches condition (9.3) played no rôle in guaranteeing the existence of a mean value for g; only in showing it to be non-zero.

(ii) If the series (9.2) converge and g belongs to $\mathcal{L}^\alpha$, then the local mean of g satisfies the oscillation condition (10.3), and the representation (10.4) is valid.

(iii) If the first two series at (9.2) converge and $\beta = \exp((\sigma-1)^{-1})$, then

$$\sum_{p\le\beta} p^{-1}(g(p)-1) = \sum p^{-\sigma}(g(p)-1) + o(1), \quad \sigma \to 1+.$$

(iv) Let $\theta > 0$, and let δ be a positive number sufficiently small that $(1+\delta)^\theta \le 3/2$. If we modify g_4 by tightening the condition $|g(q)-1| \le 1/2$ to $|g(q)-1| \le \delta$, then $|g_4(n)|^\theta$ satisfies the conditions of Theorem 9.1, and so has a mean value.

Moreover, if (in an obvious notation) $|g|^\theta = |g_4|^\theta * g_6$ and $\theta \le \alpha$, then the argument involving functions g_4, g_5 in the Third Approach modifies readily to give the convergence of $\sum |g_6(q)|q^{-1}$ and so of $\sum |g_6(m)|m^{-1}$. Hence $|g|^\theta$ has a mean value for $\theta \le \alpha$.

In particular, for multiplicative functions which belong to the class $\mathcal{L}^\alpha$, the existence of a non-zero mean value improves

$$\limsup_{x\to\infty} x^{-1} \sum_{n\le x} |g(n)|^\alpha < \infty \quad \text{to} \quad \lim_{x\to\infty} x^{-1} \sum_{n\le x} |g(n)|^\alpha = B < \infty.$$

(v) Continuing in the notation of the previous remark, let t be the Dirichlet convolution inverse of g_6. Formally

$$\sum_{m=1}^\infty t(p^m)p^{-m\sigma} = (1+\rho)^{-1} - 1, \quad \rho = \sum_{m=1}^\infty g_6(p^m)p^{-m\sigma}.$$

If p is sufficiently large, then the series sum

$$\rho_0 = \sum_{m=1}^\infty |g_6(p^m)|p^{-m}$$

is not more than $1/2$, and

$$\sum_{m=1}^\infty |t(p^m)|p^{-m} \le \sum_{j=1}^\infty \rho_0^j \le 2\rho_0.$$

For a suitable positive integer D,

$$\sum_{(q,D)=1} |t(q)|q^{-1} \le 2\sum_q |g_6(q)|q^{-1} < \infty.$$

We employ the representation $|g_4|^\theta = |g(r)|^\theta * t$ on the integers prime to D to show that whenever g belongs to $\mathcal{L}^\theta$

$$x^{-1}\sum_{\substack{n\le x\\(n,D)=1}} |g_4(n)|^\theta \ll \sum_{(k,D)=1} |t(k)|k^{-1} \ll 1.$$

Every integer n may be decomposed into the form n_1n_2 where $(n_2, D) = 1$ and n_1 is constructed entirely from powers of the primes dividing D. Since the series $\sum |g_4(p^m)|^\theta p^{-m}$, taken over the powers of a single prime, converges,

$$x^{-1}\sum_{n\le x} |g_4(n)|^\theta \ll \sum |g_4(n_1)|^\theta n_1^{-1} < \infty.$$

For functions which satisfy condition (9.2) of Theorem 9.1, g_4 belongs to $\mathcal{L}^\alpha$ if and only if g belongs to $\mathcal{L}^\alpha$.

This ends the remarks.

Possessed of a method, we push on a little.

Theorem 11.1. *Let $\alpha > 1$. Let g be a multiplicative function for which the series*

$$\sum_{|g(p)|\le 3/2} p^{-1}|g(p)-1|^2, \qquad \sum_{|g(p)|>3/2} p^{-1}|g(p)|^\alpha, \qquad \sum\sum_{p,k\ge 2} p^{-k}|g(p^k)|^\alpha$$

converge. Define

$$A(f,x) = \sum_{\substack{q\le x\\|g(q)-1|\le 1/2}} q^{-1}\log g(q)$$

where the principal value of the logarithm is taken. Then

$$\lim_{x\to\infty} e^{-A(f,x)}x^{-1}\sum_{n\le x} g(n)$$

exists finitely, and there is a representation

$$x^{-1}\sum_{n\le x} g(n) = \prod_{p\le x}\left(1-\frac{1}{p}\right)\left(1+\frac{g(p)}{p}+\frac{g(p^2)}{p^2}+\cdots\right)+$$

$$+\,o\left(\exp\left(\operatorname{Re}\sum_{p\le x}\frac{g(p)-1}{p}\right)\right)$$

as $x \to \infty$.

Note. It is not assumed that g belongs to any space $\mathcal{L}^\alpha$.

Proof. For those functions which satisfy $|g(q)-1| \le 1/2$ on the prime powers these assertions may be justified using the method at the beginning of the Third Approach to Theorem 9.1, replacing $g(n)$ by $g(n)\exp(-A(f,x))$, and $g_\tau(n)$ by $g_\tau(n)\exp(-A(f_\tau,x))$, $1 \le n \le x$.

An application of Lemma 2.2 shows that

$$x^{-1}\sum_{n\le x}|g_\tau(n)|^2 \ll \exp\left(\sum_{p\le\tau}p^{-1}(|g(p)|^2-1)\right),$$

which from the convergence of the first series at (9.2) is

$$\ll \exp\left(\sum_{p\le\tau}p^{-1}2(|g(p)|-1)\right).$$

Morever, from this same condition the series $\sum p^{-1}(|g(p)| - \operatorname{Re} g(p))$, taken over those primes for which $|g(p)| \le 3/2$, converges. It follows readily that

$$\exp(-2\operatorname{Re} A(f_\tau,x))x^{-1}\sum_{n\le x}|g_\tau(n)|^2 \ll 1$$

uniformly in $x \ge \tau \ge 4$. Then

$$\lim_{\tau\to\infty}\sup_{x\ge\tau}\left|\exp(-A(f,x))x^{-1}\sum_{n\le x}g(n) - \exp(-A(f_\tau,x))x^{-1}\sum_{n\le x}g_\tau(n)\right| = 0.$$

It is easy to see that

$$E(\tau) = \lim_{y\to\infty}\exp(-A(f_\tau,y))y^{-1}\sum_{n\le y}g_\tau(n)$$

exists, and has the value

$$\prod_{p\le\tau}\left(1-\frac{1}{p}\right)\left(1+\frac{g(p)}{p}+\frac{g(p^2)}{p^2}+\cdots\right)\exp\left(-\sum_{m=1}^{\infty}\frac{\log g(p^m)}{p^m}\right).$$

Since $E = \lim E(\tau)$, $\tau \to \infty$ exists, our first assertion is justified.

Moreover,

$$e^{-A(f,x)}x^{-1}\sum_{n\le x} g(n) = E + o(1) = E(x) + o(1),$$

so that

$$x^{-1}\sum_{n\le x} g(n) = \prod_{p\le x}\left(1-\frac{1}{p}\right)\left(1+\frac{g(p)}{p}+\cdots\right) + o\left(\exp\left(\operatorname{Re}\sum_{p\le x}\frac{g(p)-1}{p}\right)\right)$$

as $x \to \infty$. The second assertion is justified.

More generally, still in the notation of the Third Approach, if g belongs to $\mathcal{L}^\alpha$ for some $\alpha > 1$, then so does g_4, and the local means of g_4 satisfy the oscillation condition (10.3). Since $g = g_5 * g_4$,

$$\begin{aligned}\sum_{n\le x} g(n) &= \sum_{m\le \log x} g_5(m) \sum_{r\le x/m} g_4(r) + O\left(x \sum_{\log x < m \le x} |g_5(m)|m^{-1}\right)\\ &= \sum_{m=1}^{\infty} g_5(m)m^{-1}\sum_{r\le x} g_4(r) + o(x), \quad x\to\infty.\end{aligned}$$

However, the example $g(q) = 1 - (\log\log q)^{-1}$ shows that the precision of this asymptotic representation need not suffice to remove the condition $|g(q)-1| \le 1/2$; the function $\exp(\operatorname{Re} A(f,x))$ may approach zero as $x \to \infty$. The complex valued $\exp(A(f,e^u))$ is a slowly varying function of u in a sense like that of Karamata. Slowly varying factors can be introduced into the method of the stable dual, but the results obtained do not apply in the present situation without further argument. Rather than pursue this complication, I modify the treatment of the special case.

Define multiplicative functions γ, δ by

$$\gamma(q) = \begin{cases} g(q) & \text{if } |g(q)-1| \le \frac{1}{2}\,, \\ 1 & \text{otherwise} \end{cases} \qquad \delta(q) = \begin{cases} g(q) & \text{if } |g(q)-1| > \frac{1}{2}, \\ 1 & \text{otherwise.} \end{cases}$$

We represent g as a product $\gamma\delta$, not as a convolution.

For each $\beta > 1$

$$\begin{aligned}&\|e^{-A(f,x)}g - e^{-A(f_\tau,x)}\gamma_\tau\delta\|_1\\ &\le \|f - f_\tau - \{A(f,x) - A(f_\tau,x)\}\|_{\beta'}(\|e^{-A(f,x)}g\|_\beta + \|e^{-A(f_\tau,x)}\gamma_\tau\delta\|_\beta),\end{aligned}$$

with the understanding that the additive function f is now defined in terms of γ rather than g. Let $\beta = \min(\sqrt{\alpha}, 2)$, so that $\beta' \geq 2$. From Theorem 2.1 the β' norm is

$$\ll \left(\sum_{\substack{q \leq x \\ q_0 > \tau}} |f(q)|^2 q^{-1} \right)^{1/2} + \left(\sum_{\substack{q \leq x \\ q_0 > \tau}} |f(q)|^{\beta'} q^{-1} \right)^{1/\beta'},$$

uniformly in $x \geq 2$. Since the $f(q)$ are uniformly bounded, these sums are $o(1)$ as $\tau \to \infty$.

Choose u so that $\beta u = \alpha$. Then

$$\|e^{-A(f,x)} g\|_\beta \leq \|e^{-A(f,x)} \gamma\|_{u'\beta} \|\delta\|_\alpha.$$

Here $\|\delta\|_\alpha^\alpha$ is

$$\ll x^{-1} \sum_{n \leq x} |\delta(n)|^\alpha \ll \sum_{n \leq x} |\delta(n)|^\alpha n^{-1} \ll \prod_{p \leq x} (1 + \delta(p)^\alpha p^{-1} + \cdots)$$

$$\ll \exp\left(\sum_{|g(q)-1|>1/2} |g(q)|^\alpha q^{-1} \right) \ll 1,$$

uniformly in $x \geq 2$. Moreover, with $v = u'\beta$, Lemma 2.2 shows that

$$\|e^{-A(f,x)} \gamma\|_v^v \ll \exp\left(\operatorname{Re} \left\{ -\sum_{q \leq x} v q^{-1} \log \gamma(q) + \sum_{q \leq x} q^{-1} (|\gamma(q)|^v - 1) \right\} \right),$$

which after Lemma 10.3 is also bounded uniformly for $x \geq 2$.

Similar estimates hold for the norm involving γ_τ. The argument continues as before save that $E(\tau)$ is modified by a factor

$$\prod_{p > \tau} \left(1 - \frac{1}{p}\right) \left(1 + \sum_{\substack{m=1 \\ |g(p^m)-1| \leq 1/2}}^{\infty} \frac{1}{p^m} + \sum_{\substack{m=1 \\ |g(p^m)-1| > 1/2}}^{\infty} \frac{g(p^m)}{p^m} \right),$$

a factor which approaches 1 as $\tau \to \infty$.

Theorem 11.1 follows.

Since the method of the stable dual adapts to allow the introduction of slowly varying factors, there is the possibility of formulating and establishing for Theorem 11.1 a version analogous to Theorem 9.1.

The argument employed by Delange to characterise multiplicative functions which have values in the complex unit disc, and a non-zero mean value, is quite different from the method of the stable dual. To establish the sufficiency of the conditions of Delange, from primes to integers, by applying the (standard) Turán–Kubilius inequality, is an idea of Rényi, [143]. I have adapted his idea during the Third Approach to Theorem 9.1, introducing complex logarithms and applying the Turán–Kubilius inequality for complex functions; rather than considering $|g(n)|$ and $\arg g(n)$ separately, and applying the Turán–Kubilius inequality for real functions, as he did. Within the aesthetic of stable duality, the proof of Theorem 9.1 implicit in the proof of Theorem 11.1 is perhaps the most satisfactory.

Variant proofs of Theorem 9.1 in the case $\alpha = 2$ were given by a number of authors, including Daboussi and Delange, [17], Schwarz and Spilker, [153]. A survey of these and related results may be found in Schwarz, [152].

Theorem 9.1 was generalised from $\alpha = 2$ to the cases $\alpha > 1$ independently by myself, [43], and using a different method, by Daboussi, [15]. The proof which I gave in my paper employed Turán–Kubilius duals and Fourier analysis, including results from the probabilistic theory of numbers.

I have given an elaborate treatment of Theorem 9.1, but the insight gained will enable me to deal with multiplicative functions in $\mathcal{L}^\alpha$ which have mean value zero. Before doing so it will be convenient to apply the method of the stable dual to give rapid proofs of the original theorems of Wirsing and Halász.

12

Exercises: Why the form?

A little topology. Let G be an abelian topological group. For a character χ_1 in $\widehat{G}$, a compact set K in G and $\varepsilon > 0$, consider the set $S(\chi_1, K, \varepsilon)$ of characters χ on G which satisfy $|\chi(y) - \chi_1(y)| < \varepsilon$ on every y in K. We may regard each $S(\chi_1, K, \varepsilon)$ as an analogue of a sphere in a metric space. Note that it contains χ_1 itself. The unions of these various sets comprise the open sets in the standard topology on $\widehat{G}$.

1. Prove that if χ_2 belongs to $S(\chi_1, K, \varepsilon)$, then there is an $\eta > 0$ for which $S(\chi_2, K, \eta)$ is a subset of $S(\chi_1, K, \varepsilon)$.

Hint: Prove that

$$\sup_{y \in K} |\chi_2(y) - \chi_1(y)| \leq \varepsilon - \theta$$

for some positive θ.

2. Prove that a set E in $\widehat{G}$ is open if and only if each character χ_1 in $\widehat{G}$ is contained in an $S(\chi_1, K, \varepsilon)$ contained in E.

3. Give $\mathbb{Z}$, the additive group of integers, the discrete topology. We identify a typical character on $\mathbb{Z}$ with a real $\beta (\text{mod}\, 1)$,

$$\chi_\beta : n \mapsto e^{2\pi i \beta n}.$$

Define a (translation invariant) metric ρ on $\widehat{\mathbb{Z}}$ by $\rho(\chi_\alpha, \chi_\beta) = \|\alpha - \beta\|$. Prove that this metric induces the standard topology on $\widehat{\mathbb{Z}}$.

4. As mentioned in the introductory background, $(Q^*)^\wedge$ is the direct product of denumerably many copies of $\mathbb{R}/\mathbb{Z}$. If p_j denotes the jth rational prime, then (in a usual way) we may define a metric δ on $(Q^*)^\wedge$ by

$$\delta(g, h) = \sum_{j=1}^{\infty} 2^{-j} \left(\frac{\|\alpha_j - \beta_j\|}{1 + \|\alpha_j - \beta_j\|} \right),$$

where $g(p_j) = e^{2\pi i \alpha_j}$, $h(p_j) = e^{2\pi i \beta_j}$, $j = 1, 2, \ldots$ and g, h are typical completely multiplicative functions into the complex unit circle. Prove that this metric induces the standard topology on $(Q^*)^\wedge$.

5. In actual practice it is convenient to use a family of metrics

$$\delta_\sigma(g,h) = \left(\sum_{j=1}^{\infty} p_j^{-\sigma}|g(p_j) - h(p_j)|^2\right)^{1/2},$$

on $(Q^*)^\wedge$, with $\sigma > 1$, varied according to the needs of the argument. Prove that each such metric also induces the standard topology.

In many problems we may confine ourselves to multiplicative functions g which are 1 on all but finitely many primes. The metric

$$\rho(g,h) = \left(\sum_{j=1}^{\infty} p_j^{-1}|g(p_j) - h(p_j)|^2\right)^{1/2}$$

then proves convenient. This metric may be used to topologise the direct sum of countably many copies of $\mathbb{R}/\mathbb{Z}$, one for each prime p_j. Unlike the direct product under the regime of the metric of exercise 4, the direct sum does not then lie within a sphere of finite radius.

Note that all these metrics are translation invariant. Moreover, each induces an equivalence relation on the set of all multiplicative functions.

6. Prove that a character χ_α in $\widehat{\mathbb{Z}}$ has order q if and only if α has the form $a/q(\bmod 1)$, $1 \le a \le q$, $(a,q) = 1$. Prove that the characters of prime order are dense in $\widehat{\mathbb{Z}}$.

The following exercise assumes some algebraic number theory.

7. Strictly speaking, a Dirichlet character cannot belong to $(Q^*)^\wedge$. We may allow it to if, according to the introductory background but contrary to classical practice, we define the character to be 1 on the prime factors of its associated modulus. Prove that with this understanding Dirichlet characters of prime order to prime moduli are dense in $(Q^*)^\wedge$.

Hint: Identify Dirichlet characters of a fixed order with power residue symbols and employ a reciprocity law; cf. [28], [29].

By extension, Theorem 9.1 asserts that to possess a non-zero mean value, a multiplicative function in $\mathcal{L}^\alpha$, $\alpha > 1$, needs to be near the identity of $(Q^*)^\wedge$. A characterisation of multiplicative functions which are similarly near to a Dirichlet character follows.

Almost periodic functions

Combinations of characters on $\mathbb{Z}$:

$$P(n) = \sum_{j=1}^{t} c_j e^{2\pi i \beta_j n}, \quad \beta_j \text{ in } \mathbb{R},\ c_j \text{ in } \mathbb{C},$$

form a linear space over $\mathbb{C}$. For $\alpha \geq 1$, let B^α denote the (Bohr–Besicovitch) space of arithmetic functions h with the property that for every $\varepsilon > 0$, some P satisfies $\|h - P\|_\alpha < \varepsilon$.

8. Prove that if h belongs to B^α, then h belongs to $\mathcal{L}^\alpha$. Moreover,

$$\lim_{x\to\infty} x^{-1} \sum_{n\leq x} h(n)\chi(n)$$

exists whenever χ is a character on $\mathbb{Z}$, or a Dirichlet character.

The real β for which $\chi(n) = e^{-2\pi i\beta n}$ yields a non-zero limit comprise the Fourier–Bohr spectrum of h.

9. Prove that if h belongs to B^α, then so does $|h|$.

Hint: If h is well approximated by P, and $|P(n)| \leq M$ on all positive integers n, then consider a polynomial approximation to $|y|$ on the interval $[-M, M]$.

10. Prove that if h belongs to B^α, then so does $|h|^{1/2}$.

11. Prove that if a multiplicative function g belongs to B^α and satisfies $\|g\|_1 > 0$, then the series $\sum p^{-1}$, taken over those primes for which $|g(p)| > 3/2$, converges.

Hint: Prove that

$$\lim_{M\to\infty} \limsup_{x\to\infty} x^{-1} \sum_{\substack{n\leq x \\ |g(n)|>M}} |g(n)| = 0.$$

12. Let g be a multiplicative function in $\mathcal{L}^\alpha$, $1 < \alpha \leq 2$. Let S be a finite collection of primes for which $|g(p)| \leq 3/2$, and Λ the corresponding sum of their reciprocals. Prove that

$$\Lambda \sum_{n\leq x} g(n)e^{2\pi i\beta n} = \sum_{\substack{pm\leq x \\ p\in S}} g(p)g(m)e^{2\pi i\beta pm} + O(x(\Lambda^{1/2} + \Lambda^{1-1/\alpha})),$$

the implied constant depending at most upon α. We have replaced a sum by (the specialisation of) a bilinear form. Moreover, one of the (implicit) variables in the form is supported on a set of primes over which we have considerable control.

Hint: Theorem 3.1.

The notation continues for the next two exercises.

13. Prove that for irrational β,

$$x^{-1} \sum_{m\leq x} \left| \sum_{\substack{p\leq x/m \\ p\in S}} g(p)e^{2\pi i\beta pm} \right|^2 \to \sum_{p\in S} p^{-1}|g(p)|^2, \quad x \to \infty.$$

14. Prove that

$$\limsup_{x\to\infty} x^{-1}\left|\sum_{n\le x} g(n)e^{2\pi i\beta n}\right| \ll \Lambda^{-1/\alpha} + \Lambda^{-1/\alpha'},$$

where $\alpha^{-1} + (\alpha')^{-1} = 1$.

15. Prove that a multiplicative function in B^α, $\alpha \ge 1$, cannot have irrationals in its Fourier–Bohr spectrum. Note that the case $\alpha = 1$ is included.

16. Let g belong to B^α and have a non-empty Fourier–Bohr spectrum. Prove that the limit in exercise 8 is non-zero for at least one Dirichlet character χ.

17. Establish the result of Daboussi, [14]: A multiplicative function g belongs to B^α and has a non-empty Fourier–Bohr spectrum if and only if for some Dirichlet character χ the series

$$\sum_p p^{-1}(g(p)\chi(p) - 1), \qquad \sum_{|g(p)|\le 3/2} p^{-1}|g(p)\chi(p) - 1|^2,$$

$$\sum_{|g(p)|>3/2} p^{-1}|g(p)|^\alpha, \quad \sum_p \sum_{k=2}^{\infty} p^{-k}|g(p^k)|^\alpha.$$

converge.

There is much to spare in this treatment. The classification of multiplicative functions from the viewpoint of almost periodic functions continues in the exercises of Chapter 18.

13

Theorems of Wirsing and Halász

Theorem 13.1. *Let g be a real multiplicative function, satisfying $|g(n)| \le 1$ on the positive integers, and for which the series $\sum p^{-1}(1-g(p))$ diverges. Then*

$$x^{-1}\sum_{n\le x} g(n) \to 0, \quad x\to\infty.$$

This result, variously ascribed as a conjecture of Wintner or Erdős, was first proved by Wirsing, using a different method, [170].

Proof. Let $m(x)$ denote the local mean of g. It follows readily from Lemma 2.2 that

$$m(x) \ll \exp\left(-\sum_{p\le x} p^{-1}(1-|g(p)|)\right).$$

We may therefore assume the convergence of the series $\sum p^{-1}(1-|g(p)|)$. Since $(1-g(p))^2 = 2(1-g(p)) - (1-|g(p)|^2)$, this ensures the divergence of the series $\sum p^{-1}(1-g(p))^2$.

The theorem may now be established directly from Theorem 6.6, or by various short cuts. I shall deduce it from Theorem 6.1.

Setting $a_n = g(n)$ in the dual of a classical Turán–Kubilius inequality, the case $\alpha = 2$ of Theorem 3.1, we obtain the approximate functional equation

$$\sum_{p\le x^{1/2}} \frac{1}{p}\left|g(p)m\left(\frac{x}{p}\right) - m(x)\right|^2 \ll 1.$$

Since $||u|-|v|| \le |u-v|$, this inequality continues to hold with $|g(p)|$ in place of $g(p)$, $|m(x)|$ in place of $m(x)$. The convergence of the series $\sum p^{-1}(1-|g(p)|)$ enables $|g(p)|$ to be replaced by 1. Applying the Cauchy–Schwarz inequality and appealing to an estimate from elementary number theory

$$\sum_{p\le x^{1/2}} \frac{1}{p}\left|\left|m\left(\frac{x}{p}\right)\right| - |m(x)|\right| \ll (\log\log x)^{1/2}. \tag{13.1}$$

This is an inequality of the type (6.2) considered in Theorem 6.1, with $f(p) = 0$.

We set $P = (\log\log x)^3$, $P^K = x^{1/3}$, $\tau = K^{1/2}$, $\psi(w) = \phi(w) = |m(x/w)|$ in Theorem 6.1. Clearly $\psi(w) \ll 1$. Moreover, again since $||u|-|v|| \le |u-v|$,

$$\sup_{w \le z \le w(1+P^{-1/3})} |\psi(w) - \psi(z)| \ll P^{-1/3}$$

uniformly for $P^{-\tau} \le w \le P^{\tau}$. We may therefore take a constant multiple of $(\log\log x)^{-1/2}$ for ε. From Theorem 6.1, $\phi(w) - \phi(1) = D\log w + O(\varepsilon)$ where $D\tau \log P \ll 1$. Over the range $1 \le w \le \exp((\log x)^{1/3})$ we have

$$|m(x/w)| - |m(x)| \ll \frac{(\log x)^{1/3}}{\tau \log P} + \varepsilon \ll (\log\log x)^{-1/2}.$$

If $|m(x)|$ exceeds a certain (absolute) multiple of $(\log\log x)^{-1/2}$, then $m(x/w)$ will have a constant sign, and

$$m(x/w) - m(x) \ll (\log\log x)^{-1/2}. \tag{13.2}$$

If the condition on $|m(x)|$ fails, then a similar estimate holds anyway.

With $T = \exp((\log x)^{1/3})$ we apply this estimate to the initial functional inequality:

$$|m(x)|^2 \sum_{p \le T} p^{-1}|g(p) - 1|^2 \ll 1.$$

The proof is complete.

Remark. Note that the large primes (in the range $(\log\log x)^3 < p \le x^{1/2}$) are used to estimate the oscillation of the local means $m(x)$, and the small primes (in the range $p \le \exp((\log x)^{1/3})$) are used to complete the proof.

Theorem 13.2. *Let g be a multiplicative function with values in the complex unit disc. Suppose that for each real λ the series $\sum p^{-1}(1 - \operatorname{Re} g(p)p^{i\lambda})$ diverges. Then*

$$x^{-1} \sum_{n \le x} g(n) \to 0, \quad x \to \infty.$$

This proposition was first proved by Halász, [92], using Fourier analysis in the complex plane.

Remarks. Since the summands are non-negative it is an easy exercise to show that the series over the primes diverges uniformly on every compact set of λ-values.

As for Theorem 13.1, an application of Lemma 2.2 shows that the series $\sum p^{-1}(1-|g(p)|)$ and so the series $\sum p^{-1}(1-|g(p)|^2)$ may without loss of generality be assumed convergent. Since

$$|1-g(p)p^{i\lambda}|^2 = 2(1-\operatorname{Re} g(p)p^{i\lambda}) - (1-|g(p)|^2),$$

we may assume the series $\sum p^{-1}|1-g(p)p^{i\lambda}|^2$ to diverge uniformly on every compact set of λ values.

Proof of Theorem 13.2. We begin with formally the same approximate functional equation as in the proof of Theorem 13.1. In the present case $g(p)$ and $m(x)$ are treated directly. The convergence of the series $\sum p^{-1}(1-|g(p)|)$ enables $g(p)$ to be replaced by $g(p)/|g(p)|$ whenever $g(p) \neq 0$, and to be replaced by 1 otherwise. The analogue of the equation (13.1) is

$$\sum_{p\le x^{1/2}} p^{-1}\left|z_p m\left(\frac{x}{p}\right) - m(x)\right| \ll (\log\log x)^{1/2}$$

with certain unimodular z_p. This is an equation of the type considered in Theorem 6.6.

We make the same choices for P, K, τ as before, but set $\psi(w) = \phi(w) = m(x/w)$. Then the analogue

$$m(x/w) = w^{i\lambda}m(x) + O((\log\log x)^{-1/2})$$

of the oscillation estimate (13.2) follows directly from Theorem 6.6. With once again $T = \exp((\log x)^{1/3})$

$$|m(x)|^2 \sum_{p\le T} \frac{1}{p}|g(p)p^{i\lambda} - 1|^2 \ll 1.$$

Moreover, if $|\lambda| > 1$, then arguing directly (cf. the initial application of Theorem 6.1 in Chapter 10)

$$\phi\left(1+\frac{1}{2|\lambda|}\right) - \phi(1) \ll |\lambda|^{-1} + x^{-1}.$$

Since $\lambda \ll P^{1/3}$, $\lambda m(x) \ll 1$ in every case. Altogether

$$|m(x)|^2\left(|\lambda|^2 + \sum_{p\le T} p^{-1}|g(p)p^{i\lambda} - 1|^2\right) \ll 1,$$

and the proof is complete.

Remark. This proof of Halász' theorem shows that if

$$\limsup_{x\to\infty} x^{-1} \left| \sum_{n\le x} g(n) \right| \ge c > 0,$$

then there are c_1, c_2 depending only upon c, and a real τ, $|\tau| \le c_1$, so that

$$\sum \frac{1}{p}(1 - \operatorname{Re} g(p)p^{-i\tau}) \le c_2.$$

I shall use this remark in Chapter 16.

For a multiplicative function g with values in the complex unit disc, the asymptotic behaviour of the local mean depends upon whether the series $\sum p^{-1}(1 - \operatorname{Re} g(p)p^{i\tau})$, taken over the primes, converges for some real τ. Could there be more than one such τ? I shall show that there cannot.

For characters on Q^* we may argue in terms of the family of translation invariant metrics introduced in exercise 5 of Chapter 12. Convergence of the series $\sum p^{-1}(1 - \operatorname{Re} g(p)p^{i\tau})$ amounts to existence of a uniform bound on the distances $\delta_\sigma(n^{i\tau}, g)$, $\sigma > 1$. Here I have employed a clear but corrupt notation. If the assertion is to hold for values τ_1 and τ_2 of τ, then the $\delta_\sigma(n^{i\tau_1}, n^{i\tau_2})$ and so $\delta_\sigma(n^{i(\tau_1-\tau_2)}, 1)$ will also be uniformly bounded. We are reduced to showing that the function $n^{i\tau}$ cannot be near to 1 unless $\tau = 0$. For increased generality and not to apply an exercise, I argue directly.

The representation(s)

$$\zeta(s) = s\int_1^\infty y^{-s-1}[y]dy = s(s-1)^{-1} + s\int_1^\infty y^{-s-1}([y]-y)dy, \tag{13.3}$$

are obtained using integration by parts. The second of the integrals is well defined in the punctured half-plane $\sigma = \operatorname{Re}(s) > 0$, $s \ne 1$, and defines a continuous function there. In particular, for real $\tau \ne 0$, $\lim \zeta(\sigma + i\tau)$, $\sigma \to 1+$ exists. Clearly $(\sigma - 1)\zeta(\sigma) \to 1$ as $\sigma \to 1+$.

Lemma 13.3. *The series* $\sum p^{-1}(1 - \operatorname{Re} p^{i\tau})$ *diverges for every real* $\tau \ne 0$.

Proof. If for some non-zero τ the series converges, then

$$\lim_{\sigma\to 1+} \sum p^{-\sigma}(1 - \operatorname{Re} p^{i\tau})$$

exists. From the Euler product representation of $\zeta(s)$

$$\zeta(\sigma)\zeta(s)^{-1} = \prod_p \left(\frac{1-p^{-s}}{1-p^{-\sigma}}\right) = \exp\left(\sum_{m=1}^\infty \frac{1}{m}\left(\frac{1}{p^{m\sigma}} - \frac{1}{p^{ms}}\right)\right).$$

In the exponential the terms with $m > 1$ comprise a series uniformly absolutely convergent in every half-plane $\sigma \ge 1/2 + \delta > 1/2$. Our temporary hypothesis ensures that $\lim |\zeta(\sigma)\zeta(s)^{-1}|$ and therefore $\lim \zeta(\sigma)$ exists as $\sigma \to 1+$. This we have shown not to be.

To reach the general case I employ the following result.

Lemma 13.4. *The inequality*

$$1 - \operatorname{Re} z_1\bar{z}_2 \le 2(1 - \operatorname{Re} w_1 z_1) + 2(1 - \operatorname{Re} w_2 z_2).$$

holds uniformly for z_1, z_2, w_1, w_2 in the complex unit disc and satisfying $w_1\bar{w}_2$ real.

Remark. In the complex unit disc $|1 - z|^2 \le 2(1 - \operatorname{Re} z)$ with equality if and only if z is unimodular.

Proof of Lemma 13.4. If a w_j vanishes, then the asserted inequality is trivially valid. For $w_j \neq 0$ we may replace z_j by $z_j \exp(i \arg w_j)$ and reduce ourselves to the case that w_j is real and positive. If some $-\operatorname{Re} z_j$ is non-negative, then the inequality is again immediate. Without loss of generality each $-\operatorname{Re} z_j$ is negative, and $w_1 = 1 = w_2$.

Set $z_j = \rho_j \exp(i\theta_j)$, $0 \le \rho_j \le 1$. We are to prove

$$2\sum_{j=1}^{2}(1 - \rho_j \cos\theta_j) - (1 - \rho_1\rho_2 \cos((\theta_1 - \theta_2))$$

non-negative. From the alternative representation

$$2(1 - \rho_1 \cos\theta_1)(1 - \rho_2\cos\theta_2) + 1 - \rho_1\rho_2\cos(\theta_1 + \theta_2)$$

this is evident.

Suppose now that the series $\sum p^{-1}(1 - \operatorname{Re} g(p)p^{i\tau_j})$, $j = 1, 2$, converge. From Lemma 13.4 so will the series $\sum p^{-1}(1 - p^{i(\tau_1 - \tau_2)})$, and from Lemma 13.3, $\tau_1 = \tau_2$.

Remarks. (i) The inequality

$$1 - \operatorname{Re} z_1 \cdots z_r \le r\sum_{j=1}^{r}(1 - \operatorname{Re} z_j)$$

is valid for $r \ge 1$ and the z_j in the complex unit disc. For $r = 2$ it is a particular case of Lemma 13.4. Let $r \ge 3$. If any $\operatorname{Re} z_j < 1/r$, then the corresponding term $r(1 - \operatorname{Re} z_j)$ guarantees the result. Assuming $\operatorname{Re}(z_j) \ge 1/r$, partial differentiation with respect to $|z_j|$ shows that without loss of generality z_j may be taken on the unit circle: $e^{i\theta_j}$ with θ_j real. The expression to be dominated then has the alternative representation

$$2\left(\sin\frac{1}{2}(\theta_1 + \cdots + \theta_r)\right)^2.$$

Application of the functional equation for sine together with the Cauchy–Schwarz inequality now suffices.

(ii) An appeal to its Euler product representation shows that $K(s)$, the Dirichlet series corresponding to a multiplicative function k with values in the complex unit disc, and vanishing on powers of 2, satisfies

$$e^{-1} \le |K(s)| \exp\left(-\sum_{p\ge 3}\sum_{m=2}^{\infty} \operatorname{Re} k(p^m)p^{-ms} \right) \le e,$$

uniformly in the half-plane $\sigma = \operatorname{Re}(s) > 1$. Combined with Lemma 13.4 this yields

Lemma 13.5.

$$\prod_{j=1}^{2} \left| \sum_{n=1}^{\infty} g(n)h_j(n)n^{-\sigma_j} \right|^2 \le 2e^{10}\zeta(\sigma)^3 \left| \sum_{(n,2)=1} h_1(n)\overline{h_2(n)}n^{-\sigma} \right|$$

uniformly for multiplicative functions g, h_1, h_2 with values in the complex unit disc, real $\sigma_j \ge \sigma > 1$, $j = 1, 2$.

If $h_j(n) = \chi_j(n)n^{-i\tau_j}$, $s_j = \sigma_j + i\tau_j$, $\sigma_j \ge \sigma > 1$, $j = 1, 2$, then the Dirichlet series $G(s_j, \chi_j)$ at s_j, corresponding to the function $g\chi_j$, g braided with the Dirichlet character χ_j, satisfies

$$|G(s_1, \chi_1)G(s_2, \chi_2)|^2 \le 3e^{10}\zeta(\sigma)^3 |L(\sigma + i(\tau_1 - \tau_2), \chi_1\bar{\chi}_2)|,$$

with a Dirichlet L-series in the upper bound; cf. [58], [75]. In particular, when g is real-valued,

$$|G(\sigma + i\tau)|^4 \le 3e^{10}\zeta(\sigma)^3 |\zeta(\sigma + 2i\tau)|$$

uniformly for real $\sigma > 1, \tau$.

(iii) Let g be a real valued multiplicative function which satisfies $|g(n)| \le 1$ on the positive integers. If the series $\sum p^{-1}(1 - g(p))$ diverges, then by Wirsing's theorem g has asymptotic mean value zero. If the same series over the primes converges, then by remark (i) at the end of Chapter 11, or by Theorem 11.1,

$$\lim_{x\to\infty} x^{-1} \sum_{n\le x} g(n)$$

exists. This limit always exists, a result due to Wirsing, [170].

(iv) If we allow g to assume values in the complex unit disc, then either $\sum p^{-1}(1 - \operatorname{Re} g(p)p^{i\tau})$ diverges for all real τ, or not. In the first case Halász' theorem shows that g has asymptotic mean value zero. In the second case, Theorem 11.1 guarantees the existence of a real $\omega(x)$ so that

$$\lim_{x\to\infty} x^{-1}\exp(-i\omega(x))\sum_{n\leq x} g(n)$$

exists. In fact there are always constants A, real τ, and a slowly oscillating function L, from the positive reals to the complex unit circle, so that

$$\sum_{n\leq x} g(n) = Ax^{1+i\tau}L(\log x) + o(x), \qquad x\to\infty.$$

Conjectured by Wirsing in the above paper, this was established by Halász, [92]. We do not need this refinement, which may be readily deduced from what we have obtained so far. However, it is convenient to note that for multiplicative functions with values in the complex unit disc,

$$\lim_{x\to\infty} x^{-1}\left|\sum_{n\leq x} g(n)\right|$$

always exists.

14

Again Wirsing's theorem

An attraction of the theorem of Wirsing established in Chapter 13 is that applied to the Möbius function $\mu(n)$ it furnishes a proof of the prime number theorem. However, since the method of the stable dual as exposed in this monograph employs the good distribution of prime numbers over short intervals guaranteed by Lemma 6.2, a sharp version of the prime number theorem has been assumed. In this chapter I show how experience gained in the method of the stable dual together with a little Fourier analysis, yields a reasonably simple proof of Wirsing's theorem that does not appeal to the prime number theorem.

To begin with g will denote a multiplicative function with values in the complex unit disc and associated Dirichlet series $G(s)$. In the notation of Theorem 13.2, $m(x)$ will denote the local mean $x^{-1}\sum g(n)$, $1 \le n \le x$.

Theorem 14.1. *The following propositions are equivalent:*

(i) $\lim_{x\to\infty} m(x) = 0$,

(ii) $\lim_{x\to\infty} \frac{1}{\log x}\int_1^x \frac{|m(y)|^2}{y}\,dy = 0$,

(iii) $\lim_{\sigma\to 1+}(\sigma-1)\int_{-\infty}^{\infty}\left|\frac{G(\sigma+i\tau)}{\sigma+i\tau}\right|^2 d\tau = 0$.

Proof. That (i) implies (ii) is clear. In the other direction, let $0 < \varepsilon < 1/2$, and let S denote the sum of the reciprocals of the primes in the interval $(x^\varepsilon, x^{1/2}]$. Setting $a_n = g(n)$ in the mean-square case of Theorem 3.1, the dual of a classical Turán–Kubilius inequality, we obtain

$$\sum_{x^\varepsilon < p \le x^{1/2}} \frac{1}{p}\left|g(p)m\left(\frac{x}{p}\right) - m(x)\right|^2 \ll 1,$$

and after an application of the Cauchy–Schwarz inequality,

$$m(x) = S^{-1}\sum_{x^\varepsilon < p \le x^{1/2}} \frac{g(p)}{p} m\left(\frac{x}{p}\right) + O(S^{-1/2}).$$

For $1 \le u \le v$, the function $m(x)$ satisfies

$$m(v) - m(u) = v^{-1}(u - v)m(u) + v^{-1} \sum_{u<m\le v} g(m).$$

Denote by I a typical interval $[y, y(1+\varepsilon)]$, $0 < y \le x\varepsilon^{-1}(1+\varepsilon)^{-1}$. Uniformly for p, w in I

$$\left|\left|m\left(\frac{x}{p}\right)\right| - \left|m\left(\frac{x}{w}\right)\right|\right| \le \left|m\left(\frac{x}{p}\right) - m\left(\frac{x}{w}\right)\right| \le 4\varepsilon.$$

We average the outside terms of these inequalities over I using the measure $w^{-1}dw$:

$$\left|m\left(\frac{x}{p}\right)\right| \le \frac{1}{\log(1+\varepsilon)} \int_I \left|m\left(\frac{x}{w}\right)\right| \frac{dw}{w} + 4\varepsilon.$$

Cover the interval $(x^\varepsilon, x^{1/2}]$ by adjoining intervals I. Bearing in mind the elementary estimate

$$\sum_{p\le z} \frac{1}{p} = \log\log z + \text{constant} + O\left(\frac{1}{\log z}\right), \qquad z \ge 2,$$

we may sum over the primes p in I, then over the intervals I, to reach

$$\sum_{x^\varepsilon<p\le x^{1/2}} \frac{1}{p}\left|m\left(\frac{x}{p}\right)\right| \ll \frac{1}{\log(1+\varepsilon)\log x^\varepsilon} \int_1^x \left|m\left(\frac{x}{w}\right)\right| \frac{dw}{w} + \varepsilon S.$$

Together with the approximate functional equation, this yields

$$m(x) \ll \frac{1}{\varepsilon^2 \log(1/\varepsilon)} \cdot \frac{1}{\log x} \int_1^x \frac{|m(y)|}{y}\, dy + \varepsilon,$$

the implied constant absolute for x sufficiently large in terms of ε. That (ii) implies (i) follows rapidly from an application of the Cauchy–Schwarz inequality.

We may seem not to have gone far. However, experience in analytic number theory shows that an average of an arithmetic function may often be more readily dealt with than the function itself. In Fourier analytic terms we are 'smoothing' the function to be transformed. For the present theorem the derivation of the approximate functional equation will contain our entire appeal to the number theoretical nature of g. For the equivalence of propositions (ii) and (iii) we need only that the values of g not be 'large'.

The change of variables $u = e^w$ in the integral representation

$$s^{-1}G(s) = \int_1^\infty u^{-s}m(u)du, \quad s = \sigma + i\tau,$$

shows $(s\sqrt{2\pi})^{-1}G(s)$ as a function of τ to be the Fourier transform of the function which is $m(e^w)e^{-w(\sigma-1)}$ when $w > 0$, and zero otherwise. Since $s^{-1}G(s)$ and its transform belong to the Lebesgue class $L^2(\mathbb{R})$, Plancherel's theorem gives

$$\int_{-\infty}^\infty \left|\frac{G(s)}{s}\right|^2 d\tau = 2\pi \int_0^\infty |m(e^w)e^{-w(\sigma-1)}|^2 dw = 2\pi \int_1^\infty |m(u)|^2 u^{-2\sigma+1} du.$$

With $\varepsilon > 0$, $\sigma = 1 + (\log x)^{-1}$, the hypothesis (ii) ensures that

$$(\sigma - 1)\int_1^{x^{1/\varepsilon}} \frac{|m(u)|^2}{u^{2\sigma-1}} du \to 0, \quad x \to \infty.$$

Moreover, from the trivial estimate $|m(u)| \le 1$ the corresponding weighted integral over the range $(x^{1/\varepsilon}, \infty]$ does not exceed $\frac{1}{2}e^{-2/\varepsilon}$. That (ii) implies (iii) follows rapidly. Arguing the converse is simpler.

The proof of Theorem 14.1 is complete.

The argument employing Plancherel's relation during the proof of the Theorem 14.1 shows that at $\zeta(s)$ the integral of $|G(s)|^2|s|^{-2}$ over $\mathbb{R}$ is extremal. The analyst enquires whether this remains true if $|G(s)|^2$ is replaced by some other power of $|G(s)|$? It is worthwhile to pursue this question.

Lemma 14.2. *Let $\delta > 0$. Then*

$$\int_{\mathbb{R}} |G(s)|^\delta |s|^{-2} d\tau \le e^\delta \int_{\mathbb{R}} |\zeta(s)|^\delta |s|^{-2} d\tau,$$

uniformly for g with values in the complex unit disc, $\sigma > 1$.

Proof. We make a further appeal to the number theoretic nature of g. For completely multiplicative g, the Euler product of $G(s)$ affords the representation

$$G(s)^{\delta/2} = \exp\left(\frac{\delta}{2}\sum_p \sum_{m=1}^\infty \frac{1}{m}\left(\frac{g(p)}{p^s}\right)^m\right), \quad \sigma > 1,$$

corresponding to which the integral involving $|G(s)|^\delta$ has the representation

$$\int_{\mathbb{R}} \left|\exp\left(\sum_p \sum_m \frac{\delta g(p^m)}{2mp^{ms}}\right)\right|^2 \frac{d\tau}{|s|^2}.$$

Interpreting the exponential as a Dirichlet series and appealing to Plancherel's relation as in the proof of Theorem 14.1 shows this integral not to exceed the similar one with every $g(p)$ replaced by 1. For completely multiplicative functions, in particular for characters on Q^*, the inequality of Lemma 14.2 holds with e^δ replaced by 1.

Otherwise, for each multiplicative g define the completely multiplicative function g_1 by $g_1(p) = g(p)$, and the function h by Dirichlet convolutions $g = h * g_1$. Then h is multiplicative, $h(p) = 0$, $h(p^k) = g(p^k) - g(p^{k-1})g(p)$, $k \geq 2$. In particular

$$\left|\sum_{n=1}^{\infty} h(n)n^{-s}\right| \leq \prod_p \left(1 + 2\sum_{k=2}^{\infty} \frac{1}{p^k}\right) \leq \exp\left(\sum \frac{2}{p(p-1)}\right) < e,$$

so that

$$|G(s)| \leq e \left|\sum_{n=1}^{\infty} g_1(n)n^{-s}\right|$$

in the half-plane $\operatorname{Re}(s) > 1$. The complete lemma is apparent.

We may appreciate Lemma 14.2, using a bound on $\zeta(s)$. For each positive integer N, again integrating by parts,

$$\zeta(s) = \sum_{n=1}^{N} n^{-s} + \frac{N^{1-s}}{s-1} + s\int_N^\infty \frac{[y]-y}{y^{s+1}}\, dy.$$

The choice $N = 2 + [|s|]$ gives

$$\zeta(s) \ll \sum_{n=1}^{N} \frac{1}{n} + \frac{1}{|s-1|} + |s|\int_N^\infty \frac{dy}{y^2} \ll |s-1|^{-1} + \log(2+|s|).$$

With this bound, for each fixed $\delta > 1$,

$$\int_{\mathbb{R}} |G(s)|^\delta |s|^{-2} d\tau \ll (\sigma - 1)^{1-\delta}$$

uniformly in $\sigma > 1$, g with values in the complex unit disc.

We are almost ready for Wirsing's theorem.

Lemma 14.3. *Assume that for the real valued g the series $\sum p^{-1}(1-g(p))$ diverges. Then*

$$\sup_{|\tau|\le\alpha} \zeta(\sigma)^{-1}|G(s)| \to 0, \ \sigma \to 1+,$$

where $\alpha = \exp((\sigma-1)^{-1/2})$.

Proof. Employing Euler products, as in the treatment of Theorem 9.1 towards the end of Chapter 9, we see that

$$\zeta(\sigma)^{-1}|G(s)| \ll \exp\left(- \sum_{3\le p\le\beta} \frac{1}{p}(1-g(p)) - \sum_{3\le p\le\beta} g(p)\left(\frac{1}{p} - \frac{1}{p^\sigma}\right) \right.$$
$$\left. + \operatorname{Re} \sum_{3\le p\le\beta} \frac{g(p)}{p^\sigma}\left(\frac{1}{p^{i\tau}} - 1\right)\right)$$

with $\beta = \exp((\sigma-1)^{-1})$, uniformly in $\sigma = \operatorname{Re}(s) > 1$. Let $0 < \varepsilon < 1$. On $\operatorname{Re}(s) = \sigma$, $|\tau| \le \varepsilon^{-4}(\sigma-1)$, the second explicit sum is bounded, the third

$$\ll |\tau| \sum_{p\le\beta} \frac{\log p}{p} \ll |\tau| \log\beta \ll \varepsilon^{-4}.$$

For σ sufficiently near to 1, $\zeta(\sigma)^{-1}|G(\sigma)| < \varepsilon$.

On $\operatorname{Re}(s) = \sigma$, $\varepsilon^{-4}(\sigma-1) < |\tau| \le \alpha$, Lemma 13.5 together with the above bound for $|\zeta(s)|$ gives

$$\zeta(\sigma)^{-1}|G(s)| \ll (\zeta(\sigma)^{-1}|\zeta(\sigma+2i\tau)|)^{1/4}$$
$$\ll \left((\sigma-1)\left(\frac{1}{\sigma-1+|\tau|} + \log(2+|\tau|)\right)\right)^{1/4},$$

less than a constant multiple of ε when σ is sufficiently near to 1.

Lemma 14.3 is proved. The choice of α is not particularly significant.

Second proof of Theorem 13.1. We establish proposition (iii) of Theorem 14.1. The trivial bound $|G(s)| \le \zeta(\sigma)$ shows the contribution of the tail $|\tau| > \alpha$ towards the integral not to exceed $2\zeta(\sigma)^2\alpha^{-1} = o(1)$, as $\sigma \to 1+$. Appeals to Lemma 14.3 and Lemma 14.2, respectively, show that over the complementary range of τ-values, $(\zeta(\sigma)^{-1}|G(s)|)^{1/2}$ is uniformly $o(1)$ as $\sigma \to 1+$, and the integrals of $\zeta(\sigma)^{-1/2}|s|^{-2}|G(s)|^{3/2}$ are uniformly bounded.

15

Exercises: The prime number theorem

In classical notation, $\pi(x)$ denotes the number of primes not exceeding x. Hadamard and de la Vallée Poussin established the prime number theorem

$$\pi(x)\frac{\log x}{x} \to 1, \quad x \to \infty,$$

in 1896. Their proofs combined Hadamard's theory of integral functions with Riemann's analytic continuation of $\zeta(s)$ to the punctured complex plane, $s \neq 1$.

1. Prove that

$$\zeta(\sigma)^3|\zeta(\sigma + i(\tau_1 - \tau_2))\zeta(\sigma + i\tau_1)^2\zeta(\sigma + i\tau_2)^2| \geq 1$$

is valid for real τ_1, τ_2 and $\sigma > 1$.

Subsequent proofs were in part motivated by a will to reduce the prominent rôle of the theory of complex variables. Landau, at the beginning of the twentieth century, showed that the proposition

$$x^{-1}\sum_{n \leq x} \mu(n) \to 0, \quad x \to \infty,$$

and the prime number theorem could be derived, the one from the other, by elementary means, [116].

Define the arithmetic function r by $r(n) = \log n - \tau(n) + 2\gamma$, where $\tau(n)$ denotes Dirichlet's divisor function, γ Euler's constant.

2. Prove that $\mu * r$ at n is $\Lambda(n) - 1$ if $n > 1$, $\Lambda(n) - 1 + 2\gamma$ if $n = 1$. Moreover,

$$\sum_{n \leq y} r(n) \ll y^{1/2}$$

uniformly for $y \geq 1$.

It is easier to consider the weighted sum

$$\psi(x) = \sum_{n \le x} \Lambda(n)$$

than the prime counting function itself.

3. Prove that $\psi(x) \ll x$, $x \ge 2$. Moreover, if μ has asymptotic mean value zero, then $x^{-1}\psi(x) \to 1$, $x \to \infty$.

Integration by parts shows that the first of these estimates amounts to the upper bound of Chebyshev, the second to the prime number theorem.

4. Use the relation $\mu \log = \mu * \Lambda$ and the fact that

$$\sum_{r \le x} \mu(r) \left[\frac{x}{r}\right] = 1, \quad x \ge 1,$$

to argue from the prime number theorem to the proposition that μ has asymptotic mean value zero, in an elementary fashion.

Introducing ideas of Borel and Carathéodory, Landau obviated the need to analytically continue $(s-1)\zeta(s)$ over the whole plane. It would suffice to reach some half-plane $\sigma > \alpha$ with $\alpha < 1$. This enabled him to give the first proof of the prime ideal theorem; the corresponding Dedekind zeta function had at that time only the more limited continuation, [115].

With his tauberian theorem, Wiener, [168], [169], showed that not only would it suffice to know the behaviour of $(s-1)\zeta(s)$ in the half-plane $\sigma \ge 1$; for large values of τ, $\zeta(1+i\tau)$ need not be bounded explicitly in terms of $|\tau|$. Let us remove the need for such bounds in our second proof of Wirsing's theorem.

5. In the notation of Theorem 14.1, prove that

$$\int_{|\tau| \le 1} |G(s)|^2 d\tau \ll (\sigma - 1)^{-1},$$

the implied constant absolute.

6. Prove that for $T \ge 1$, and with the same uniformities,

$$\int_{|\tau| \ge T} \left|\frac{G(s)}{s}\right|^2 d\tau \ll \frac{1}{T(\sigma - 1)}.$$

Note that these inequalities remain valid even if g is not multiplicative.

7. Modify the proof of Wirsing's theorem, given in Chapter 14, to employ only that $\zeta(s) - (s-1)^{-1}$ is uniformly bounded in each box $1 < \sigma \le 2$, $|\tau| \le T$.

An elementary proof of the prime number theorem was achieved by Erdős and Selberg in 1949, [83], [155], respectively. The foundation for their proofs was an approximate formula of Selberg:

$$\sum_{n\le x}\Lambda(n)\log n+\sum_{mn\le x}\Lambda(m)\Lambda(n)=2x\log x+O(x),\quad x\ge 1.$$

8. In the notation of exercise 2, prove that $\mu\log=-\Lambda*\mu$, $(\Lambda-\mathbf{1})\log=(\mu\log)*r+\mu*(r\log)$, $(\Lambda-\mathbf{1})\log+\Lambda*(\Lambda-\mathbf{1})=\mu*(r\log)$ and establish Selberg's formula.

Once a completely elementary proof of the prime number theorem was achieved, ambition moved towards a theory both strong enough to deliver the prime number theorem and applicable to the study of multiplicative functions allowed arbitrary values in the complex unit disc. This was achieved in two stages, by Wirsing, [170], and Halász, [92], in papers surely to be classics.

9. Modify the proof of Wirsing's theorem, given in Chapter 14, to deliver a proof of Halász' theorem.

Hint: Exercise 6.

In his original paper, Halász estimates $m(x)$ by an L^1-Mellin inversion in the complex plane. The use of $g\log$ rather than g allows a factorisation $(-G'/G)G$ of the corresponding Dirichlet series, preparatory to an appeal to Plancherel's relation through the agency of the Cauchy–Schwarz inequality.

Amplify the notation of Theorem 14.1 with

$$m_1(x)=x^{-1}\sum_{n\le x}g(n)\log n.$$

10. Prove the following propositions equivalent.

(i) $\lim_{x\to\infty}m(x)=0$,

(ii) $\lim_{x\to\infty}\frac{1}{\log x}\,m_1(x)=0$,

(iii) $\lim_{x\to\infty}\frac{1}{(\log x)^3}\int_1^x\frac{|m_1(y)|^2}{y}\,dy=0$,

(iv) $\lim_{\sigma\to 1+}(\sigma-1)^3\int_{\mathbb{R}}\left|\frac{G'(\sigma+i\tau)}{\sigma+i\tau}\right|^2d\tau=0$.

11. Apply a Chebyshev bound for $\psi(x)$ to show that for completely multiplicative g,

$$\int_{\mathbb{R}}\left|\frac{G'(s)}{sG(s)}\right|^2d\tau\ll\frac{1}{\sigma-1},$$

and give another proof of Halasz' theorem.

During a conversation with H.-E. Richert I mentioned that curiously enough the application of Plancherel's relation to Dirichlet series occurs already in a 1956 paper of Chudakov, [12]. "It occurs already in my Ph.D.", he rapidly retorted.

Whilst the reduction of the local mean $m(x)$ to its logarithmic average, given in Theorem 14.1, is quantitatively weak, it allows complete control over the primes p appearing in the approximate functional equation involving $m(x)$ and S. At the price of weakening this control, the following method, still elementary, increases precision by applying results from the theory of sieves.

12. The arithmetic function $\eta(n)$ is defined to be 1 if $n = 1$, and 0 if $n > 1$. Prove that

$$\eta(n) = \sum_{\substack{d|n \\ \omega(d)\le 2r}} \mu(d) + O\left(\sum_{\substack{d|n \\ \omega(d)=2r}} |\mu(d)| \right)$$

where $\omega(d)$ denotes the number of distinct prime factors of n, is valid uniformly for all positive integers n and r.

13. Prove that for any arithmetic function h and positive integer D

$$\sum_{n\le x} h(n) = \sum_{\substack{n\le x \\ (n,D)=1}} h(n) - \sum_{\substack{d|D \\ d>1 \\ \omega(d)\le 2r}} \mu(d) \sum_{\substack{n\le x \\ n\equiv 0 (\bmod d)}} h(n)$$

$$+ O\left(\sum_{n\le x} |h(n)| \sum_{\substack{d|n \\ \omega(d)=2r}} |\mu(d)| \right),$$

the implied constant absolute.

14. Let $x \ge 2$, $0 < \varepsilon < A\varepsilon \le 1$. Let D be the product of the primes p in the interval $(x^\varepsilon, x^{A\varepsilon}]$. Show that if $r \ge e^2 \log A + c_1$ for a certain positive absolute constant c_1, then the number of integers not exceeding x, which are divisible by some divisor d of D with $\omega(d) = 2r$, is $\ll A^{-1}x$.

15. In the notation of exercise 14, show that if $d \mid D$ and $\omega(d) \le 2r$, then $d \ll x^\gamma$ with $\gamma = 2e^2\varepsilon A \log A$, and an absolute implied constant.

16. Prove that any multiplicative function with values in the complex unit disc satisfies

$$\sum_{n\le x} g(n) \ll \sum_{\substack{d|D \\ x^\varepsilon < d \le x^\gamma}} |\mu(d)| \left| \sum_{n\le x/d} g(n) \right| + A^{-1}x$$

for all x sufficiently large in terms of ε and A.

17. Let $B > 0$. Prove that if $x^\varepsilon < U \le x^\gamma$, then the number of divisors of D which lie in the interval $(U, U + U(\log x)^{-B})$ is $\ll \varepsilon^{-1} U (\log x)^{-B-1}$, uniformly for $x^\varepsilon \ge 2$.

18. Let L denote $\log x$. Prove that

$$|m(x/d)|^2 \ll L^B \int_U^{U+UL^{-B}} |m(x/y)|^2 y^{-1} dy + L^{-B}$$

uniformly for $U < d \le U + UL^{-B}$.

19. Prove that

$$\sum_{\substack{d|D \\ x^\varepsilon < d \le x^\gamma}} \frac{1}{d} \left| m\left(\frac{x}{d}\right)\right|^2 \ll \frac{1}{\varepsilon \log x} \int_{x^{1-\gamma}}^{x^{1-\varepsilon}} \frac{|m(u)|^2}{u} du + \frac{1}{\varepsilon L^{B+1}}.$$

20. Prove that

$$m(x) \to 0 \quad \text{if and only if} \quad \frac{1}{\log x} \int_{x^{1/2}}^{x} \frac{|m(u)|^2}{u} du \to 0, \quad x \to \infty.$$

Whereas the proof of Theorem 14.1 allows a power of $\log \log x$ to be saved over the trivial estimate for $m(x)$, the above argument allows a power of $\log x$ to be saved, as the following can show.

21. Prove that there is a positive constant c so that

$$x^{-1} \sum_{n \le x} g(n) \ll \left((\sigma - 1) \int_{\mathbb{R}} \left| \frac{G(s)}{s} \right|^2 d\tau + (\log x)^{-1} \right)^c,$$

where $\sigma = 1 + (\log x)^{-1}$, $s = \sigma + i\tau$, and $G(s)$ is the Dirichlet series associated with g.

An application of the above reduction argument to the study of multiplicative functions on residue classes, and so to the distribution of primes in such classes, may be found in [75].

We may abandon all control over the interpolating primes, view the logarithmic function as $\mathbf{1} * \Lambda$ and consider directly the local mean of $g \log$, to obtain

$$m(x) \log x = \sum_{p \le x} \frac{g(p) \log p}{p} m\left(\frac{x}{p}\right) + O(x).$$

This equation, too, is effective in the pursuit of the prime number theorem through the agency of the Möbius function, but the presence of primes as large as x complicates its application to the problems considered in [75].

Time has passed. What the elementary proof of the prime number theorem perhaps provoked, can now be used to set it.

22. Let $0 < \varepsilon < 1$. Deduce from Selberg's formula that $\psi(x(1+\varepsilon)) - \psi(x) \le 3\varepsilon x$ for all x sufficiently large in terms of ε.

23. Deduce from Selberg's formula that $E(x) = \psi(x) - x$ satisfies

$$E(x) \log x \ll \sum_{p \le x} \log p \left| E\left(\frac{x}{p}\right)\right| + O(x), \quad x \ge 2.$$

By the argument of Theorem 14.1 show that the prime number theorem will follow if

$$\lim_{\sigma \to 1+} (\sigma - 1) \int_{\mathbb{R}} \left| \frac{\zeta'(s)}{\zeta(s)} + \zeta(s) \right|^2 \frac{d\tau}{|s|^2} = 0,$$

and establish the latter using only that $\zeta(s) - (s-1)^{-1}$ is analytic in some disc $0 < |s-1| < c$, $\zeta'(s)$ is continuous on every notched box $\sigma \ge 1$, $|\tau| \le T$, $|s-1| \ge c$, and $\zeta(s)$ does not vanish on $\operatorname{Re}(s) = 1$, $s \ne 1$.

24. Suppose that $G(s)$ can be analytically continued into a connected open set containing $\operatorname{Re}(s) \ge 1$, and that $G(s)$ has (simple) zeros ρ_j, $j = 1, 2, \ldots$, on $\operatorname{Re}(s) = 1$. Prove that

$$\lim_{\sigma \to 1+} (\sigma - 1) \int_{\mathbb{R}} \frac{d\tau}{|sG(s)|^2} = \sum_j \frac{\pi}{|\rho_j G'(\rho_j)|^2}.$$

To establish the prime number theorem via the Plancherel relation is to demonstrate that $\zeta(1 + i\tau) \ne 0$, $\tau \ne 0$.

16

Finitely distributed additive functions

In a 1946 paper, Erdős isolated the following notion. A real valued additive function f is *finitely distributed* if there are positive constants c_1 and c_2, and an unbounded sequence of real x, so that each interval $[1, x]$ contains a subset of integers n_j, $1 \le j \le k$, $k \ge c_1 x$, on which $|f(n_j) - f(n_k)| \le c_2$ is satisfied. These awkwardly defined objects proved most convenient in the study of the value distribution of additive functions. As I shall show in the next chapter, they also occur naturally in the study of multiplicative functions. They gain their convenience from a characterisation by Erdős, [82].

Theorem 16.1. *An additive function f is finitely distributed if and only if there is a constant c so that the series*

$$\sum_{|f(p)-c\log p|>1} \frac{1}{p}, \qquad \sum_{|f(p)-c\log p|\le 1} \frac{|f(p) - c\log p|^2}{p}$$

converge.

I give two proofs of this theorem, one with integration and one without.

Proof: integers to primes, with integration. For any real u,

$$\int_{-1}^{1} (1 - |t|)e^{itu} dt = \left(\frac{\sin(u/2)}{u/2}\right)^2.$$

The integral is at least $4\pi^{-2}$ when $|u| \le 1$.

For real t define the multiplicative function $g(n) = \exp(itc_2^{-1} f(n))$, and set

$$m(x) = x^{-1} \sum_{n \le x} g(n),$$

as usual. Then

$$\int_{-1}^{1} (1 - |t|)|m(x)|^2 dt \ge 4c_1^2\pi^{-2}$$

for an unbounded sequence of x-values.

We showed in Chapter 13 that

$$\gamma(t) = \lim_{x\to\infty} |m(x)|$$

always exists. Let $0 < d \le c_1/\pi$. Let E denote the set of t-values in $[-1,1]$ for which $\gamma(t) \ge d$, and $|E|$ its Lebesgue measure. Then $4c_1^2\pi^{-2} \le d^2(2-|E|)+|E|$. We fix d at c_1/π, so that $|E| > 2c_1^2\pi^{-2}$. Note that if t belongs to E, then so does $-t$.

If t is a real for which $\gamma(t)$ is positive, then it is a corollary of Halász' theorem that there is a unique real τ for which the series $\sum p^{-1}(1-\operatorname{Re} g(p)p^{-i\tau})$ converges. The function τ of t is defined and uniformly bounded on the set E of positive Lebesgue measure.

If τ is defined on t_1, t_2, then the series

$$\sum p^{-1}(1-\operatorname{Re}\exp(i\{t_j c_2^{-1} f(p) - \tau(t_j)\log p\})), \quad j = 1, 2,$$

converge. From Lemma 13.4 the series

$$\sum p^{-1}(1-\operatorname{Re}\exp(i\{(t_1-t_2)c_2^{-1} f(p) - (\tau(t_1)-\tau(t_2))\log p\}))$$

also converges. Hence $\tau(t_1-t_2)$ exists and has the value $\tau(t_1)-\tau(t_2)$. Moreover, if τ is defined on t, then it is also defined on $-t$, and $\tau(-t)$ coincides with $-\tau(t)$.

A theorem of Steinhaus asserts that the difference set of a set of positive Lebesgue measure contains a proper interval about the origin. A proof is given in [41] Lemma 1.1. Together with what we have established this shows that τ is defined on a proper interval about the origin and so on the whole real line. It satisfies Cauchy's functional equation: $\tau(x+y) = \tau(x)+\tau(y)$. Moreover, it is uniformly bounded on some (and therefore each) bounded proper interval about the origin. It follows that τ is continuous at $t = 0$.

Indeed, suppose that $|\tau(t)| \le M$ for $|t| < \delta$. Given $\varepsilon > 0$, choose an integer $n > M/\varepsilon$. Then for $|t| < n^{-1}\delta$, $|\tau(t)| = |n^{-1}\tau(nt)| \le n^{-1}M < \varepsilon$. By translation τ is everywhere continuous. A result of Cauchy now guarantees $\tau(t)$ the form $c_2^{-1}ct$ for some constant c.

We have demonstrated the convergence of the series

$$\sum p^{-1}(1-\operatorname{Re}\exp(it\{f(p) - c\log p\}))$$

for each real t. Moreover, the sum function is uniformly bounded on each bounded set of t values.

To complete the proof in this direction, we employ the (in)equalities

$$1 - \operatorname{Re} e^{2iv} = 2(\sin v)^2 \geq 8v^2\pi^{-2}$$

if $|v| \leq 1$, and

$$\int_0^1 (1 - \operatorname{Re} e^{2ivt})dt = 1 - \frac{\sin 2v}{2v} \geq \frac{1}{2}$$

if $|v| > 1$. Together these give

$$\min(1, |v|^2) \leq 2\int_0^1 (1 - \operatorname{Re} e^{2ivt})dt + \frac{\pi^2}{8}(1 - \operatorname{Re} e^{2iv}).$$

Setting $v = f(p) - c\log p$ for each prime p in turn, we obtain the convergence of the series $\sum p^{-1}\min(1, |f(p) - c\log p|^2)$.

On $\mathbb{R}$, $\left(\frac{2}{u}\sin\frac{u}{2}\right)^2$ approximates the function which is 1 if $u = 0$, zero otherwise. Restricted to $[-\pi/2, \pi/2]$ this function of Dirac type is exactly represented by $\lim(\cos u)^{2m}$, $m \to \infty$.

Proof: integers to primes, without integration. For real θ, $e^{i\theta} + e^{-i\theta} = 2\cos\theta = 2 - 4\left(\sin\frac{\theta}{2}\right)^2 \geq 2 - \theta^2$. In the above notation there is a unimodular ω such that

$$g(n_j)\omega + \overline{g(n_j)\omega} \geq 2 - (c_2|t|)^2, \quad 1 \leq j \leq k.$$

For $|t| \leq \sqrt{2}\,c_2^{-1}$ and every positive integer m

$$\sum_{n\leq x}(g(n)\omega + \overline{g(n)\omega})^{2m} \geq (2 - c_2^2t^2)^{2m}c_1x.$$

The sum has the alternative representation

$$\sum_{\substack{r=0\\r\neq m}}^{2m}\binom{2m}{r}\sum_{n\leq x}(g(n)\omega)^{2(m-r)} + \binom{2m}{m}[x].$$

There is an integer w, $2 \leq w \leq 2m$, for which

$$\begin{aligned} x^{-1}\left|\sum_{n\leq x} g(n)^w\right| &\geq \left\{c_1(2 - c_2^2t^2)^{2m} - \binom{2m}{m}\right\}\left(2^{2m} - \binom{2m}{m}\right)^{-1} \\ &= c_1\left(1 - \frac{1}{2}\,c_2^2t^2\right)^{2m} + O(m^{-1/2}), \end{aligned}$$

the implied constant absolute. Further restrict t to $c_2|t|m \le 1$ and fix m at a value so large that the lower bound exceeds $c_1/2$. Without loss of generality the same w occurs for an unbounded sequence of x-values. There exists γ, depending at most upon c_1, and a real τ so that

$$\sum \frac{1}{p}|1 - g(p)^w p^{i\tau}|^2 \le \gamma.$$

Here w may vary with t. In view of Remark (i) following Lemma 13.4, we may replace w by $(2m)!$ provided we multiply τ by $w^{-1}(2m)!$ and γ by $((2m)!)^2$. The proof now proceeds as before, but without appeal to the result of Steinhaus.

Proof: primes to integers. The proposition in this direction is not applied in the present volume, so I sketch a proof.

Set $g(n) = \exp(itf(n))$, t real. Assume the convergence of the series over the primes, and define

$$\omega(x) = \sum_{\substack{p \le x \\ |f(p) - c\log p| \le 1}} p^{-1}(f(p) - c\log p).$$

It follows from Theorem 11.1 that

$$\phi(t) = \lim_{x\to\infty} [x]^{-1} \sum_{n\le x} g(n)\exp(-it\omega(x)) \tag{16.1}$$

exists. An examination of the infinite product representation for ϕ shows that it is a continuous function of t. In particular it is continuous at $t = 0$.

The average of $g(n)\exp(-it\omega(x))$ considered at (16.1) is the characteristic function of the frequency

$$F_x(z) = [x]^{-1} \sum_{\substack{n\le x \\ f(n) - \omega(x) \le z}} 1.$$

By a classical theorem from the theory of probability, the weak limit of the $F_x(z)$ exists. Denote this limit by $F(z)$.

Choose z_0 so that $F(z_0) - F(-z_0) > 1/2$. For all sufficiently large values of x, there are at least $(F(z_0) - F(-z_0) - 1/4)x > x/4$ integers n, not exceeding x, for which $|f(n) - \omega(x)| \le z_0$. In particular, f is finitely distributed.

A detailed treatment of finitely distributed additive functions, within the discipline of probabilistic number theory, is given in [41] Vol. 1, Chapter 7.

The convergence of the two series in Theorem 16.1 is clearly the necessary and sufficient condition that the frequencies

$$[x]^{-1} \sum_{\substack{n \le x \\ f(n)-\alpha(x) \le z}} 1, \qquad x \to \infty,$$

possess a limiting distribution for *some* $\alpha(x)$, not necessarily the $\omega(x)$ chosen during the proof of Theorem 16.1.

Theorem 16.1 is related to, and indeed rapidly establishes, a celebrated theorem of Erdős and Wintner [86], which asserts that the frequencies

$$[x]^{-1} \sum_{\substack{n \le x \\ f(n) \le z}} 1$$

possess a limiting distribution as $x \to \infty$ if and only if the three series

$$\sum_{|f(p)|>1} \frac{1}{p}, \quad \sum_{|f(p)| \le 1} \frac{f(p)^2}{p}, \quad \sum_{|f(p)| \le 1} \frac{f(p)}{p}$$

converge. A direct proof of this theorem within the philosophy of the method of the stable dual is given on pages 9–12 of the introduction to my book, [56].

The following consequence of Theorem 16.1 will be applied in the next chapter.

Theorem 16.2. *If there are positive constants c_1, c_2 so that for an unbounded sequence of x values, $|f(n)| \le c_2$ for at least $c_1 x$ integers n, not exceeding x, then the series*

$$\sum_{|f(p)|>1} p^{-1}, \qquad \sum_{|f(p)| \le 1} p^{-1} |f(p)|^2$$

converge.

Proof. Granted the condition on the $f(n)$, Theorem 16.1 guarantees the convergence of $\sum p^{-1} \min(1, |f(p) - c \log p|^2)$ for some c. It will suffice to show that $c = 0$.

Define the additive function h by $h(p) = f(p) - c \log p$ if $|f(p) - c \log p| \le 1$, $h(p) = 0$ for the remaining primes, and $h(p^k) = 0$ if $k \ge 2$.

Given $r > 0$, those integers n, up to x, which are divisible by the square of some prime $q > r$, are in number at most

$$\sum_{q > r} \left[\frac{x}{q^2} \right] \ll \frac{x}{r},$$

and so at most $c_1x/8$ if r is fixed at a sufficiently large value. Those n up to x which are divisible by a power q^k of the prime q, with $k \geq m$, are in number at most

$$\sum_{k=m}^{\infty} \left[\frac{x}{q^k}\right] \leq \frac{x}{q^m}\left(1-\frac{1}{q}\right)^{-1} \leq \frac{x}{2^{m-1}}.$$

It follows that if we fix r, m at sufficiently large (absolute) values, then amongst the integers n ($\leq x$) on which $|f(n)| \leq c_2$, there are at least $2c_1x/3$ which have the form $n'n''$ where n' is squarefree, $(n', n'') = 1$, and n'' does not exceed a bound depending only upon c_1.

Moreover, at the loss of at most $c_1x/6$ further integers n up to x, we may assume that the n' are not divisible by any prime p for which $|f(p) - c\log p| > 1$. Here we employ the convergence of the sum $\sum p^{-1}$ taken over such primes. We are left with at least $c_1x/2$ integers n_j on which $f(n) - c\log n - h(n)$ is uniformly bounded.

From the Turán–Kubilius inequality we see that

$$\sum_{n_j \leq x} \left| h(n_j) - \sum_{p\leq x} p^{-1}h(p) \right|^2 \ll x \sum_{p \leq x} p^{-1}|h(p)|^2 \ll x,$$

having employed the convergence of the second sum in Theorem 16.1. An application of the Cauchy–Schwarz inequality gives

$$\sum_{p\leq x} p^{-1}|h(p)| \leq \left(\sum_{p\leq x} p^{-1} \sum_{p \leq x} p^{-1}(h(p))^2 \right)^{1/2} \ll (\log\log x)^{1/2},$$

and together

$$\sum_{n_j \leq x} (f(n_j) - c\log n_j)^2 \ll \sum_{n_j\leq x} h(n_j)^2 + x \ll x \log\log x.$$

By hypothesis $|f(n_j)| \leq c_2$, so that

$$c^2 \sum_{n_j \leq x} (\log n_j)^2 \ll x \log\log x.$$

Moreover

$$\sum_{n_j\leq x} (\log n_j)^2 \geq \sum_{m \leq c_1x/2} (\log m)^2 > \sum_{\sqrt{x} < m \leq c_1 x/2} (\log x^{1/2})^2 > c_1 x(\log x)^2/9$$

for x sufficiently large. Thus $c \ll \log\log x(\log x)^{-2}$ for an unbounded sequence of x-values, and must have the value zero.

It is not difficult to give a characterisation, in terms of their values on the primes, of the additive functions f which satisfy the initial conditions of Theorem 16.2.

17

Multiplicative functions of the class $\mathcal{L}^\alpha$. Mean value zero

Given that a multiplicative function belongs to the class $\mathcal{L}^\alpha$, the extra information that it possesses a mean value zero is not particularly helpful. It might vanish on every integer greater than 1. In seeking analogues of the theorems of Wirsing and Halász it is more fruitful to consider the multiplicative functions in $\mathcal{L}^\alpha$ which do not have a mean value zero. These may be characterised. Indeed, I shall characterise those multiplicative g in $\mathcal{L}^\alpha$ for which $\|g\|_1 > 0$.

Theorem 17.1. *In order that the real non-negative multiplicative function g belong to $\mathcal{L}^\alpha$, $\alpha > 1$, and satisfy*

$$\liminf_{j\to\infty} x_j^{-1} \sum_{n\le x_j} g(n) > 0$$

on an unbounded sequence x_j, it is necessary and sufficient that:

The partial sums

$$\sum_{p\le x} p^{-1}(g(p)-1) \tag{17.1}$$

be bounded above uniformly in x, and below uniformly on the sequence x_j;

The series

$$\sum_{g(p)\le 3/2} p^{-1}(g(p)-1)^2, \quad \sum_{g(p)>3/2} p^{-1}g(p)^\alpha, \quad \sum\sum_{p,m\ge 2} p^{-m}g(p^m)^\alpha \tag{17.2}$$

converge.

Proof of Theorem 17.1. Let g belong to $\mathcal{L}^\alpha$ and satisfy

$$x_j^{-1}\sum_{n\le x_j} g(n) \ge c > 0, \quad j = 1, 2, \ldots$$

For these same values $x = x_j$,

$$x^{-1} \sum_{\substack{n \le x \\ g(n) > c/4}} g(n) \ge c/2.$$

Moreover, for a suitably large τ

$$x^{-1} \sum_{\substack{n \le x \\ g(n) > \tau}} g(n) \le \tau^{(\alpha-1)} x^{-1} \sum_{n \le x} g(n)^\alpha < c/8.$$

Hence

$$x^{-1} \sum_{\substack{n \le x \\ c/4 < g(n) \le \tau}} g(n) > c/4. \tag{17.3}$$

Define the additive function f by $f(q) = \log g(q)$ if $g(q) > 0$, and $f(q) = 1$ otherwise. For those integers counted in the sum (17.3) $|f(n)|$ does not exceed a number independent of x. From Theorem 16.2 the series $\sum p^{-1} \min(1, |f(p)|)^2$ converges. Since $\log w = w - 1 + O((w-1)^2)$ uniformly for real w in the interval $[1/2,\ 3/2]$, this condition amounts to the convergence of

$$\sum_{|g(p)-1| \le 1/2} p^{-1}(g(p)-1)^2, \qquad \sum_{|g(p)-1| > 1/2} p^{-1}$$

which in turn is equivalent to the convergence of

$$\sum_{g(p) \le 3/2} p^{-1}(g(p)-1)^2, \qquad \sum_{g(p) > 3/2} p^{-1}. \tag{17.4}$$

We can now bring to bear the machinery developed during the proof of Theorem 9.1. What follows is a gazeteer for a journey around Theorem 9.1.

Since g belongs to $\mathcal{L}^\alpha$, the inequalities (9.5) and (9.6) are valid. As in the Second Approach to Theorem 9.1 we obtain the functional inequality (10.1). The convergence of the series (17.4) enables us to replace $g(p)$ in (10.1) by 1. We obtain the slow oscillation condition (10.2), and so (10.3): the local means $m(x)$ of g satisfy $m(x/w) - m(x) \ll (\log\log x)^{-1/2}$ uniformly for $1 \le w \le \exp(\sqrt{\log x})$.

Employing the slow oscillation of $m(x)$, and the hypothesis $\|g\|_1 > 0$, in conjunction with the functional inequalities (9.5) and (9.6), we obtain the convergence of the series at (17.2) save possibly for the convergence of $\sum p^{-m}g(p^m)^\alpha$, $m \geq 2$, for finitely many primes p. However, this last convergence may be obtained by deriving a uniform lower bound

$$wx^{-1} \sum_{n \leq x/w} g_1(n) \geq c_1 > 0, \quad 1 \leq w \leq \log x,$$

for all sufficiently large members $x = x_j$ of the unbounded sequence of values in the hypothesis of the present theorem. Here g_1 is the function defined immediately preceding (9.7) in the proof of Theorem 9.1. As in the Second Approach to Theorem 9.1, in the section following (10.3), we can then bound

$$\sum_{p^m \leq \log x} g(p^m)^\alpha \sum_{n \leq x/p^m} g_1(n)^\alpha$$

from above and below. This argument depends heavily upon the slow oscillation of the local means of g_1, a result obtained by stable duality. The application of Euler products to establish the convergence of individual $\sum p^{-m}g(p^m)^\alpha$, as given in the First Approach to Theorem 9.1, is no longer available.

By Hölder's inequality g belongs to $\mathcal{L}^1$, and an integration by parts shows that $\zeta(\sigma)^{-1} \sum g(n)n^{-\sigma}$ is bounded above uniformly for $1 < \sigma \leq 2$. From the Euler product representations of these Dirichlet series, $\sum p^{-\sigma}(g(p) - 1)$ is also bounded above, uniformly for $1 < \sigma \leq 2$. In view of Remark (iii) in Chapter 11, the partial sums $\sum p^{-1}(g(p) - 1)$, $p \leq x$, are uniformly bounded above.

A short way to obtain the lower bound on the partial sums (17.1) is to appeal to Theorem 11.1. A more elementary argument goes as follows.

In the notation of the Second Approach to Theorem 9.1, $\sum g_3(n)n^{-1}$ converges. Since the value of g does not exceed that of $g_3 * g_2$ on the same integer,

$$x^{-1} \sum_{n \leq x} g(n) \leq x^{-1} \sum_{m \leq x} g_3(m) \sum_{r \leq x/m} g_2(r).$$

Again the mean values of g_2 may be estimated using Lemma 2.2. This time

$$x^{-1} \sum_{n \leq x} g_2(n) \ll \exp\left(\sum_{\substack{p \leq x \\ g(p) \leq 3/2}} p^{-1}(g(p) - 1) \right), \quad x \geq 2.$$

By what we have established so far, g_2 belongs to $\mathcal{L}^1$. The contribution towards the local mean of $g_3 * g_2$ which arises from those $m > x^{1/2}$ in the above representation is $\ll \sum g_3(m)m^{-1}$, $m > x^{1/2}$, which is $o(1)$ as $x \to \infty$. The contribution from the m not exceeding $x^{1/2}$ is

$$\ll \sum_{m \le x^{1/2}} g_3(m)m^{-1} \exp\left(\sum_{\substack{p \le x/m \\ g(p) \le 3/2}} p^{-1}(g(p)-1) \right).$$

The convergence of the first series at (17.2) allows the range $p \le x/m$ to be extended to $p \le x$ at the expense of a factor uniformly bounded for $x \ge 2$. With $x = x_j$ the desired lower bound on the partial sums (17.1) follows now from the lower bound on the local mean of g.

In the other direction, let δ be a positive real satisfying $(1+\delta)^\alpha \le 3/2$. Let t_1 be the multiplicative function which coincides with g on the prime powers q for which $1-\delta \le g(q) \le 1+\delta$, and which is zero on the remaining prime powers. Application of Lemma 2.2 shows that t_1 belongs to $\mathcal{L}^\alpha$. Moreover, if $g^\alpha = t_1^\alpha * t_2$, then the argument involving g_4, g_5 in the Third Approach to Theorem 9.1 readily modifies to give the convergence of $\sum |t_2(m)|m^{-1}$. The usual argument with convolutions shows that g belongs to $\mathcal{L}^\alpha$. A related argument involving the mean value of $|g|^\alpha$ is made as Remark (iv) in Chapter 11.

Since the series (17.2) converge, we may appeal to Theorem 11.1. The partial sums $\sum p^{-1}(g(p)-1)$, $p \le x$ are bounded above, and

$$x^{-1}\sum_{n \le x} g(n) = \prod_{p \le x}\left(1-\frac{1}{p}\right)\left(1+\frac{g(p)}{p}+\cdots\right) + o(1), \quad x \to \infty.$$

A uniform lower bound for the local means $m(x_j)$, with j sufficiently large, follows rapidly.

Theorem 17.2. *In order that the real multiplicative function g belong to $\mathcal{L}^\alpha$, $\alpha > 1$, and satisfy*

$$\limsup_{j \to \infty} x_j^{-1} \left| \sum_{n \le x_j} g(n) \right| > 0 \tag{17.5}$$

on an unbounded sequence x_j, it is necessary and sufficient that

The partial sums

$$\sum_{p \le x} p^{-1}(g(p)-1) \tag{17.6}$$

be bounded above uniformly in x, and below uniformly on the sequence x_j;

The series:

$$(17.7) \quad \sum_{|g(p)|\le 3/2} p^{-1}(g(p)-1)^2, \quad \sum_{|g(p)|>3/2} p^{-1}|g(p)|^\alpha, \quad \sum\sum_{p,m\ge 2} p^{-m}|g(p^m)|^\alpha$$

converge, and that for each prime p

$$(17.8) \qquad \sum_{k=1}^{\infty} p^{-k} g(p^k) \ne -1.$$

Corollary (A generalisation of Wirsing's theorem). *If the real valued multiplicative* g *belongs to* $\mathcal{L}^\alpha$, *and*

$$\sum_{p\le x} \frac{1}{p}(1-g(p)) \to \infty, \ x \to \infty,$$

then g *has mean value zero.*

Proof of Theorem 17.2. Assume that g belongs to $\mathcal{L}^\alpha$ and satisfies condition (17.5). Then $|g|$ belongs to $\mathcal{L}^\alpha$, and $\||g|\|_1 > 0$. Conditions (17.1) and (17.2) of Theorem 17.1 are satisfied with $|g|$ in place of g. In particular the convergence of the second and third series at (17.7) and a uniform upper bound for the partial sums (17.6) is assured.

Following the argument in the proof of Theorem 17.1, we obtain the slow oscillation of the local means $m(x)$, first for $|m(x)|$ and then for $m(x)$ itself: $m(x/w) - m(x) \ll (\log\log x)^{-1/2}$ uniformly for $1 \le w \le \exp(\sqrt{\log x})$. This condition together with the inequalities (9.5) and (9.6) gives the convergence of the first series at (17.7).

The validity of (17.7) guarantees the asymptotic representation (10.4) of Lemma 10.2. If $1 + \sum p^{-k} g(p^k)$ vanishes for some prime p, then arguing by way of Euler products, as in the proof of Theorem 9.1, during the First Approach, $\zeta(\sigma)^{-1} \sum g(n) n^{-\sigma} \to 0$, $\sigma \to 1+$. The slow oscillation of $\sum h(d) d^{-1}$, $d \le x$, enables us to apply the Hardy–Littlewood tauberian theorem, and $\sum h(d) d^{-1}$ sums to zero. Hence g has mean value zero, contrary to assumption. Condition (17.8) is satisfied.

Alternatively, we may employ the representation

$$(17.9) \quad x^{-1} \sum_{n\le x} g(n) = \prod_{p\le x} \left(1 - \frac{1}{p}\right)\left(1 + \frac{g(p)}{p} + \cdots\right) + o(1), \quad x \to \infty,$$

guaranteed by Theorem 11.1. This same representation now yields the asserted lower bound on the partial sums (17.6).

In the opposite direction we note that if g satisfies conditions (17.6)–(17.8), then so does $|g|$. By Theorem 17.1, $|g|$ and so g belongs to $\mathcal{L}^\alpha$. Straightaway the representation (17.9) becomes available from Theorem 11.1, and with it a uniform lower bound on the $|m(x_j)|$, j sufficiently large.

Theorem 17.3. *In order that a complex valued multiplicative function g belong to $\mathcal{L}^\alpha$, $\alpha > 1$, and satisfy*

$$\limsup_{j\to\infty} x_j^{-1} \left| \sum_{n\le x_j} g(n) \right| > 0$$

on an unbounded sequence x_j, it is necessary and sufficient that for some real τ:

The partial sums

$$\sum_{p\le x} p^{-1}(Re\, g(p)p^{i\tau} - 1) \tag{17.10}$$

be bounded above uniformly in x, and below uniformly on the x_j;

The series (17.11):

$$\sum_{|g(p)|\le 3/2} p^{-1}|g(p)p^{i\tau} - 1|^2, \quad \sum_{|g(p)|>3/2} p^{-1}|g(p)|^\alpha, \quad \sum_{p,m\ge 2}\sum p^{-m}|g(p^m)|^\alpha$$

converge, and that for each prime p,

$$\sum_{m=1}^{\infty} p^{-m(1-i\tau)} g(p^m) \neq -1. \tag{17.12}$$

Corollary (A generalisation of Halász' theorem). *If the complex multiplicative g belongs to $\mathcal{L}^\alpha$, and for each real τ,*

$$\sum_{p\le x} \frac{1}{p}(1 - Re\, g(p)p^{i\tau}) \to \infty, \qquad x \to \infty,$$

then g has mean value zero.

Proof of Theorem 17.3. We modify the proof of Theorem 17.2. I identify the main changes.

Let g belong to $\mathcal{L}^\alpha$ and not have mean value zero. Again $|g|$ satisfies the conditions of Theorem 17.1. The convergence of the series $\sum p^{-1}(1-|g(p)|)^2$ taken over those primes with $|g(p)| \le 3/2$, and that of $\sum p^{-1}$ taken over those primes p with $|g(p)| > 3/2$, enable us to replace g in the approximate functional equation (10.1) by $g(p)/|g(p)|$ when $g(p)$ is non-zero, by 1 otherwise.

In place of the slow oscillation of the local mean $m(x)$, Theorem 6.6 gives

$$m(x/w) = w^{i\lambda}m(x) + O((\log\log x)^{-1/2})$$

for some real λ, uniformly for $1 \le w \le \exp(\sqrt{\log x})$. Moreover, λ is constrained by $|\lambda|^{1/\alpha'}|m(x)| \ll 1$.

For x running through the sequence x_j we obtain a uniform bound

$$|\lambda_x|^{1/\alpha'} + \sum_{\substack{p\le\log x\\ |g(p)|\le 3/2}} p^{-1}|g(p)p^{i\lambda_x} - 1|^2 \le c_2, \quad \lambda_x \text{ real}.$$

Without loss of generality we may assume that $\lambda_x \to \tau$ as $x\ (= x_j) \to \infty$. Then for each fixed $P > 0$

$$\sum_{\substack{p\le P\\ |g(p)|\le 3/2}} p^{-1}|g(p)p^{i\tau} - 1|^2 \le c_2.$$

The convergence of all the series at (17.11) can now be obtained.

We continue to follow the proof of Theorem 17.2, but with $g(n)n^{i\tau}$ in place of g. If condition (17.12) fails, then $g(n)n^{i\tau}$ has mean value zero, and after an integration by parts, so does g. Thus no sum $1+\sum p^{-m(1-i\tau)}g(p^m)$ vanishes.

The remaining conditions on the partial sums (17.10) may be obtained by following the corresponding argument in the proof of Theorem 17.2, replacing $g(n)$ there by $g(n)n^{i\tau}$. Note that the analogue of (17.9) established then gives

$$x_j^{-1}\left|\sum_{n\le x_j} g(n)n^{i\tau}\right| \ge \delta > 0$$

for all j sufficiently large. However, the factor $n^{i\tau}$ may be stripped off using an integration by parts, bearing in mind the oscillation condition (10.3) satisfied by the local means of g.

In the other direction the validity of conditions (17.10), (17.11) for g, shows that $|g|$ satisfies the conditions of Theorem 17.1. In particular g belongs to $\mathcal{L}^\alpha$. The representation (17.9) is available with $g(n)n^{i\tau}$ in place of $g(n)$, and from it a uniform lower bound for

$$x_j^{-1}\left|\sum_{n\le x_j} g(n)n^{i\tau}\right|, \quad j \text{ sufficiently large}.$$

Again the local means of g oscillate subject to (10.3), and an integration by parts allows the removal of the factors $n^{i\tau}$. This completes the proof.

I cannot deny that the proofs of Theorems 17.2 and 17.3 give the impression that they should accompany a physical exercise routine. Bouncing around between propositions, however, can be an effective method in the study of projective geometry.

In the treatment of multiplicative functions that I have given in this chapter and Chapters 9–11 and 13, I have eschewed all application of the Fourier analytic method of Halász. Indeed the treatment is almost hermetic within the method of the stable dual. As I mentioned at the end of the Second Approach to Theorem 9.1, my original treatment of Theorem 9.1 in the case $\alpha = 2$ combined the use of a Turán–Kubilius dual with the method of Halász, [35]. In particular, I showed that whenever the behaviour of a multiplicative function g on the prime powers is under sufficient control, the local means of g oscillate slowly, even if g is not known to belong to the class $\mathcal{L}^1$. This was then applied to gain the convergence of the series $\sum |g(p^m)|^\alpha p^{-m}$ for individual primes p. My whole argument had much in common with the Second Approach to Theorem 9.1.

A variant exposition of this approach may be found in Chapter 10 of my book, [41]. I showed in Theorem 10.8 of that chapter, and Lemma 8 of [43] that if g satisfies condition (9.2) of Theorem 9.1, then

$$\lim_{x\to\infty} (xL(\log x))^{-1} \sum_{n\le x} g(n) \tag{17.13}$$

exists and is finite, with

$$L(u) = \exp\left(\sum_p p^{-1-u^{-1}}(g(p)-1)\right).$$

Here $L(u)$ is a slowly varying function of u, and the value of the limit is

$$\prod_p \left(1-\frac{1}{p}\right)\left(1+\frac{g(p)}{p}+\cdots\right)\exp\left(\frac{1-g(p)}{p}\right).$$

As the example $g(q) = 1+(\log\log q)^{-1}$ shows, such a multiplicative function need not belong to $\mathcal{L}^\alpha$ for any $\alpha \ge 1$. The quality of this result, obtained with the full panoply of Halász' Fourier analytic method, may be compared with that of Theorem 11.1, whose derivation may be viewed elementary.

The oscillation estimate (10.3): $m(x) - m(x/w) \ll (\log\log x)^{-1/2}$ has a quantitative aspect. One may ask how sharp it is, and whether the result might in some sense be made uniform in g. For functions with values in the complex unit disc the strongest estimates of this type currently available are derived by Fourier analysis, [69]. Estimates of this type are useful in treating generalised character sums.

A necessary and sufficient criterion that a multiplicative function which belongs to some $\mathcal{L}^\alpha$, $\alpha > 1$, may possess a mean value zero, I give in the paper, [43]. It is the negation of the union of the conditions (17.10)–(17.12) of the present Theorem 17.3. In the treatment of multiplicative functions with mean value zero given in that paper the duals of Turán–Kubilius inequalities play essential rôles.

For a treatment of multiplicative functions which includes Theorem 9.1 but does not consider theorems involving a mean value zero, and which purposefully avoids the introduction of inequalities of Turán–Kubilius type, see Mauclaire, [123].

An historical overview of the general theory of multiplicative functions may be found in my review, [65]. Bearing that review in mind, it is interesting to compare the treatment accorded multiplicative functions in the present volume with that in the paper of Atkinson and Cherwell, [3], carried out under an aesthetic derived from Cherwell's statistical studies of the gaps between prime numbers.

18

Exercises: Including logarithmic weights

1. Prove that for any irrational β and multiplicative function g in $\mathcal{L}^\tau$, $\tau > 1$,

$$\lim_{x\to\infty} x^{-1} \sum_{n\le x} g(n) e^{-2\pi i\beta n} = 0.$$

Hint: Use Theorem 17.1 to control the size of $g(p)$. In their original proof of this result, given in detail only for $\tau = 2$, Daboussi and Delange, [18], see also Delange, [22], argue somewhat differently.

Let D be a non-zero integer. Then every positive integer n can be uniquely expressed as a product $n_1 n_2$ where n_1 is made up of powers of the primes which divide D, and $(n_2, D) = 1$.

2. Show that if an arithmetic function h belongs to $\mathcal{L}^\tau$, $\tau > 1$, and satisfies

$$\limsup_{x\to\infty} x^{-1} \left| \sum_{n\le x} h(n) \right| > 0,$$

then there is a positive K so that

$$\limsup_{x\to\infty} x^{-1} \left| \sum_{\substack{n\le x \\ n_1 \le K}} h(n) \right| > 0.$$

3. Let g be a multiplicative function which belongs to $\mathcal{L}^\tau$, $\tau > 1$, and satisfies

$$\limsup_{x\to\infty} x^{-1} \left| \sum_{n\le x} g(n) e^{2\pi i\beta n} \right| > 0$$

for a rational number $\beta = u/v$ with $(u, v) = 1$. Prove that there is a Dirichlet character $\chi(\operatorname{mod} v)$ so that

$$\limsup_{x\to\infty} x^{-1} \left| \sum_{n\le x} g(n)\chi(n) \right| > 0.$$

Is there a result in the other direction?

4. Prove that in the circumstances of exercise 3, there is a unique primitive Dirichlet character χ for which the series

$$\sum_{|g(p)|\le 3/2} p^{-1}|g(p)\chi(p)-1|^2$$

converges.

5. Characterise, in terms of their behaviour on the primes, those multiplicative functions g which belong to $\mathcal{L}^\tau$, $\tau > 1$, and satisfy

$$\limsup_{x\to\infty} x^{-1}\left|\sum_{n\le x} g(n)e^{2\pi i\beta n}\right| > 0$$

for some real β.

How does this characterisation compare with the result of Daboussi established by exercise 17 of Chapter 12?

6. Characterise those real non-negative multiplicative functions g which satisfy

$$\liminf_{x\to\infty} x^{-1} \sum_{c_1<g(n)\le c_2} 1 > 0$$

for some $0 < c_1 < c_2 < \infty$.

7. Characterise those real non-negative multiplicative functions g which satisfy

$$\limsup_{x\to\infty} x^{-1} \sum_{c_1<g(n)\le c_2} 1 > 0$$

for some $0 < c_1 < c_2 < \infty$.

Weighted sums; Logarithmic density

If g is a multiplicative function with values in the complex unit disc, then the weighted sum

$$S(x) = \sum_{n\le x} g(n)n^{-1}$$

satisfies $S(x) - S(x/y) \ll \log y$ uniformly for $2 \le y \le x$. In a sense the function $S(x)/\log x$ is already slowly oscillating. The following series of exercises show that $S(x)/\log x$ is more manageable than the mean of g itself.

8. Prove that

$$\sum_{n=1}^{\infty} n^{-\sigma}\left|f(n) - \sum_{q} q^{-\sigma}(1-q_0^{-\sigma})f(q)\right|^2 \le \zeta(\sigma)\sum_{q} q^{-\sigma}(1-q_0^{-\sigma})|f(q)|^2$$

uniformly for $\sigma > 1$ and all complex valued additive functions for which the upper bound sum, taken over the prime powers, converges.

9. In the notation of Chapter 2, prove that

$$\sum_{n\le x} n^{-1}|f(n)-E(f)|^2 \ll D^2 \log x,$$

uniformly for $x \ge 2$ and additive functions f.

10. Prove that

$$\sum_{q\le x} q \left| \sum_{\substack{n\le x\\ n\cong 0 (\mathrm{mod}\, q)}} \frac{a_n}{n} - \frac{1}{q}\left(1-\frac{1}{q_0}\right)\sum_{n\le x}\frac{a_n}{n}\right|^2 \ll \log x \sum_{n\le x} \frac{|a_n|^2}{n}$$

uniformly for $x \ge 2$ and all complex numbers a_n.

11. Prove that

$$|S(x)|^2 \sum_{q\le x} q^{-1}|g(q)-1|^2 \ll (\log x)^2,$$

uniformly for $x \ge 2$.

This inequality may be compared with the inequality bounding the local mean $m(x)$ of g in Theorem 13.2. The divergence of the series

$$\sum q^{-1}|g(q)-1|^2$$

alone forces $S(x)/\log x$ to approach zero as $x \to \infty$.

The next exercise characterises the natural analogue of finitely distributed additive functions with logarithmic weights.

12. Without appealing to the theorem of Halász, prove that a real additive function f satisfies

$$\limsup_{x\to\infty} \frac{1}{\log x} \sum_{\substack{n\le x\\ |f(n)-\alpha(x)|\le c}} \frac{1}{n} > 0$$

for some $\alpha(x)$, $x \ge 1$, and positive c, if and only if the series

$$\sum_{|f(p)|>1} \frac{1}{p}, \qquad \sum_{|f(p)|\le 1} \frac{f(p)^2}{p},$$

converge.

19

Encounters with Ramanujan's function $\tau(n)$

In January of 1980 my wife and I arrived in London for a visit of six months. The English pound was artificially high against the American dollar, but that is a different story. I was a guest at Imperial College whilst on a John Simon Guggenheim Fellowship. My immediate mathematical aim was to devise and develop a method for treating in wide generality the differences of additive functions: $f_1(an+b) - f_2(An+B)$, $aB \neq Ab$, $n = 1, 2, \ldots$. In this I was successful, the outcome forming part of a volume on arithmetic functions and integer products, [56]. Further developments of those results may be found in the memoir [77], and in Chapters 31 and 32 of the present work.

I was at the time interested in several other mathematical topics. In particular, if a multiplicative function belonged to a class $\mathcal{L}^\alpha$ for some $\alpha > 1$, when could it have a mean value zero? I had in mind the function τ of Ramanujan. Unfortunately the use of the symbol τ for both Dirichlet's divisor function and the function of Ramanujan is standard.

Ramanujan's function $\tau(n)$ is defined by the identity

$$\sum_{n=1}^{\infty} \tau(n) x^n = x \prod_{j=1}^{\infty} (1 - x^j)^{24}.$$

It was conjectured by Ramanujan that τ is a multiplicative function, satisfies $\tau(p^{r+1}) = \tau(p)\tau(p^r) - p^{11}\tau(p^{r-1})$, $r \geq 1$, on the prime powers, and $|\tau(p)| < 2p^{11/2}$, [136]. The first two conjectures were established by Mordell, [129].

Hardyproved that

$$c_1 x^{12} \leq \sum_{n \leq x} \tau(n)^2 \leq c_2 x^{12}$$

for certain positive constants c_1, c_2, and all sufficiently large x, [94]. Hardy's bounds were decisively improved by Rankin, [137], who obtained positive constants A, δ so that

$$\sum_{n \leq x} \tau(n)^2 = A x^{12} + O(x^{12-\delta}).$$

Rankin effected his asymptotic estimate by analytically continuing the Dirichlet series $\sum \tau(n)^2 n^{-11-s}$ to the left of the line $\operatorname{Re} s = 1$, except for the simple pole at $s = 1$. It was not until 1974 that Deligne established the analogue of the Riemann hypothesis for local zeta functions, [23], and thereby settled the second conjecture of Ramanujan affirmatively.

If we set $g(n) = |\tau(n)|n^{-11/2}$, then the multiplicative function g belongs to the class $\mathcal{L}^2$. I asked myself whether the asymptotic behaviour of the mean

$$x^{-1} \sum_{n \le x} |\tau(n)| n^{-11/2}, \qquad x \to \infty,$$

could be investigated using only characteristic properties of multiplicative functions. It seemed unlikely that an analytic continuation of the Dirichlet series $\sum |\tau(n)|n^{-11/2-s}$ would be soon forthcoming.

On Wednesday, March the fifth of that year, I gave a talk at King's College, London. By that time matters had reached the stage that I could prove that one of the propositions

$$x^{-1} \sum_{n \le x} |\tau(n)| n^{-11/2} \to 0, \qquad x \to \infty,$$

$$\frac{1}{\log x} \sum_{p \le x} ||\tau(p)|p^{-11/2} - 1| \frac{\log p}{p} \to 0, \qquad x \to \infty,$$

held, and not both. Loosely: Either $|\tau(n)|n^{-11/2}$ is generally small, or $|\tau(p)|p^{-11/2}$ is generally near to 1. Note that the second of the two propositions is consistent with the estimate $|\tau(p)|p^{-11/2} < 2$ of Deligne.

During the question period at the end of the lecture, another visitor from the United States stood up from the large and distinguished audience and said that he didn't believe in either of the propositions.

Later in the evening, in a whispered aside, the objection was withdrawn. The sceptic had been convinced by a colleague that between the two propositions there was a gap.

I was scheduled to give a talk in Cardiff, Friday, May the second, and had suggested for it the same topic as my lecture at King's. When the time came I had made progress.

Theorem 19.1. *Let g be a non-negative multiplicative arithmetic function which has a mean value. Then g^δ has a mean value for each δ, $0 < \delta < 1$. Moreover, if any of these further mean values is non-zero, then the series*

$$\sum p^{-1} (g(p)^{1/2} - 1)^2,$$

taken over the prime numbers, converges.

With what we have presently to hand, this is not difficult.

Proof of Theorem 19.1. Suppose that g^δ, $0 < \delta < 1$, does not have mean value zero. Then by Theorem 17.1 with $\alpha = 1/\delta$, the series

$$(19.1) \qquad \sum_{g(p)\le 3/2} p^{-1}(g(p)^\delta - 1)^2, \quad \sum_{g(p)>3/2} p^{-1}g(p), \quad \sum_{p,m\ge 2} p^{-m}g(p^m)$$

converge.

For this we need only that g^δ belongs to $\mathcal{L}^{1/\delta}$, but we know more. Since g has a mean value,

$$\lim_{\sigma\to 1+} \zeta(\sigma)^{-1}\sum_{n=1}^{\infty} g(n)n^{-\sigma}$$

exists. Moreover, in view of our extra hypothesis the value of this limit must be non-zero. By considering the Euler product representation of $\sum g(n)n^{-\sigma}$, as in Chapter 9, we see that

$$\lim_{\sigma\to 1+} \sum p^{-\sigma}(g(p) - 1)$$

exists, and that the series $\sum p^{-1}(g(p) - 1)$ converges.

In view of the second part of condition (19.1), the series

$$(19.2) \qquad \sum_{g(p)\le 3/2} p^{-1}(g(p) - 1)$$

converges, and after Lemma 10.3 so does the series

$$\sum_{g(p)\le 3/2} p^{-1}(g(p)^\delta - 1),$$

even if the condition $g(p) \le 3/2$ be removed.

An application of Theorem 9.1 then shows that g^δ has a (non-zero) mean value.

Moreover, not only does the first series at (19.1) converge, but also the series

$$\sum_{g(p)\le 3/2} p^{-1}(g(p)^\theta - 1)^2$$

for each fixed θ, $0 < \theta < 1$. In particular

$$\sum p^{-1}(g(p)^{1/2} - 1)^2 \le \sum_{g(p)\le 3/2} p^{-1}(g(p)^{1/2} - 1)^2 + \sum_{g(p)>3/2} p^{-1}g(p) < \infty.$$

Armed with this result I established something stronger than I had at King's.

Theorem 19.2. *Let $0 < \delta \leq 2$. Then*

$$\lim_{x \to \infty} x^{-1} \sum_{n \leq x} \left(\frac{|\tau(n)|}{n^{11/2}} \right)^{\delta} = A_\delta$$

exists. In particular

$$\lim_{x \to \infty} x^{-13/2} \sum_{n \leq x} |\tau(n)| = 2A_1/13$$

exists. Moreover, either every A_δ with $0 < \delta < 2$ is zero, or the series

$$\sum_p \frac{1}{p} \left(\frac{|\tau(p)|}{p^{11/2}} - 1 \right)^2,$$

taken over the prime numbers, converges.

In view of the result of Deligne, we can write $\tau(p)p^{-11/2} = 2\cos\theta_p$, where θ_p is real and may be taken in the interval $0 \leq \theta_p \leq \pi$. There is a conjecture of Serre, ascribed to Sato and Tate, that as p varies the θ_p are distributed over this interval with a probability density $2(\sin\theta)^2/\pi$. As a consequence we would have

$$\sum_{p \leq x} \frac{1}{p} \left(\frac{|\tau(p)|}{p^{11/2}} - 1 \right)^2 = (c + o(1)) \log\log x, \qquad x \to \infty,$$

where

$$c = \frac{2}{\pi} \int_0^{\pi} (2|\cos\theta| - 1)^2 (\sin\theta)^2 = 2 - \frac{16}{3\pi}.$$

Accordingly, I conjectured that in Theorem 19.2 every limit A_δ with $0 < \delta < 2$ had the value zero.

This conjecture may be found in the account of a survey talk which I gave in the Journée held at Orsay, Paris, June 2-3, 1980, [45]. A detailed proof of Theorems 19.1 and 19.2 I gave in the paper, [46], and included a number of other conjectures concerning Ramanujan's function. In particular, the (finite) existence of A_δ for $\delta > 2$ seemed unlikely, since the convergence of the series $\sum p^{-1}((p^{-11/2}|\tau(p)|)^{\delta/2} - 1)^2$ would again be inconsistent with the Sato–Tate conjecture.

These conjectures generated interest, e.g., Murty, [130], but further progress appeared to need new information concerning τ.

In January 1983, the annual general meeting of the American Mathematical Society was held in Denver, Colorado. Paul Bateman, Wolfgang Schmidt and I organised a number theory section. It was very well attended. One of the speakers was Carlos Moreno. He informed me that he and Shahidi had proved that the function $\sum \tau(n)^4 n^{-22-s}$, analytically continued over the punctured line $\operatorname{Re}(s) = 1$, $s \neq 1$, by Shahidi, had a double pole at $s = 1$. From what I already knew of the distributional properties of τ, I was certain that this settled my first conjecture concerning the A_δ.

From the result of Rankin

$$\lim_{\sigma \to 1+} \zeta(\sigma)^{-1} \sum_{n=1}^{\infty} g(n)^2 n^{-\sigma}, \qquad g(n) = |\tau(n)| n^{-11/2},$$

exists and is non-zero. Considering Euler products

$$\lim_{\sigma \to 1+} \sum_p p^{-\sigma}(g(p)^2 - 1) \tag{19.3}$$

exists. A tauberian theorem, e.g., Lemma 10.1, then gives the convergence of the series $\sum p^{-1}(g(p)^2 - 1)$.

Similarly, from the result of Moreno and Shahidi

$$\lim_{\sigma \to 1+} \zeta(\sigma)^{-2} \sum_{n=1}^{\infty} g(n)^4 n^{-\sigma}$$

exists and is non-zero. In this case the series $\sum p^{-1}(g(p)^4 - 2)$ converges.

For real u, $u^2 - 1 = (|u| - 1)(|u| + 1)$. Since $|g(p)| < 2$, the convergence of the series $\sum p^{-1}(|g(p)| - 1)^2$ would imply that of $\sum p^{-1}(g(p)^2 - 1)^2$. However, (19.4):

$$\sum_{p \le x} p^{-1}(g(p)^2 - 1)^2 = \sum_{p \le x} p^{-1}(g(p)^4 - 2) - 2\sum_{p \le x} p^{-1}(g(p)^2 - 1) + \sum_{p \le x} p^{-1},$$

which clearly becomes unbounded with x. The A_δ with $0 < \delta < 2$ are indeed all zero. Oh what a dinner in forfeit was here missed!

In fact I felt that the approximation of the function $|u|$ by a quadratic polynomial over the interval $0 \le u \le 2$ would give something.

Lemma 19.3. *For each δ, $0 < \delta < 1$,*

$$w(\delta) = \inf_{-1 \le y \le 3} y^{-2}(1 + \delta y - (1 + y)^\delta)$$

is positive.

Proof. For a fixed real $y > -1$ the function $h(z) = 1 + zy - (1 + y)^z$ satisfies $h(0) = h(1) = 0$, $h''(z) = -(1 + y)^z(\log(1 + y))^2 < 0$, so that $h(z)$ is positive for $0 < z < 1$ and $y \ge -1$.

The following result was established by Elliott, Moreno and Shahidi, [81].

Theorem 19.4. *For each δ, $0 < \delta < 1$, there is a constant c, depending upon δ, so that*

$$\sum_{n \leq x} (|\tau(n)|n^{-11/2})^{2\delta} \leq cx(\log x)^{-w(\delta)}$$

for all $x \geq 2$.

Proof. Apply Lemma 2.2 to the function $g(n)^2$, where $g(n) = |\tau(n)|n^{-11/2}$. After Deligne we may write $\tau(p)p^{-11/2} = \alpha + \bar{\alpha}$ where $|\alpha| = 1$. Then from the second of Ramanujan's conjectures established by Mordell, argument by induction gives

$$\tau(p^r)p^{-11r/2} = \frac{\alpha^{r+1} - (\bar{\alpha})^{r+1}}{\alpha - \bar{\alpha}}.$$

In particular

$$|g(p^r)| = |\alpha^r + \alpha^{r-1}\bar{\alpha} + \cdots + (\bar{\alpha})^r| \leq r + 1.$$

In the notation of Lemma 2.2 we have

$$\Delta \ll \sup_{y \geq 2} y^{-1} \sum_{p^r \leq y} (r+1)^2 \log p^r \ll 1,$$

using the classical bound of Chebyshev. From Lemma 2.2

$$\sum_{n \leq x} g(n)^{2\delta} \ll x \exp\left(\sum_{q \leq x} \frac{g(q)^{2\delta} - 1}{q} \right).$$

The contribution to the sum in the exponential from the prime powers $q = p^r$ with $r \geq 2$ is

$$\ll \sum_p \sum_{r=2}^{\infty} r^2 p^{-r} \ll \sum p^{-2} \ll 1,$$

uniformly for $0 \leq \delta \leq 1$. Since $0 \leq g(p)^2 < 4$, by Lemma 19.3

$$g(p)^{2\delta} - 1 \leq \delta(g(p)^2 - 1) - w(\delta)(g(p)^2 - 1)^2.$$

From (19.3) and (19.4)

$$\sum_{p \leq x} p^{-1}(g(p)^{2\delta} - 1) \leq -w(\delta) \log\log x + O(1).$$

Theorem 19.4 is established.

In Chapter 9 we could avoid appeal to a tauberian theorem by using integration by parts and the bound of Chebyshev. We can do likewise here. With $\sigma = 1 + (\log x)^{-1}$,

$$\sum_{p \le x} p^{-1}(g(p)^2 - 1) = \sum_p p^{-\sigma}(g(p)^2 - 1) + O(1).$$

Similarly for the sum involving $g(p)^4 - 2$.

The power of $\log x$ in the upper bound estimate of Theorem 19.4 was improved by Rankin, [139], who also implicitly settled in the affirmative my conjecture concerning the non-existence of A_δ with $\delta > 2$; see also Murty, [131]. An account of related results is given by Rankin, [140] and Murty, [132]. The finer behaviour of Ramanujan's function is far from settled. I close this chapter with a further conjecture contained in my paper, [46].

Assume for the moment the conjecture of Lehmer, [118], that $\tau(n)$ is never zero. Then the function $f(n) = \log |\tau(n)| n^{-11/2}$ is additive, and one might seek to apply to it the machinery of probabilistic number theory, e.g. as expounded in [41]. To this end it would be desirable to know when $|\tau(p)|p^{-11/2}$ is very small, say less than p^{-c} for a given positive value of c. Results in this direction are patchy; see, for example, Murty, [132]. To fix matters I assume that in the notation following Theorem 19.2,

$$\sum_{\substack{p \le x \\ |\theta_p - \frac{\pi}{2}| \le w}} \frac{1}{p} \le \frac{c \log\log x}{(-\log w)^4}$$

uniformly for $x \ge 2$, $x^{-\lambda} < w \le \pi/4$ for some fixed positive λ and c. Then continuing to assume the validity of the Sato–Tate conjecture we can derive the asymptotic estimates

$$(\log\log x)^{-1} \sum_{p \le x} p^{-1} \log g(p) \to \frac{2}{\pi} \int_0^\pi (\log 2|\cos\theta|)(\sin\theta)^2 d\theta = -\frac{1}{2},$$

and

$$(\log\log x)^{-1} \sum_{p \le x} p^{-1} (\log g(p))^2 \to \frac{2}{\pi} \int_0^\pi (\log 2|\cos\theta|)^2 (\sin\theta)^2 d\theta = \mu^2,$$

with $\mu > 0$, as $x \to \infty$. In particular, f is guaranteed to be of Kubilius class H; cf. [41], Chapter 12. Let $F_x(z)$ denote the frequency amongst the positive integers up to x of those for which

$$\frac{|\tau(n)|}{n^{11/2}} \le \frac{e^{z\mu\sqrt{\log\log x}}}{\sqrt{\log x}}$$

Our combined assumptions would imply that as $x \to \infty$ the $F_x(z)$ would converge weakly to the normal distribution, with mean zero and variance one.

Conjecture. *For each real z,*

$$F_x(z) \to \frac{1}{\sqrt{2\pi}} \int_{-\infty}^{z} e^{-u^2/2} du, \qquad x \to \infty.$$

Lehmer's conjecture might fail but a theorem of Serre, [159], asserts that those integers n for which $\tau(n) = 0$ have asymptotic density zero, i.e.,

$$\lim_{x \to \infty} x^{-1} \sum_{\substack{n \le x \\ \tau(n) \neq 0}} 1 = 1.$$

In view of the concluding remarks to Chapter 3 of [41], the conjecture should then be modified so that the frequency is restricted to those integers on which τ does not vanish.

20

The operator T on L^2

We may localise the assertion of Theorem 8.1 and deepen our knowledge of the operator underlying the Turán–Kubilius inequality. I begin with the case $\alpha = 2$. For a complex additive arithmetic function f, and real $x \geq 2$, define

$$\Delta(f) = x^{-1} \sum_{n \leq x} \left| f(n) - x^{-1} \sum_{m \leq x} f(m) \right|^2 .$$

When x is an integer, $\Delta(f)$ has the form of a variance.

Theorem 20.1. *For all x sufficiently large, $\Delta(f)$ lies between positive absolute multiples of*

$$\min_{\lambda} \left(|\lambda|^2 + \sum_{q \leq x} q^{-1} |f(q) - \lambda \log q|^2 \right),$$

uniformly in f.

In this chapter q continues to denote a power of the prime q_0. To emphasise its dependence upon the function f, I shall denote the local mean

$$x^{-1} \sum_{n \leq x} f(n)$$

by $N(f)$, rather than the $m(x)$ of Chapter 8.

The upper bound in Theorem 20.1 is straightforward. The local mean has the representation(s)

$$x^{-1} \sum_{q \leq x} f(q) \left\{ \left[\frac{x}{q} \right] - \left[\frac{x}{q q_0} \right] \right\} = \sum_{q \leq x} \frac{f(q)}{q} \left(1 - \frac{1}{q_0} \right) + O\left(x^{-1} \sum_{q \leq x} |f(q)| \right).$$

An appeal to the Cauchy–Schwarz inequality, employing Chebyshev's inequality for the number of primes up to a given bound, shows that this error term is

$$\ll \left(\frac{1}{\log x} \sum_{q \le x} \frac{|f(q)|^2}{q} \right)^{\frac{1}{2}}.$$

It follows from the Turán–Kubilius inequality that

$$\Delta(f) \ll \sum_{q \le x} q^{-1} |f(q)|^2.$$

Integration by parts gives for the local mean of $\log n$ the estimate $\log x + O(1)$, and shows that $\Delta(\log) \ll 1$ uniformly in $x \ge 2$. By the Cauchy–Schwarz inequality $\Delta(f) \le 2\Delta(f - \lambda \log) + 2\Delta(\lambda \log)$ for any λ, and we have the asserted upper bound. Note that $\Delta(\log) \gg 1$ for all $x \ge 2$.

Much more difficult is the lower bound, first established by Ruzsa [150]. He deduced it from a concentration function estimate for real additive functions, which he had derived by an elaboration of the Fourier analytic method of Halász. As a proof of Theorem 20.1 his argument is formidable. I show that by operating within the background philosophy and functional analytic format of the method of the stable dual, a rapid proof of quite different nature may be given. It is a variant of the method lightly touched upon during the proof of Theorem 8.1.

I recall the (finite-dimensional) Hilbert spaces defined in Chapter 3. Again let s denote the number of prime powers not exceeding x. We endow the space $L^2(\mathbb{C}^s)$ with the inner product

$$(f, h) = \sum_{q \le x} q^{-1}(1 - q_0^{-1}) f(q) \overline{h(q)}.$$

The space $M^2(\mathbb{C}^t)$, $t = [x]$, is given an analogue of the classical inner product:

$$[a, b] = [x]^{-1} \sum_{n \le x} a_n \bar{b}_n,$$

where a, b are vectors with n^{th} coordinates a_n, b_n respectively. These inner products are consistent with the norms defined on the spaces in Chapter 3. The Turán–Kubilius inequality asserts that the operator $A : L^2(\mathbb{C}^s) \to M^2(\mathbb{C}^t)$ given by

$$(Af)(n) = f(n) - \sum_{q \le x} f(q) q^{-1}(1 - q_0^{-1})$$

has norm bounded independently of x. So therefore does its adjoint A^*. Hence $\|A^* A f\| \ll \|Af\|$. We see from this that in pursuit of a lower bound for $\Delta(f)$, which is essentially a lower bound for $\|Af\|^2$, we might investigate the operator $A^*A : \mathbb{C}^s \to \mathbb{C}^s$, self-adjoint with respect to the first inner product. In the event I begin with the following approximate functional equation.

Lemma 20.2. *The inequality*

$$\sum_{q\le x} \frac{1}{q} \left| f(q) - \alpha\left(\frac{x}{q}\right) + \alpha(x) \right|^2 \ll \Delta(f) + \frac{\log\log x}{\log x} \sum_{q\le x} \frac{|f(q)|^2}{q}$$

with

$$\alpha(y) = \sum_{q\le y} \frac{f(q)}{q}\left(1 - \frac{1}{q_0}\right),$$

is valid for all additive f, real $x \ge 2$.

Proof. For completely multiplicative f this result follows rapidly from the example (5.2) given in Chapter 5. I have simply replaced the local mean of f by a sum over prime powers. For a general f, I note that the adjoint operator of A is represented by

$$(A^*a)(q) = \frac{q}{(1-q_0^{-1})[x]}\left(\sum_{\substack{n\le x \\ n\cong 0 (\mathrm{mod}\, q)}} a_n - \frac{1}{q}\left(1 - \frac{1}{q_0}\right) \sum_{n\le x} a_n \right).$$

If we introduce the condition $(n,q) = 1$ into the final sum, then we translate A^* by an operator whose action on a is in $L^2(\mathbb{C}^s)$ norm

$$\ll \left(\sum_{q\le x} \frac{1}{qx^2} \Big(\sum_{\substack{n\le x \\ (n,q)>1}} |a_n| \Big)^2 \right)^{1/2} \ll \left(x^{-1} \sum_{n\le x} |a_n|^2 \sum \frac{1}{qq_0} \right)^{1/2} \ll \|a\|,$$

this last norm in $M^2(\mathbb{C}^t)$. Arguing in this way we obtain a uniform bound for the operator $(\check{A}a) : \mathbb{C}^t \to \mathbb{C}^s$ defined by

$$(\check{A}a)(q) = \frac{1}{[x/q] - [x/(qq_0)]}\left(\sum_{\substack{n\le x \\ n\cong 0 (\mathrm{mod}\, q)}} a_n - \frac{1}{q} \sum_{\substack{n\le x \\ (n,q)=1}} a_n \right).$$

In the early outline argument we replace A^* by $\check{A}$.

Set $a_n = f(n) - N(f)$ in the inequality

$$\sum_{q\le x} \frac{1}{q}\left(\left[\frac{x}{q}\right]-\left[\frac{x}{qq_0}\right]\right)^{-2}\left|\sum_{\substack{n\le x\\ n\cong 0(\mathrm{mod}\, q)}} a_n - \frac{1}{q}\sum_{\substack{n\le x\\ (n,q)=1}} a_n\right|^2 \ll x^{-1}\sum_{n\le x}|a_n|^2.$$

The upper bound is $\ll \Delta(f)$. For each prime power q

$$N(f)\left(\sum_{\substack{n\le x\\ n\cong 0(\mathrm{mod}\, q)}} 1 - \frac{1}{q}\sum_{\substack{n\le x\\ (n,q)=1}} 1\right) \ll N(f) \ll (\log\log x)^{1/2}\|f\|,$$

in terms of the norm on $L^2(\mathbb{C}^s)$. We may then proceed as in Chapter 8. The modified weight $[x/q]-[x/(qq_0)]$ has been chosen to balance the coefficient of $f(q)$ in the representation

$$\sum_{\substack{n\le x\\ n\cong 0(\mathrm{mod}\, q)}} f(n) = f(q)\left(\left[\frac{x}{q}\right]-\left[\frac{x}{qq_0}\right]\right) + \sum_{\substack{m\le x/q\\ (m,q)=1}} f(m).$$

This completes the proof of Lemma 20.2.

Instead of solving for the function $\alpha(y)$ we define an operator $T:\mathbb{C}^s \to \mathbb{C}^s$ by

$$(Tf)(q) = \sum_{x/q<\ell\le x} \frac{f(\ell)}{\ell}\left(1-\frac{1}{\ell_0}\right),$$

where ℓ denotes a prime power. T is self-adjoint with respect to the inner product on $L^2(\mathbb{C}^s)$, and in terms of the identity operator on $\mathbb{C}^s$, Lemma 20.2 asserts that

$$\|(I-T)f\|^2 \ll \Delta(f) + \frac{\log\log x}{\log x}\|f\|^2.$$

In particular, $\|T\|$ is bounded uniformly for $x \ge 2$.

From what we have so far established concerning approximate functional equations, including Theorem 6.1, we expect $\alpha(y)$ to be close to a multiple of $\log y$ as y varies, and $f(q)$ to be near the same multiple of $\log q$. We take advantage of this.

Integration by parts shows that uniformly for $1 \le r \le x$,

$$\sum_{\substack{n\le x\\ n\equiv 0(\mathrm{mod}\, r)}} (\log n - N(\log)) \ll 1 + \log(x/r) + r^{-1}\log x.$$

As a consequence $\|\check{A}(\log - N(\log))\| \ll \log\log x/(\log x)^{1/2}$. Since $\|\check{A}(f-\lambda\log - N(f-\lambda\log))\| \le \|\check{A}(f-N(f))\| + |\lambda|\|\check{A}(\log - N(\log))\|$, the argument of Lemma 20.2 gives the following result.

Lemma 20.3.

$$\|(I-T)(f-\lambda\log)\|^2 \ll \Delta(f) + (\|f-\lambda\log\|^2 + |\lambda|^2)(\log x)^{-\frac{1}{2}}$$

uniformly in f, λ and $x \geq 2$.

It would be possible to consider the functional inequality of Lemma 20.2 directly. However, the form of the inequality in Lemma 20.3 brings advantages when Theorem 20.1 is generalised to other spaces. A direct computation shows that A^*A differs from $I-T$ by an operator which would introduce complications into the following argument without appreciably affecting the outcome.

To treat the inequality of Lemma 20.3 I employ a classical spectral decomposition theorem for self-adjoint operators.

Lemma 20.4. *Let B be an operator self-adjoint with respect to an inner product over a complex finite dimensional space. Then there is an orthogonal basis for the space consisting of eigenvectors of B.*

Proof. The proof goes by induction on the dimension of the space. For dimension 1 the lemma asserts that B has an eigenvector. For spaces of dimension greater than 1, let w be an eigenvector of B, corresponding to an eigenvalue λ. Let S be the orthogonal complement of w in the parent space. Then B takes S into S. For in terms of the inner product $(w, Bz) = (Bw, z) = (\lambda w, z) = \lambda(w, z) = 0$ for all vectors z in S. By the induction hypothesis S has an orthogonal basis made up of eigenvectors of B, and by adjoining w to it, so does the parent space.

The next lemma asserts that approximate eigenvalues and eigenvectors are near to genuine eigenvalues and eigenvectors.

Lemma 20.5. *In the notation of Lemma 20.4*

(i) *For any vector y and complex number β, there is an eigenvalue λ of B such that*

$$|\lambda-\beta|\|y\| \leq \|By-\beta y\|.$$

(ii) *For any eigenvalue μ of B which is at a positive distance δ from the rest of the spectrum, and any vector y, there is a further vector z, belonging to the eigenspace corresponding to μ, so that*

$$\delta\|y-z\| \leq \|By-\mu y\|.$$

Proof. Let w_j, $j = 1, \ldots, s$ be an orthonormal basis for the parent space, made up of eigenvectors of B. If

$$y = \sum_{j=1}^{s} c_j w_j,$$

then

$$\|By - \beta y\|^2 = \sum_{j=1}^{s} |c_j|^2 |\lambda_j - \beta|^2,$$

where λ_j is the eigenvalue associated with w_j. Choose for λ a value λ_j that minimises $|\lambda_j - \beta|^2$, and note that $\sum |c_j|^2 = \|y\|^2$. This gives (i).

For (ii), define z to be the vector $\sum c_j w_j$ summed over all those w_j with associated eigenvalue μ. Then

$$\|By - \mu y\|^2 \geq \sum_{\lambda_j \neq \mu} |c_j|^2 |\lambda_j - \mu|^2 \geq \delta^2 \|y - z\|^2.$$

We shall largely determine the spectrum of our particular operator T by applying Lemma 20.5. The following argument will provide candidates for the approximate eigenvectors y.

Let W be the vector space of polynomials with real coefficients, constant coefficient zero. We define an operator $D : W \to W$ by

$$\psi(u) \to \int_{1-u}^{1} \psi(v) v^{-1} dv.$$

Strictly speaking D operates on the functions $[0,1] \to \mathbb{R}$ which are specialisations of polynomials. We may then appeal to the fact that if a polynomial vanishes when specialised to the values of the unit interval, then it vanishes identically. Alternatively, D may be defined formally by $u^j \to j^{-1}(1 - (1-u)^j)$ for each $j \geq 1$, and linearity.

For each positive integer k, let W_k denote the subspace of W comprised of polynomials of degree at most k. D takes W_k into itself. Moreover, any polynomial with leading term u^k is transformed into a polynomial with leading term $(-1)^{k+1} u^k k^{-1}$. Let J denote the identity operator on W. Then $D + (-1)^k k^{-1} J$ takes W_k into a subspace of itself of smaller dimension. Viewed from W_k into W_k, this operator has a non-trivial kernel.

Lemma 20.6. *For each $k \geq 1$, the operator D has an eigenvalue $(-1)^{k+1} k^{-1}$, with an associated eigenvector which is a polynomial of degree k over the real numbers, constant coefficient zero. This polynomial is determined to within a scalar factor.*

Proof. Only the final assertion needs justification. We define on W the inner product

$$(\phi, \psi) = \int_0^1 \phi(u) \overline{\psi(u)} du,$$

with respect to which D is self-adjoint. The difference of any two monic polynomials ψ_k, Λ_k which satisfy the conditions of Lemma 20.6 will have a representation $\sum t_j\psi_j$ with reals t_j, $1 \le j \le k-1$. Since $\psi_k - \Lambda_k$ and the ψ_j, $1 \le j \le k-1$, have distinct corresponding eigenvalues, they are mutually orthogonal. For each $j < k$, $t_j(\psi_j, \psi_j) = (\psi_k - \Lambda_k, \psi_j) = 0$. Thus $\psi_k = \Lambda_k$.

Lemma 20.7. *T has spectrum $(-1)^{k+1}k^{-1} + O((\log x)^{-\frac{1}{9}})$, $k = 1, \dots, s$; uniformly for $x \ge 2$. For $1 \le k \le (\log x)^{1/10}$ and all sufficiently large x, these eigenvalues are distinct and simple.*

During the proof of this lemma I shall many times employ the elementary estimate

$$\sum_{q \le x} \frac{1}{q}\left(1 - \frac{1}{q_0}\right) = \log\log x + c + O\left(\frac{1}{\log x}\right),$$

valid for some constant c and all $x \ge 2$. It will also be sometimes convenient to denote $\log x$ by L.

Proof of Lemma 20.7. Let ψ_k be the monic polynomial in Lemma 20.6 associated with the eigenvalue $(-1)^{k+1}k^{-1}$ of D, $k \ge 1$. For a particular value of k let y be the vector of $\mathbb{C}^s$ with q-coordinate $\psi_k(\log q/\log x)$. If ℓ is a prime power not exceeding x, then the ℓ component of Ty is

$$\sum_{x/\ell < q \le x} \frac{1}{q}\left(1 - \frac{1}{q_0}\right)\psi_k\left(\frac{\log q}{\log x}\right).$$

For $\ell \le xe^{-\sqrt{L}}$ we express this sum in the form

$$\int_{1-\frac{\log \ell}{\log x}}^{1} \psi_k(u)d\left(\sum_{\frac{\log q}{\log x} \le u} \frac{1}{q}\left(1 - \frac{1}{q_0}\right)\right)$$

and integrate by parts, appealing to the elementary estimate. We so obtain

$$\frac{(-1)^{k+1}}{k}\psi_k\left(\frac{\log \ell}{\log x}\right) + O\Big((\log x)^{-\frac{1}{2}}\Big).$$

Truncating the range $x/\ell < q$ to $e^{\sqrt{L}} < qq$ allows a similar estimate when $xe^{-\sqrt{L}} < \ell \le x$. Motivating this procedure is the notion that for large x the discrete operators T well approximates the continuous operator D. In this step we apply that ψ_k is an eigenvector of D.

Clearly

$$\|(T + (-1)^k k^{-1} I)y\|^2 \ll (\log x)^{-1} \sum_{\ell \le x} \ell^{-1} \ll L^{-1}\log L.$$

Moreover, $\|y\| \gg 1$ for all x sufficiently large. It follows from part (i) of Lemma 20.5 that T has an eigenvalue ρ_k within $O((L^{-1}\log L)^{1/2})$ of $(-1)^{k+1}k^{-1}$. The eigenvalues might be multiple.

Let ν_k, $k = 1, \dots, s$, denote the totality of eigenvalues of T, counted multiply with the dimension of their eigenspaces. Then

$$\nu_1^2 + \cdots + \nu_s^2 = \operatorname{Trace} T^2.$$

The operator T may be represented by the s-by-s matrix with ℓ^{-1} in place (q, ℓ) if the prime powers q, ℓ satisfy $q\ell > x$, and with zero entry if $q\ell \leq x$. The trace of T^2 is therefore

$$\sum_{q \leq x} \frac{1}{q} \sum_{x/q < \ell \leq x} \frac{1}{\ell}.$$

The inner sum here is $-\log(1 - \log q/\log x) + O((\log(x/q))^{-1})$ for $q \leq xe^{-\sqrt{L}}$, and $\ll \log L$ otherwise. The double sum may be treated much as was the sum defining Ty at ℓ. It has the estimate

$$\int_{L^{-1}\log 2}^{1-L^{-\frac{1}{2}}} -\log(1-v)v^{-1}dv + O(L^{-\frac{1}{2}}\log L).$$

Expanding $\log(1-v)$ as a power series in v and integrating term by term, as we may inside the radius of convergence, we see that

$$\operatorname{Trace} T^2 = \sum_{j=1}^{\infty} \frac{1}{j^2}((1-L^{-\frac{1}{2}})^j - (L^{-1}\log 2)^j) + O\left(\frac{\log L}{L^{\frac{1}{2}}}\right).$$

For $|w| \leq 1$, $|w^j - 1| \leq j|w|$. The terms in the sum with $j \leq L^{\frac{1}{2}}$ contribute $\zeta(2) + O(L^{-\frac{1}{2}}\log L)$. The terms with $j > L^{\frac{1}{2}}$ contribute $O(L^{-\frac{1}{2}})$. Thus

$$\nu_1^2 + \cdots + \nu_s^2 = \zeta(2) + O(L^{-\frac{1}{2}}\log L).$$

From what we have already proved, if x is large enough and j does not exceed a certain absolute multiple of $(L(\log L)^{-1})^{\frac{1}{4}}$, then the ρ_j are distinct. For these we have

$$\sum_j \rho_j^2 \geq \sum_j j^{-2}\left(1 - O\left(j(L^{-1}\log L)^{\frac{1}{2}}\right)\right) = \zeta(2) + O((L^{-1}\log L)^{\frac{1}{4}}).$$

Those ν_j not amongst these early ρ_j are uniformly $\ll (L^{-1}\log L)^{\frac{1}{8}}$. The validity of Lemma 20.7 is now clear.

No attempt has been made to sharpen the estimates for the eigenvalues of T. We shall only need that T has a simple eigenvalue near to 1 which is well separated from the rest of the spectrum.

I continue for the moment with the notation used during the proof of Lemma 20.7.

Since T is self-adjoint, its eigenvalues are real. We order them to satisfy $|\nu_1| \geq \cdots \geq |\nu_s|$. Let d_j be an eigenvector of T corresponding to ν_j, so that the d_j, $1 \leq j \leq s$, form a basis for $\mathbb{C}^s$, orthonormal with respect to the inner product (,). For $1 \leq j \leq (\log x)^{1/10}$ and all x large the d_j are determined to within a unimodular scalar multiple.

Let h_k be the vector of $\mathbb{C}^s$ given q-coordinate $\psi_k(\log q/\log x)$, and then rescaled so that in $L^2(\mathbb{C}^s)$ it has norm 1. Thus $h_1 = \|\log\|^{-1}\log$.

From what we have established during the proof of Lemma 20.7,

$$\|(T+(-1)^k k^{-1}I)h_k\| \ll (\log x)^{-\frac{1}{4}},$$

and therefore

$$\|(T-\nu_k I)h_k\| \ll (\log x)^{-\frac{1}{9}}.$$

For $1 \leq k \leq (\log x)^{1/10}$ and all x large, it follows from part (ii) of Lemma 20.5 that

$$\|h_k - (h_k, d_k)d_k\| \ll (\log x)^{-\frac{1}{9}}.$$

In particular, $|(h_k, d_k)| = 1 + O((\log x)^{-\frac{1}{9}})$. By rescaling d_k, if necessary, we may assume the inner product (h_k, d_k) to be real and non-negative. Then

$$\|h_k - d_k\| \ll (\log x)^{-\frac{1}{9}}$$

for all $x \geq 2$, the implied constant depending upon k.

Proof of Theorem 20.1. Define operators S_1, T_1 on $L^2(\mathbb{C}^s)$ to itself by

$$S_1 f = Tf - (f, h_1)h_1, \qquad T_1 f = Tf - \nu_1(f, d_1)d_1.$$

Then

$$(S_1 - T_1)f = (\nu_1 - 1)(f, d_1)d_1 + (f, d_1)(d_1 - h_1) + (f, d_1 - h_1)h_1.$$

Each of the three inner product terms is $\ll (\log x)^{-\frac{1}{9}}\|f\|$ in $L^2(\mathbb{C}^s)$ norm. For example, $\|d_1 - h_1\| \ll (\log x)^{-\frac{1}{9}}$ and by the Cauchy–Schwarz inequality $|(f, d_1)| \leq \|f\|\|d_1\| = \|f\|$.

Moreover, if z in $\mathbb{C}^s$ has spectral representation $\sum c_j d_j$, $1 \leq j \leq s$, then

$$\|T_1 z\|^2 = \sum_{j=2}^{s} \nu_j^2 |c_j|^2 \leq \nu_2^2 \sum_{j=2}^{s} |c_j|^2 \leq \nu_2^2 \|z\|^2.$$

Thus $\|T_1\| \leq \frac{1}{2} + O((\log x)^{-\frac{1}{9}})$, with a like estimate for $\|S_1\|$. As a consequence the resolvent operator $(I - S_1)^{-1}$ has a norm that is bounded uniformly for all sufficiently large x.

The choice $\lambda = (f, \log)\| \log \|^{-2} = (f, h_1)\| \log \|^{-1}$ allows

$$(I - S_1)^{-1}(I - T)(f - \lambda \log) = f - \lambda \log,$$

and from the inequality of Lemma 20.3

$$\|f - \lambda \log \|^2 \ll \Delta(f) + |\lambda|^2 (\log x)^{-\frac{1}{9}}.$$

To complete the proof of Theorem 20.1 we note that for $x \geq 2$

$$|\lambda|^2 \ll \Delta(\lambda \log) \ll \Delta(\lambda \log - f) + \Delta(f).$$

From what was said at the beginning of the chapter $\Delta(\lambda \log - f) \ll \|f - \lambda \log \|^2$, and to this we may apply the bound just obtained.

I have elaborated this proof of Theorem 20.1 slightly so that it may be readily generalised.

For each positive integer m, and the same underlying value of x, define operators T_m, S_m from $\mathbb{C}^s$ to $\mathbb{C}^s$ by

$$T_m f = Tf - \sum_{k=1}^{m} \nu_k (f, d_k) d_k,$$

$$S_m f = Tf - \sum_{k=1}^{m} (-1)^{k+1} k^{-1} (f, h_k) h_k.$$

For each fixed k, $\|d_k\|$ is bounded uniformly in $x \geq 2$. From what we have established concerning the approximation of d_k by h_k, it follows readily that

$$\|S_m - T_m\| \ll (\log x)^{-1/9}$$

uniformly in $x \geq 2$. The operators S_m, T_m are bounded uniformly in $x \geq 2$. In particular, for each positive integer r,

$$\|S_m^r\| \leq (m+1)^{-r} + O((\log x)^{-1/9})$$

holds uniformly for $x \geq 2$.

These results are concerned with the operator T and its projections acting on the space $L^2(\mathbb{C}^s)$. In the next chapter I show that similar results are valid when T acts on $L^\alpha(\mathbb{C}^s)$, $(L^2 \cap L^{\alpha'}(\mathbb{C}^s))'$, and so on.

21

The operator T on L^α and other spaces

In this chapter I extend the results of Chapter 20 to other spaces. For $\alpha > 1$, $x \geq 2$, define

$$\Delta_\alpha(f) = \left(x^{-1} \sum_{n \leq x} \left| f(n) - x^{-1} \sum_{m \leq x} f(m) \right|^\alpha \right)^{1/\alpha}.$$

This notation is slightly at odds with that of Chapter 20, since $\Delta(f) = \Delta_2(f)^2$. Whereas it was convenient to regard $\Delta(f)$ as a variance, I shall view $\Delta_\alpha(f)$ as a norm.

Theorem 21.1. *Let $\alpha > 1$. For all x sufficiently large, $\Delta_\alpha(f)$ lies between positive multiples of*

$$\min_\lambda (|\lambda| + \| f - \lambda \log \|),$$

uniformly in f, where the norm is that on $(L^2 \cap L^{\alpha'}(\mathbb{C}^s))'$ if $1 < \alpha < 2$, that on $L^2 \cap L^\alpha(\mathbb{C}^s)$ if $\alpha \geq 2$.

I shall show that the minimum over λ may be removed in favour of the choice $\lambda = (f, \log)\| \log \|_2^{-2}$, defined in terms of the inner product on $L^2(\mathbb{C}^s)$ considered in Chapter 20. This choice for λ is independent of α.

Although not necessary to the proof of Theorem 21.1, it is instructive to first consider the operator

$$D : \psi(u) \to \int_{1-u}^{1} \psi(v) \frac{dv}{v}$$

on the space L^α with respect to the measure $v^{-1}dv$ on $0 \leq v \leq 1$. For this the following generalisation of Lemma 20.5 is helpful.

Lemma 21.2. *Let T be a self-adjoint operator on a complex Hilbert space. For any vector x and complex number α, there is a λ in the closure of the spectrum of T so that*

$$|\lambda - \alpha| \|x\| \leq \|Tx - \alpha x\|.$$

Let μ be an eigenvalue of T which is at a positive distance δ from the rest of the spectrum. Then for any vector x there is a further vector z, belonging to the eigenspace of T corresponding to μ, so that

$$\delta \|x - z\| \leq \|Tx - \mu x\|.$$

Proof of Lemma 21.2. The first of these two propositions is equivalent to the assertion that for a self-adjoint operator T, the norm of the resolvent operator $R(\alpha; T)$ does not exceed the reciprocal of the distance of α from the spectrum of T. If the family of projections $E(\lambda)$ is the resolution of the identity associated with T, then it suffices to consider the equation

$$\|(T - \alpha I)x\|^2 = \int_{-\infty}^{\infty} |\lambda - \alpha|^2 d\|E(\lambda)x\|^2;$$

Yosida, [175], Chapter XI, §§5,6. Corollary 2 on page 312 of that reference is readily extended to cover the case of the complex valued $f(\lambda) = \lambda - \alpha$.

The second proposition may be deduced from the same equation, with μ in place of α. If (u, v) on the real line doesn't contain any point of the spectrum of T, then $E(\beta) = E(\gamma)$ for whenever $u < \beta < \gamma < v$. Let $\delta > \varepsilon > 0$. From the hypothesis on μ

$$\|(T - \mu I)x\|^2 \geq (\delta - \varepsilon)^2 \int 1 d\|E(\lambda)x\|^2$$

where the integral is taken over the real line with the interval $[\mu - \delta + \varepsilon, \mu + \delta - \varepsilon]$ removed. This last integral has the alternative representation

$$\left\| x - \int_{\mu-\delta+\varepsilon}^{\mu+\delta-\varepsilon} 1 dE(\lambda)x \right\|^2,$$

and

$$\int_{\mu-\delta+\varepsilon}^{\mu+\delta-\varepsilon} 1 dE(\lambda)x = \lim_{\nu \to 0+} (E(\mu) - E(\mu - \nu))x$$

belongs to the eigenspace of T associated with μ; Yosida, [175], Chapter XI, §8.

Formally this proof runs on the lines of that for Lemma 20.5.

I shall apply Lemma 21.2 only for compact operators, and then only to motivate a number theoretic argument.

An application of Hölder's inequality shows that into L^α, D is bounded, indeed compact.

The map D is one-to-one, but its image in L^α contains only continuous functions, and so doesn't cover the whole of L^α. Thus 0 is the sole member of its continuous spectrum.

The dual space of L^α may be identified with $L^{\alpha'}$, $\alpha^{-1} + (\alpha')^{-1} = 1$, and the dual operator D' is formally the same as D. Moreover, since D is compact, D and D' have the same spectra, even down to the finite multiplicity of each eigenvalue. Without loss of generality we may confine ourselves to the cases $\alpha \geq 2$. This allows us to drop from L^α down to L^2.

Since $v^{-1}dv$ does not give a (finite) measure to the whole interval (0,1), it is advantageous to consider the functions g of $L^\alpha(0,1)$ restricted to the interval $\delta \leq u \leq 1-\delta$, and so belonging to $L^2(\delta, 1-\delta)$, $0 < \delta < 1/2$. I shall denote by g_1 both a typical function g on $(0,1)$ restricted to $(\delta, 1-\delta)$, and the function

$$h(u) = \begin{cases} g_1(u), & \delta \leq u \leq 1-\delta, \\ 0, & 0 \leq u < \delta \text{ or } 1-\delta < u \leq 1 \end{cases}$$

which partially recovers g.

On $L^2(\delta, 1-\delta)$ the operator

$$G : w \rightarrow \int_{1-u}^{1-\delta} w(v) \frac{dv}{v}$$

is self-adjoint. As demonstrated by the exercises in the following chapter, the eigenfunctions of G satisfy a second order differential equation, but they can no longer be a set of polynomials orthogonal with respect to the measure $v^{-1}dv$, and there does not seem to be a simple explicit representation for them. It is more convenient to think of δ as small, and regard $L^2(\delta, 1-\delta)$ as an approximation to $L^2(0,1)$.

Let g be a unit eigenfunction of D on $L^\alpha(0,1)$, with associated eigenvalue $\lambda \neq 0$. On the interval $\delta \leq u \leq 1$ we have

$$Dg = Dg_1 + \int_{1-\delta}^{1} g(v) \frac{dv}{v}.$$

We see that in $L^2(0,1)$

$$\|Dg_1 - \lambda g_1\|^2 = \int_{\delta}^{1-\delta} \left| \int_{1-\delta}^{1} g(v) \frac{dv}{v} \right|^2 \frac{du}{u} + \int_{1-\delta}^{1} |Tg_1|^2 \frac{du}{u}.$$

Bearing in mind that on $L^\alpha(0,1)$, $\|g_1\| \le 1$, we may estimate $|Tg_1|$ by means of Hölder's inequality, and obtain for the last of these integrals the upper bound $2\delta(\log 1/\delta)^{2/\alpha'}$. Similarly

$$\left|\int_{1-\delta}^{1} g(v)\frac{dv}{v}\right| \le (-\log(1-\delta))^{2/\alpha'},$$

and the penultimate integral is not more than $(2\delta)^{2/\alpha'}\log 1/\delta$. Altogether

$$\|Dg_1 - \lambda g_1\| \ll \delta^c$$

for some positive constant c. Moreover, since g is not almost surely zero, $\|g_1\| \ge c_0 > 0$ holds in $L^2(0,1)$ for all sufficiently small δ.

It follows from Lemma 21.2 that for some eigenvalue $\lambda_j = (-1)^{j+1}j^{-1}$ of D on $L^2(0,1)$ we have $|\lambda - \lambda_j| \ll \delta^{c/2}$, the implied constant depending at most upon α. Letting $\delta \to 0^+$ we see that λ belongs to the closure of the spectrum of D on $L^2(0,1)$, and so to that spectrum itself. Moreover, the polynomials $\psi_k(u)$ applied in the proof of Lemma 20.7 belong to $L^\alpha(0,1)$ for every $\alpha > 0$, and are still eigenfunctions of D on L^α, $\alpha > 1$. Disregarding multiplicities, the spectrum of D on L^α, $\alpha > 1$, is the same as that of D on L^2.

In fact every eigenvalue of D is still simple. For example, when $\alpha > 1$ we can define the map

$$D_1 : g \mapsto Dg - (g, \rho_1)\rho_1$$

where ρ_1 is the eigenvector $\psi_1(u)$ of D, scaled to satisfy $\|\rho_1\|_2 = 1$, and the inner product of g and ρ_1 is taken in $L^2(0,1)$. Since ρ_1 belongs to every $L^\beta(0,1)$, $\beta \ge 1$, D_1 is well defined. D_1 is also compact and formally self-dual. The above argument shows that its spectrum is that of D with the eigenvalue 1 omitted. It has spectral radius 1/2. Suppose now that g is an eigenvalue of D with corresponding eigenvalue 1, and fix an integer k so that $\|D_1^k\| < 3/4$. A simple check shows that $D_1^k g = g - (g, \rho_1)\rho_1$, from which

$$g = (I - D_1^k)^{-1}(g, \rho_1)\rho_1 = (g, \rho_1)\rho_1.$$

Apart from its normalisation, g is uniquely determined. In particular, the eigenvalue 1 is simple.

To establish Theorem 21.1 in this style we need an estimate for the spectral radius of T_1, an operator which depends upon x. It is not immediately clear that T is uniformly bounded from $L^\alpha(\mathbb{C}^s)$ to itself.

Lemma 21.3. *The operator*

$$T : L^\alpha(\mathbb{C}^s) \to L^\alpha(\mathbb{C}^s)$$

is bounded uniformly in $x \geq 2$.

This result may be established using the measure $d\mu$ introduced in Chapter 3. This measure assigns $q^{-1}(1-q_0^{-1})$ to each prime power q (not exceeding x). However, insight is gained by introducing further spaces on which to exhibit an underlying weak convergence of measures.

For each $x \geq 2$ define the function $F : \mathbb{R} \to \mathbb{R}$ by

$$F(u) = \sum_{q \leq x^u} \frac{1}{q}\left(1 - \frac{1}{q_0}\right).$$

On the space $L^\alpha(0,1)$ with respect to the measure dF, and into itself, we (typically) define the operator H by

$$(Hw)(u) = \int_{1-u}^{1} w(v)dF(v).$$

Unless otherwise specified, integrals involving possibly atomic measures are taken over intervals of the form $(a, b]$.

We may enclose the points $\log q/\log x$, $2 \leq q \leq x$, in non-overlapping open intervals of real numbers. If we define $h(v)$ to be $f(q)$ whenever v is contained in the interval enclosing $\log q/\log x$, $2 \leq q \leq x$, and to be zero on the unit interval otherwise, then for values $\log \ell/\log x$ of u, with ℓ again a prime power, $(Hh)(u)$ coincides with $(Tf)(\ell)$. We may view the asymptotic behaviour of the operators T as an application of the weak convergence of the measures dF as $x \to \infty$. A bound for the norm of H is already a bound for the norm of T.

It will be convenient to denote $\log x$ by L.

Lemma 21.4. *The operator H is bounded uniformly for $x \geq 2$.*

Proof. An application of Hölder's inequality shows that

$$|(Hw)(u)| \leq \|w\|_\alpha \left(\int_{1-u}^{1} 1dF\right)^{1/\alpha'}, \qquad 0 \leq u \leq 1.$$

The integral of the identity function over the interval $(1-u, 1]$ has the estimate

$$-\log(1-u) + O(((1-u)L)^{-1}),$$

certainly uniformly for $0 \le u \le 1 - L^{-1/2}$, $x \ge 2$. Hence

$$\int_0^{1-L^{-1/2}} |(Hw)(u)|^\alpha dF(u)$$

$$\ll \|w\|_\alpha^\alpha \int_0^{1-L^{-\frac{1}{2}}} (-\log(1-u))^{\alpha-1} + ((1-u)L)^{1-\alpha} dF(u).$$

Since dF gives measure $\ll \log L$ to the whole unit interval, the term involving $(1-u)L$ contributes $\ll L^{-(\alpha-1)/2} \log L$ to the bounding integral. Over the range $0 \le u \le 1/2$, $-\log(1-u) \ll u$, and the corresponding contribution to the integral is

$$\ll \sum_{q \le x^{1/2}} \frac{1}{q} \left(\frac{\log q}{\log x}\right)^{\alpha-1} \ll 1.$$

For $1/2 < u \le 1 - L^{-1/2}$ the logarithmic function needs a little more care.

There is a constant c so that the error term

$$z(y) = \sum_{q \le y} \frac{1}{q}\left(1 - \frac{1}{q_0}\right) - \log\log y - c$$

is $\ll L^{-1}$, uniformly for $2 \le x^{1/2} \le y \le x$. Then

$$\int_{1/2}^{1-L^{-1/2}} (-\log(1-u))^{\alpha-1} dF(u) =$$

$$\int_{1/2}^{1-L^{-1/2}} (-\log(1-u))^{\alpha-1} \frac{du}{u} + \int_{1/2}^{1-L^{-1/2}} (-\log(1-u))^{\alpha-1} dz(x^u).$$

For each fixed $\varepsilon > 0$, $-\log(1-u) \ll (1-u)^{-\varepsilon}$, and the second of these three integrals does not exceed a constant multiple of

$$\int_{1/2}^{1-L^{-1/2}} \frac{du}{(1-u)^{1/2}},$$

an integral which is less than $\sqrt{2}$ in value. An integration by parts gives for the third of the integrals the representation

$$\left(\frac{1}{2}\log L\right)^{\alpha-1} z(xe^{-L^{1/2}}) - (\log z)^{\alpha-1} z(x^{1/2})$$

$$- \int_{1/2}^{1-L^{-1/2}} z(x^u)(\alpha-1)(-\log(1-u))^{\alpha-2} \frac{du}{1-u}.$$

Here, the two plain terms involving z are by an earlier remark $\ll L^{-1}(\log L)^{\alpha-1}$. The integral is similarly

$$\ll L^{-1}\int_{1/2}^{1-L^{-1/2}}(-\log(1-u))^{\alpha-2}\frac{du}{1-u}\ll L^{-1}(\log L)^{\alpha-1}.$$

The integral of $|(Hw)(u)|^{\alpha}dF(u)$ over the range $(0,1-L^{-1/2}]$ is $\ll \|w\|_{\alpha}^{\alpha}$ uniformly in $x\geq 2$.

Our preliminary estimate for $(Hw)(u)$ is $\ll \|w\|_{\alpha}(\log L)^{1/\alpha'}$, uniformly for $0\leq u\leq 1$, so that

$$\int_{1-L^{-1/2}}^{1}|(Hw)(u)|^{\alpha}dF(u)\ll \|w\|_{\alpha}^{\alpha}(\log L)^{\alpha-1}\int_{1-L^{-1/2}}^{1}1dF(u).$$

The operator H is indeed bounded uniformly in $x\geq 2$.

The validity of Lemma 21.3 is immediate.

During the proof of Lemma 20.7 I introduced, in an explicit form, the integral

$$\int_{1-\frac{\log \ell}{\log x}}^{1}\psi_k(u)dF(u).$$

This representation was suggested by the knowledge that as $x\to\infty$ the measures $dF(u)$ on the interval (0,1) converge weakly to $u^{-1}du$. Nonetheless, the focus of that lemma is the operator T with its dependence on x, and not the operator D to which, in a certain sense, it converges. Difficulties in the presentation of Lemma 20.7 arise in part from there not being a simple common space on which to exhibit the operators T and D.

The treatment of Lemma 21.4 raises the measures $dF(u)$ into prominence. The limiting behaviour of the operators T may be viewed as a manifestation of the weak convergence of the $dF(u)$. Attractive in its clarity, this approach, too, has to overcome technical difficulties, as I indicate with exercises in the following chapter.

In the event I continue with the operators T themselves.

For each positive integer j (again) define the operator $S_j:\mathbb{C}^s\to\mathbb{C}^s$ by

$$S_jf=Tf-\sum_{i=1}^{j}\frac{(-1)^{i+1}}{i}(f,h_i)h_i.$$

I denote the norm of an operator U from $L^{\alpha}(\mathbb{C}^s)$ to $L^{\alpha}(\mathbb{C}^s)$ by $\|U\|_{\alpha}$.

Lemma 21.5.
$$\|S_j^k\|_\alpha \ll \|S_j^{k-2}\|_2$$
uniformly for $x \geq 2$ and integers $k \geq 3$.

Proof. In terms of the measure $d\nu$ from Chapter 3, the operator S_j has a representation

$$(S_j f)(\ell) = \int \sigma(\ell, q) f(q) d\nu(q),$$

where the kernel $\sigma(\ell, q)$ is defined to be

$$1 - \sum_{i=1}^{j} \frac{(-1)^{i+1}}{i} h_i\left(\frac{\log \ell}{\log x}\right) h_i\left(\frac{\log q}{\log x}\right)$$

if $q\ell > x$, and this expression with the leading term 1 removed if $q\ell \leq x$. For our present purpose it will suffice to note that

$$\sigma(\ell, q) \ll \begin{cases} 1 & \text{if} \quad q\ell > x, \\ L^{-2} \log q \log \ell & \text{if} \quad q\ell \leq x. \end{cases}$$

Then

$$\int \sigma(\ell, q)^2 d\nu(q) \ll \int_{x/\ell}^{x} 1 d\nu + \int L^{-4} (\log \ell \log q)^2 d\nu$$
$$\ll \min\left(\log L, -\log\left(1 - \frac{\log \ell}{\log x}\right)\right) + L^{-1/2} + L^{-2}(\log \ell)^2,$$

which is $\theta(\ell)^2$, say, with $\theta(\ell) > 0$. In particular, an argument along the lines of the proof of Lemma 21.4 shows that for each $\beta \geq 1$, $\|\theta(\ell)\|_\beta \ll 1$ uniformly for $x \geq 2$.

By an application of the Cauchy–Schwarz inequality

$$|S_j f(\ell)| \ll \|f\|_2 \theta(\ell).$$

The operator S_j^k has a representation

$$(S_j^k f)(\ell) = \int \tau(\ell, q) f(q) d\nu(q),$$

with $\tau(\ell, q)$ given by (reading right to left)

$$\int \sigma(\ell, q_{k-1}) d\nu(q_{k-1}) \cdots \int \sigma(q_2, q_1) \sigma(q_1, q) d\nu(q_1).$$

If $\sigma(\cdot, q)$ denotes the function whose value on a typical prime power ℓ (not exceeding x) is $\sigma(\ell, q)$, then $\tau(\ell, q)$ is the value of $S_j^{k-1}\sigma(\cdot, q)$ at ℓ. In particular

$$\begin{aligned}|\tau(\ell,q)| &\ll \|S_j^{k-2}\sigma(\cdot,q)\|_2\theta(\ell) \\ &\ll \|S_j^{k-2}\|_2\theta(q)\theta(\ell).\end{aligned}$$

Applications of Hölder's inequality show that

$$\begin{aligned}\|S_j^k\|_\alpha &\leq \left(\int\left(\int|\tau(\ell,q)|^{\alpha'}d\nu(q)\right)^{\alpha-1}d\nu(\ell)\right)^{1/\alpha} \\ &\ll \|S_j^{k-2}\|_2\|\theta(q)\|_{\alpha'}\|\theta(\ell)\|_\alpha.\end{aligned}$$

The proof of Lemma 21.5 is complete.

S_j differs from T by a sum of rescaled projections onto various h_i. By Hölder's inequality $|(f,h_i)| \leq \|f\|_\alpha\|h_i\|_{\alpha'}$. For any fixed $\beta \geq 1$ the norms $\|h_i\|_\beta$ are bounded uniformly in $x \geq 2$. $\|S_j\|_\alpha$ is bounded uniformly in $x \geq 2$.

From what we proved in Chapter 20, Lemma 21.5 shows that

$$\|S_j^k\|_\alpha \leq c_0(j+1)^{-k} + O(L^{-1/9})$$

where the constant c_0 is independent of k. Let $\gamma-(j+1)^{-1} = 2\varepsilon > 0$. If k is fixed at a large enough value, then $c_0(j+1)^{-k} < (\gamma-\varepsilon)^k$. By representing r in the form $tk+s$, $0 \leq s < k$, we see that

$$\|S_j^r\|_\alpha \leq c_1(\gamma-\varepsilon)^r$$

with a constant c_1 that is independent of $r \geq 1$, and all x exceeding some value x_0 depending only upon α, ε.

The resolvent operator

$$(\lambda I - S_j)^{-1} = \lambda^{-1}(I + \lambda^{-1}S_j + (\lambda^{-1}S_j)^2 + \cdots)$$

has norm bounded uniformly for $|\lambda| \geq \gamma$, $x \geq x_0$.

Again T_j will be the operator $\mathbb{C}^s \to \mathbb{C}^s$ given by

$$T_jf = Tf - \sum_{i=1}^{j}\nu_i(f,d_i)d_i.$$

Theorem 21.6. *For each $\alpha > 1$ there is a positive δ so that*

$$\|d_i - h_i\| \ll (\log x)^{-\delta}, \qquad \|T_i - S_i\|_\alpha \ll (\log x)^{-\delta},$$

for each fixed i, the inequalities uniform in $x \geq 2$.

Proof. We showed in Chapter 20 that $\nu_i = (-1)^{i+1}i^{-1} + O(L^{-1/9})$. For all sufficiently large values of x, $|\nu_i| - (i+1)^{-1} > (i+1)^{-2}$, and for each $j \geq i$, $(\nu_i I - S_j)^{-1}$ is bounded in terms of α, i, j. Then

$$d_i = \sum_{r=1}^{j} \frac{(-1)^{r+1}}{r} (d_i, h_r)(\nu_i I - S_j)^{-1} h_r.$$

The inner products (d_i, h_r), evaluated in $L^2(\mathbb{C}^s)$, are bounded uniformly in $x \geq 2$. The $\|h_r\|$ in $L^\alpha(\mathbb{C}^s)$ are also bounded uniformly. Therefore $\|d_i\|$ is bounded uniformly, by this argument if x is sufficiently large, and for elementary reasons otherwise.

Further calculation shows that if $i \neq j$, then $(h_i, h_j) \ll L^{-\delta}$ for a positive constant δ not depending upon i, j or $x \geq 2$. As a consequence $\|S_j h_i\| \ll L^{-\delta}$ for $1 \leq i \leq j$, possibly with a different value of δ. This again uses the fact that h_i is an approximate eigenvector of T.

The relation

$$\lambda(\lambda I - S)^{-1} - I = \lambda(\lambda I - S)^{-1} S$$

and our estimate for the norm of $S_j h_i$ give

$$\|\nu_i(\nu_i I - S_j)^{-1} h_i - h_i\| \ll L^{-\delta},$$

enabling

$$\left\| \lambda_i d_i - \sum_{r=1}^{j} \lambda_r (d_i, h_r) h_r \right\| \ll L^{-\delta}, \quad 1 \leq i \leq j.$$

Here $\lambda_r = (-1)^{r+1} r^{-1}$, and I have appealed to the estimate $\nu_i = \lambda_i + O(L^{-1/9})$.

I apply the (bounded) operator T, $k-1$ times to the vector inside this last norm, to obtain

$$\left\| \lambda_i^k d_i - \sum_{r=1}^{j} \lambda_r^k (d_i, h_r) h_r \right\| \ll L^{-\delta}, \quad 1 \leq k \leq j.$$

Again the estimate $\nu_i = \lambda_i + O(L^{-1/9})$, $1 \leq i \leq j$, has been appealed to.

We may treat these j inequalities involving norms on $L^\alpha(\mathbb{C}^s)$ in the manner of linear equations, using Cramer's rule. The coefficient matrix (λ_i^k),

$1 \le i,\ k \le j$, has a van der Monde determinant, not zero because the λ_i are distinct and non-zero. Thus we obtain

$$\|d_i - h_i\| \ll L^{-\delta}, \qquad (d_i, h_r) \ll L^{-\delta},$$

for each fixed $i \neq r$. The inner product estimate may also be deduced from the estimate for $\|d_i - h_i\|$, using the orthogonality of the d_j in $L^2(\mathbb{C}^s)$, and the uniform bound on $\|d_i\|$ in $L^{\alpha'}(\mathbb{C}^s)$.

It is now straightforward to obtain the bound on $\|S_i - T_i\|_\alpha$. Typically $\lambda_r(f, h_r)h_r - \nu_r(f, d_r)d_r$ has the representation

$$(21.1) \qquad (\lambda_r - \nu_r)(f, h_r)h_r + \nu_r(f, h_r - d_r)d_r + \nu_r(f, d_r)(h_r - d_r).$$

In $L^\alpha(\mathbb{C}^s)$ the norm of the middle term is

$$\ll |(f, h_r - d_r)| \ll \|f\|_\alpha \|h_r - d_r\|_{\alpha'},$$

after an appeal to Hölder's inequality. From what we have already proved, and with possibly another value of δ, $\|h_r - d_r\|_{\alpha'} \ll L^{-\delta}$. The first and third terms of (21.1) may be similarly dealt with.

This completes the proof of Theorem 21.6.

Looking back, I give a version of Lemma 21.2 for the operator T on the space $L^\alpha(\mathbb{C}^s)$, but with the propositions in the opposite order.

Theorem 21.7. *Let j and r be integers, $1 \le j \le r$. For all x sufficiently large, the eigenvalue ν_j of T and any vector g in $L^\alpha(\mathbb{C}^s)$ satisfy*

$$\|g - (g, d_j)d_j\| \ll \|Tg - \nu_j g\|,$$
$$(g, d_i)\|d_i\| \ll \|Tg - \nu_j g\|, \quad i \neq j,\ 1 \le i \le r,$$

the implied constants depending only upon α, r.

Let $\beta > 0$. For all x sufficiently large, any complex number λ, $|\lambda| \ge \beta$, and vector g in $L^\alpha(\mathbb{C}^s)$, there is an eigenvalue ν_i of T so that

$$|\nu_i - \lambda| \|g\| \ll \|Tg - \lambda g\|,$$

the implied constant depending only upon β and α.

Proof. The method is the same as that employed for Theorem 21.6, so I give only a sketch.

Let $\varepsilon = \|Tg - \nu_j g\|$. With $r \ge j$ and

$$\eta(f) = \sum_{i=1}^{r} \nu_i(f, d_i)d_i,$$

we have

$$\|(\nu_j I - T_r)g - \eta(g)\| = \varepsilon.$$

The resolvent operator $(\nu_j I - T_r)^{-1}$ is uniformly bounded, and $T_r \eta(g) = 0$; hence

$$\left\| \nu_j g - \sum_{i=1}^{r} \nu_i (g, d_i) d_i \right\| \ll \varepsilon.$$

Since T is uniformly bounded, operating on the vector estimated in this norm yields

$$\left\| \nu_j^k g - \sum_{i=1}^{r} \nu_i^k (g, d_i) d_i \right\| \ll \varepsilon$$

for each fixed k, $1 \le k \le r$. Elimination justifies the first assertion of Theorem 21.7.

The proof of the second assertion goes similarly. We reach an estimate

$$D\|g\| \ll \|Tg - \lambda g\|$$

where D is the van der Monde determinant $\det(\nu_i^j)$, $0 \le i \le r$, $1 \le j \le r+1$, with $\nu_0 = \lambda$, and r chosen so that $\beta(r+1) > 1$. This determinant satisfies

$$|D| = \left| \prod_{i=1}^{r} \nu_i \prod_{0 \le i < j < r} (\nu_i - \nu_j) \right| > \gamma \prod_{j=1}^{r} |\lambda - \nu_j|$$

for some positive γ depending at most upon r, and so β. If some term in the last product is less than $(4r^2)^{-1}$, then for all sufficiently large x all other terms are larger than $(4r^2)^{-1}$. This gives the asserted bound.

In preparation for the proof of Theorem 21.1 I note that $\|S_j^k\| \ll \|S_j^{k-2}\|_2$ on each space $L^\alpha(\mathbb{C}^s)$, and so on $L^2 \cap L^\alpha(\mathbb{C}^s)$. Since the operator S_j^k is self dual, a similar bound holds on $(L^2 \cap L^{\alpha'}(\mathbb{C}^s))'$. I shall need only the case $j = 1$ of this general version of Lemma 21.5, and then only the corollary that the resolvent operator $(\lambda I - S_1)^{-1}$ is bounded uniformly for $|\lambda| \ge \gamma > 1/2$ and all x sufficiently large in terms of γ, α.

To embrace the general situation of Theorem 21.1 I employ the chain of maps

$$L^2 \cap L^\alpha(\mathbb{C}^s) \xrightarrow{A_\alpha} M^\alpha(\mathbb{C}^t) \simeq (M^{\alpha'}(\mathbb{C}^t))' \xrightarrow{\check{A}} L^2 \cap L^\alpha(\mathbb{C}^s)$$

when $\alpha \ge 2$, and a suitably modified chain when $1 < \alpha \le 2$. It is convenient to first consider $\check{A}A_\alpha$ from $L^\alpha(\mathbb{C}^s)$ to itself.

Lemma 21.8. *For each $\alpha > 1$ there is a positive δ so that*

$$\|\check{A}A_\alpha - I + T\|_\alpha \ll (\log x)^{-\delta}$$

uniformly for $x \geq 2$.

Proof. The argument has been applied in the proof of Lemma 20.2 and in Chapter 8, so I give only salient details.

A helpful estimate is

$$\frac{1}{y}\sum_{\substack{n\leq y\\(n,q)=1}} f(n) = \left(1-\frac{1}{q_0}\right)\sum_{\substack{\ell\leq y\\(\ell,q)=1}} \frac{f(\ell)}{\ell}\left(1-\frac{1}{\ell_0}\right) + O(\|f\|_\alpha (\log x)^{-1/\alpha'})$$

uniformly in $y \geq 2$, f and prime powers q. As a consequence

$$\sum_{\substack{m\leq x/q\\(m,q)=1}} f(m) - \frac{1}{q}\sum_{\substack{n\leq x\\(n,q)=1}} f(n)$$
$$= \frac{x}{q}\left(1-\frac{1}{q_0}\right)\sum_{\substack{x/q<\ell\leq x\\(\ell,q)=1}} \frac{f(\ell)}{\ell}\left(1-\frac{1}{\ell_0}\right) + O(\|f\|_\alpha L^{-1/\alpha'}).$$

Here the condition $(\ell, q) = 1$ may be removed at the expense of an error $\ll (q^{-1}x)^{1/\alpha}\|f\|_\alpha$.

Bearing in mind that

$$\frac{x}{q}\left(1-\frac{1}{q_0}\right)\left(\left[\frac{x}{q}\right]-\left[\frac{x}{qq_0}\right]\right)^{-1} = 1 + O\left(\frac{q}{x}\right)$$

uniformly for $2 \leq q \leq x$, we see that the value of $(\check{A}A_\alpha - I + T)f$ at q is

$$\ll \|f\|_\alpha (L^{-1/\alpha'} + (qx^{-1})^{1/\alpha'}) + x^{-1}q|(Tf)(q)|.$$

Employing the estimate $\ll \|f\|(\log L)^{1/\alpha'}x^{-1}q$ for the last of these bounding terms we obtain Lemma 21.8 directly. Any fixed $\delta < \min(1/\alpha', 1/\alpha)$ is permissible.

The case $\alpha = 2$ of Lemma 21.8 largely forms the content of Lemma 20.2. The following is an analogue of Lemma 20.3.

Lemma 21.9. *For each $\alpha > 1$ there is a positive δ so that*

$$\|(I-T)(f-\lambda\log)\| \ll \Delta_\alpha(f) + (\|f-\lambda\log\| + |\lambda|)(\log x)^{-\delta}$$

uniformly in f, complex λ and real $x \geq 2$. The norms are in $L^2 \cap L^\alpha(\mathbb{C}^s)$ or $(L^2 \cap L^{\alpha'}(\mathbb{C}^s))'$ as the case may be.

Proof. I confine myself to a few remarks concerning the cases $1 < \alpha < 2$. Let $g = f - \lambda\log$. In terms of the norm on $(L^2 \cap L^{\alpha'}(\mathbb{C}^s))'$

$$\|\check{A}A_\alpha - I + T\| \ll \|\check{A}A_\alpha - I + T\|_\alpha \ll L^{-\delta}.$$

Therefore

$$\|(I-T)g\| \ll \|\check{A}A_\alpha g\| + L^{-\delta}\|g\|.$$

Here

$$\|\check{A}A_\alpha g\| \leq \|\check{A}A_\alpha f\| + |\lambda|\|\check{A}A_\alpha \log\|.$$

The first of the bounding terms is $\ll \Delta_\alpha(f)$. The second may be estimated $\ll |\lambda|L^{-\delta}$ using the good distribution of the logarithmic function over residue classes, as in the derivation of Lemma 20.3.

Proof of Theorem 21.1. The upper bound

$$\Delta_\alpha(f) \ll \|f - \lambda\log\| + |\lambda|$$

follows almost at once from Theorem 2.1.

For the lower bound set $\lambda = (f, \log)\|\log\|_2^{-2}$ in Lemma 21.9. Then $(T - S_1)(f - \lambda\log) = 0$ and

$$\|(I-S_1)g\| \ll \Delta_\alpha(f) + (\|g\| + |\lambda|)L^{-\delta},$$

where g again denotes $f - \lambda\log$. Since $(I - S_1)^{-1}$ is a uniformly bounded operator

$$\|f - \lambda\log\| \ll \Delta_\alpha(f) + |\lambda|L^{-\delta}.$$

However,

$$|\lambda| \ll \Delta_\alpha(\lambda\log) \ll \Delta_\alpha(\lambda\log - f) + \Delta_\alpha(f).$$

An application of the appropriate generalised Turán–Kubilius inequality completes the proof of Theorem 21.1. It also completes the localisation of Theorem 8.1.

22

Exercises: The operator D and differentiation; the operator T and the convergence of measures

The operator D and differentiation

Consider the operator

$$D : \psi(u) \to \int_{1-u}^{1} \psi(v) \frac{dv}{v}$$

from the space L^2 with respect to the measure $v^{-1}dv$ on $[0, 1]$, to itself.

1. Prove that each eigenvector ϕ of D satisfies the relation

$$\lambda(1-u)\phi'(u) = \phi(1-u),$$

hence the differential equation

$$u(1-u)\phi''(u) - u\phi'(u) + \lambda^{-2}\phi(u) = 0.$$

Here $'$ denotes differentiation with respect to u, and λ is the corresponding eigenvalue.

2. Prove that for each positive integer r,

$$\phi_r(u) = u \sum_{k=0}^{r-1} (-1)^{r-1-k} \binom{r}{k} \binom{r-1}{k} u^{r-k-1}(1-u)^k$$

with $\lambda = \lambda_r = (-1)^{r-1}r^{-1}$ is a solution to this differential equation. In terms of the standard notation for Jacobi polynomials, $\phi_r(u)$ is $P_{r-1}^{(1,0)}$ at $(1-2u)$. See, for example, Rainville, [135] Chapter 6.

3. Prove that

$$\int_0^1 \phi_r(v)\phi_s(v)v^{-1}dv = \begin{cases} 0 & \text{if } r \neq s, \\ \frac{1}{2r} & \text{if } r = s. \end{cases}$$

4. Prove that the coefficient of u^r in $\phi_r(u)$ is

$$(-1)^{r-1}\sum_{k=0}^{r-1}\binom{r}{k}\binom{r-1}{k} = (-1)^{r-1}\binom{2r-1}{r-1}.$$

Hint:

$$(1+z)^r\left(1+\frac{1}{z}\right)^{r-1} = z^{-(r-1)}(1+z)^{2r-1}.$$

5. Prove that in the notation of Lemma 20.7,

$$\phi_r(u) = (-1)^{r+1}\binom{2r-1}{r-1}\psi_r(u).$$

It is clear that each eigenvector y of the operator G defined in Chapter 21 also satisfies the differential equation of exercise 1; however:

6. Prove that (suitably interpreted), any polynomial eigenvector y of G is also an eigenvector of D, and with the same eigenvalue; moreover $y(1-\delta)=0$. For all but countably many values of δ, the operator G has no polynomial eigenvectors, and for no value of δ, $0<\delta<1/2$, can the space $L^2(\delta, 1-\delta)$ be spanned by such eigenvectors.

The operator T and the convergence of measures

In the following exercises function analytic ideas applied in Chapter 21, during the proof of Lemma 21.5, are reinterpreted in terms of the weak convergence of measures. I use the format and notation of the proof of Lemma 21.3. In particular for each $x \geq 2$, $F:\mathbb{R}\to\mathbb{R}$ is defined by

$$F(u) = \sum_{q\leq x^u}\frac{1}{q}\left(1-\frac{1}{q_0}\right).$$

The operator H defined by

$$(Hw)(u) = \int_{1-u<v\leq 1} w(v)dF(v)$$

is viewed: on the space $L^\alpha(0,1)$ with respect to the measure dF, into itself.

If E is any dF measurable subset of the unit interval, define $\Gamma : L^\alpha(0,1)\to L^\alpha(0,1)$ with respect to dF by

$$(\Gamma w)(u) = \int_{E\cap(1-u,1]} w(v)dF(v).$$

7. Prove that $\|\Gamma\| \le \|H\|$ uniformly in E.

8. If $0 < \zeta \le 1/2$, and the set E is the interval $(0, \zeta]$, prove

$$\|\Gamma\| \ll (\zeta + L^{-1})^{\gamma}, \quad 3\gamma = \min(\alpha^{-1}, (\alpha')^{-1}),$$

uniformly for $x \ge 2$ and ζ.

We return to the operator D_1 on the space $L^{\alpha}(0,1)$ with respect to $v^{-1}dv$, defined following Lemma 21.2. On the unit square define

$$\theta(u,v) = \begin{cases} 1 - 2uv & \text{if } u + v > 1, \\ -2uv & \text{if } u + v \le 1. \end{cases}$$

Then there is an integral representation

$$(D_1 w)(u) = \int_0^1 \theta(u,v) w(v) v^{-1} dv, \quad 0 \le u \le 1.$$

The operator D_1 is derived from D by removing a projection of D. We similarly derive from H the operator H_1 given by

$$(H_1 w)(u) = \int_0^1 \theta(u,v) w(v) dF(v).$$

We further define $Z_1 : L^{\alpha}(\mathbb{C}^s) \to L^{\alpha}(\mathbb{C}^s)$ by

$$Z_1 f = Tf - 2\left(f, \frac{\log}{\log x}\right) \frac{\log}{\log x},$$

the inner product taken in $L^2(\mathbb{C}^s)$. If k is a positive integer and h is the function on $[0,1]$ defined in terms of f following the statement of Lemma 21.3, then the value of $H_1^k h$ at $\log \ell / \log x$ coincides with the value of $Z_1^k f$ at ℓ.

9. Prove that viewed on $L^{\alpha}(\mathbb{C}^s)$ to itself

$$\|Z_1 - S_1\| \ll (\log x)^{-1}, \quad x \ge 2.$$

It is easier to study the asymptotic behaviour of the operators Z_1^k that that of the S_1^k.

Let k be a positive integer, $k \ge 2$. $(H_1^k w)(u)$ has the integral representation

$$\int_0^1 \delta(u,v) w(v) dF(v)$$

where the kernel $\delta(u,v)$ is given by (reading right to left)

$$\int_0^1 \theta(u, u_{k-1} dF(u_{k-1}) \cdots \int_0^1 \theta(u_2, u_1)\theta(u_1, v) dF(u_1).$$

It is natural to regard this last as an integral on the $k-1$ dimensional box $(0,1]^{k-1}$, and taken with respect to the product measure ${}^*dF$ induced by the marginal measures $dF(u_j)$, $1 \le j \le k-1$.

10. Prove that ${}^*dF$ gives to $(0,1]^{k-1}$ a measure $(1+o(1))(\log\log x)^{k-1}$ as $x \to \infty$.

This leads to complications when studying the weak convergence of the ${}^*dF$ as $x \to \infty$.

Let $0 < \zeta \le 1/2$. Decompose H_1 into the sum of operators $B + C$, where

$$(Cw)(u) = \int_{(0,\zeta]\cap(1-u,1]} \theta(u,v)w(v)dF(v), \quad 0 \le u \le 1.$$

11. Prove that $\|B\|$ is bounded uniformly for $x \ge 2$, $0 < \zeta \le 1/2$.

12. Prove that

$$\|C\| \ll (\zeta + L^{-1})^\gamma, \qquad 3\gamma = \min(\alpha^{-1}, (\alpha')^{-1})$$

uniformly for $x \ge 2$, $0 < \zeta \le 1/2$.

13. By expanding $(B+C)^k$ into a sum of operators of the form

$$B^{r_1} C^{r_2} B^{r_3} \cdots C^{r_{2k}}$$

where each $r_j = 0, 1, r_1 + \cdots + r_{2k} = k$, or otherwise, prove that there is a positive c_1, depending at most upon α, so that

$$\|H_1^k - B^k\| \le c_1^k (\zeta + L^{-1})^\gamma$$

uniformly for $x \ge 2$, $0 < \zeta \le 1/2$, $k \ge 1$.

It will suffice to study B.

The operator B^k has a representation

$$(B^k w)(u) = \int_\zeta^1 \beta(u,v)w(v)dF(v)$$

where $\beta(u,v)$ is defined exactly like $\delta(u,v)$ save that each of the integrals $dF(u_j)$, $1 \le j \le k-1$, is taken over the interval $(\zeta, 1]$ rather than $(0,1]$.

14. Prove that the ${}^*dF$ measure of the box $(\zeta,1]^{k-1}$ is $\ll (-\log\zeta)^{k-1}$ uniformly in $x \ge 2$, $0 < \zeta \le 1/2$.

15. Prove that as $x \to \infty$, ${}^*dF$ converges weakly to the product of $k-1$ copies of the measure $v^{-1}dv$.

16. Prove that

$$\lim_{x\to\infty} \beta(u,v)$$

exists.

We call the limit function $\beta_\infty(u,v)$. If we decompose D_1 into $B_\infty + C_\infty$, where

$$(C_\infty y)(u) = \int_{(0,\zeta]\cap(1-u,1]} \theta(u,v)y(v)v^{-1}dv, \quad 0 \le u < 1,$$

then

$$(B_\infty^k y)(u) = \int_\zeta^1 \beta_\infty(u,v)y(v)v^{-1}dv.$$

The convergence of $\beta(u,v)$ is for each fixed pair (u,v). Let $1/2 < \lambda < 1$.

17. Prove that $\beta(u,v)$ satisfies the Lipschitz condition

$$\beta(t,v) - \beta(s,v) \ll (-\log\zeta)^{k-2}(|t-s| + (\log x)^{-1})$$

uniformly for $0 \le s \le t \le \lambda$, $0 < \zeta \le 1/2$, $0 \le v \le 1$, $x \ge 2$.

The implied constant may depend upon λ and k.

18. Prove that $\beta(u,v)$ satisfies a similar Lipschitz condition with respect to the v variable.

19. Prove that the convergence of $\beta(u,v)$ to $\beta_\infty(u,v)$ is uniform on $[0,\lambda]^2$.

Define $\theta_v : [0,1] \to \mathbb{C}$ by

$$\theta_v = \begin{cases} \theta(u,v) & \text{if } \zeta < u \le 1, \\ 0 & \text{if } 0 \le u \le \zeta\,. \end{cases}$$

Inspection shows that

$$\beta(u,v) = (B^{k-1}(\theta_v))(u), \quad 0 \le u \le 1.$$

20. Prove that

$$\left(\int_0^1 |\beta(u,v) - \beta_\infty(u,v)|^{\alpha'} dF\right)^{1/\alpha'} \to 0, \quad x \to \infty,$$

uniformly for $0 < u \le \lambda$.

Define the operator $N_k : L^\alpha(0,1) \to L^\alpha(0,1)$ with respect to dF, by

$$(N_k w)(u) = \int_0^1 \beta_\infty(u,v)w(v)dF(v), \quad 0 \le u \le 1.$$

21. Prove that

$$\lim_{x\to\infty} \|B^k - N_k\| = 0.$$

The operator D_1^k has an integral representation

$$(D_1^k w)(u) = \int_0^1 d(u,v)w(v)v^{-1}dv,$$

where $d(u,v)$ is given by

$$\int_0^1 \theta(u,u_{k-1})u_{k-1}^{-1}du_{k-1} \cdots \int_0^1 \theta(u_2,u_1)\theta(u_1,v)u_1^{-1}du_1.$$

Define the operator $M_k : L^\alpha(0,1) \to L^\alpha(0,1)$ with respect to dF, by

$$(M_k w)(u) = \int_0^\infty d(u,v)w(v)dF(v), \quad 0 \le u \le 1.$$

22. Prove that

$$\lim_{x\to\infty} \|H_1^k - M_k\| = 0.$$

We have replaced the kernel $\delta(u,v)$ in H_1^k by a function $d(u,v)$ which is independent of x. In this way the operators H_1^k may be regarded as reasonably standard moments of the distribution function $F(u)$.

For a fixed v, $0 \le v < 1$, let

$$d(u,v) \sim \sum_{j=1}^{\infty} c_j \rho_j(u)$$

be the representation of $d(u,v)$ arising from the spectral decomposition of $L^2(0,1)$ by the operator D.

23. Prove that $c_1 = 0$, and that for $j \ge 2$, $c_j = \lambda_j^{k-1}\rho_j(v)$, where λ_j is the eigenvalue associated with the eigenvector ρ_j.

The value of λ_j is, of course, $(-1)^{j+1}j^{-1}$.

24. Prove that

$$-\log(1-v) = \sum_{j=1}^{\infty} \lambda_j^2 \rho_j(v)^2, \quad 0 \le v < 1.$$

25. Prove that for $k \ge 3$.

$$d(u,v) = \sum_{j=2}^{\infty} \lambda_j^{k-1} \rho_j(u)\rho_j(v), \quad 0 \le u < 1,\ 0 \le v < 1,$$

may be assumed.

26. Recover the bound of Lemma 21.3 in the case $j = 1$.

27. By considering the action of D_1^k on step functions, prove that

$$\|D_1^k\| \le \liminf_{x\to\infty} \|H_1^k\|,$$

the norms being taken with respect to the appropriate L^α spaces.

The asymptotic best value for the implied constant in the standard Turán–Kubilius inequality (in our present situation that $\|A_2\|^2 \to 3/2$ as $x \to \infty$), apparently announced by Kubilius in Budapest in 1981, was established in Kubilius [110], [111]; and in Hildebrand, [99]. In the notation of Chapter 20, Kubilius iterates the coefficient matrix attached to the quadratic form (f, Tf) and estimates the corresponding trace. The argument to sharpen estimates for individual eigenvalues of that matrix by considering approximate eigenvectors is not completely given. A version of part (i) of Lemma 20.5 is stated but, again in the notation of Chapter 20, only $(f, (T - \lambda I)f)$ is calculated and not the $((T - \lambda I)f, (T - \lambda I)f)$ needed. An appropriate calculation is carried out in Kubilius, [112], wherein the estimate of the best constant in the Turán–Kubilius inequality for additive functions on the interval $[1, n]$ is improved to $3/2 + O((\log n)^{-1})$.

Some details of some precursors to Kubilius' result may be found in [56], [59]. The second of these references also contains a commentary on the differing treatment by Hildebrand mentioned above.

None of the spaces $L^\alpha, L^\alpha \cap L^\beta$ and their duals, nor of the operators T, T_m, S_m appear in these works of Kubilius and Hildebrand. A variant discussion of the operators T_n, S_m may be found in my paper [59]. In particular, in that paper I compare Theorem 21.1 with results (for real additive functions) obtained by Hildebrand, [98]. His argument applies, again, Ruzsa's bound for the concentration function of a real additive function. Note, however, that Hildebrand also employs a version of Lemma 2.6 of the present volume.

The application of iteration to estimate the spectral radii of the operators underlying inequalities of Large Sieve type was suggested (but not carried out) in [30].

It was shown by Lee, [117], that for strongly additive functions the standard Turán–Kubilius constant is eventually strictly less than 3/2. For $\alpha > 1$, $\alpha \neq 2$, no asymptotic best value of $\|A_\alpha\|$ has been established.

23

Pause: Towards the discrete derivative

In Chapter 8 I gave a characterisation of those additive functions f which belong to the class $\mathcal{L}^\alpha$ with $\alpha > 1$. What if the hypothesis is changed to: $f(n+1) - f(n)$, the first discrete derivative of f, belongs to $\mathcal{L}^\alpha$?

This question has its origins in the 1946 paper of Erdős, [82], in which he introduced the notion of finitely distributed additive functions and gave the characterisation of them otherwise obtained in Chapter 16. As an application of his characterisation Erdős showed that any non-decreasing real valued additive function must be a constant multiple of a logarithm. He drew the same conclusion from the hypothesis that $f(n+1) - f(n) \to 0$, as $n \to \infty$.

Erdős' paper contained a number of interesting assertions and problems. In particular, he conjectured that any (real) additive function for which $f(n+1) - f(n)$ was uniformly bounded below, would differ from some constant multiple of a logarithm by a uniformly bounded amount. This conjecture was established by Wirsing, [171].

Wirsing's proof is indirect. The problem is reduced to the case of a completely additive function. The appropriate conclusion then, that f is exactly the multiple of a logarithm, is assumed false. By a highly ingenious and characteristic argument, of considerable length, Wirsing obtains a contradiction. I remember his justifiable pride in the proof, and his remark that it had taken him fourteen years.

In a later note, Ruzsa showed that a one-sided bound on the difference $f(n+1) - f(n)$ of additive functions implies a two-sided bound; Ruzsa, [148], see also [56] Chapter 14, Lemma 14.2.

In view of this result of Ruzsa, we might reasonably rephrase the theorem of Wirsing in the form: $f(n+1) - f(n)$ belongs to $\mathcal{L}^\infty$ if and only if for some constant A, $f - A\log$ belongs to $\mathcal{L}^\infty$. The question raised at the beginning of this chapter now occurs naturally.

I was interested in that question for a number of reasons. I felt that the study of the value distribution of differences of additive functions, where sharp local information was not available, would not be readily susceptible to proof by contradiction. A more direct approach would seem desirable.

Looking over the proof of Wirsing I noted that he applied his (local) $\mathcal{L}^\infty$ hypothesis many times, but the circumstances in which he applied it were of a small number of types. I wondered if the more diffused (global) methodology which I had developed to study additive arithmetic functions in unbounded renormalisations, and applied to a problem of Hardy and Ramanujan, might offer an approach.

My first success in this direction was to establish that $f(n+1) - f(n)$ belongs to $\mathcal{L}^2$ if and only if for some A, the series $\sum q^{-1}|f(q) - A\log q|^2$, taken over the prime powers q, converges, [39]. According to a result in probabilistic number theory, [41] Theorems 5.5 and 7.7, or to Theorem 8.1 here, the latter condition is equivalent to $f - A\log$ belonging to $\mathcal{L}^2$. With his usual generosity, Erdős complimented me on the result.

The problems attendant on generalising this result to cover differences $f_1(an+b) - f_2(An+B)$ of distinct additive functions, with arguments confined to arithmetic progressions, $a > 0$, $A > 0$, $aB \neq Ab$, their connection with a conjecture of Kátai as well as the representation of integers by products, I discuss in my book [56].

In the following chapters I develop a (hitherto unpublished) treatment of differences $f(n+1) - f(n)$ in $\mathcal{L}^\alpha$, where $1 < \alpha \le 2$, using the method of the stable dual. The argument naturally applies to the cases $\alpha > 2$ as well.

It is suggestive to formulate a local version of the problem, using the language of Chapter 3, and to view it from the vantage of Chapter 20.

Define a shift operator E on $\mathbb{C}^t$, $t = [x]$, by $a_n \to a_{n+1}$ if $n+1 \le x$, $a_t \to 0$. For $r \ge 2$ define $E^r = E(E^{r-1})$. Likewise E^{-1} is to take a_n to a_{n-1} if $n \ge 2$, a_1 to 0. Again inductively, $E^{-r} = E^{-1}(E^{-(r-1)})$ for $r \ge 2$. With these definitions the dual of E^r is E^{-r}.

For the local study of additive functions f in mean square, the approach in Chapter 20 was to consider the self-adjoint operator $A_2^* A_2$, or a convenient relative of it, on $L^2(\mathbb{C}^s)$. The first step appealed to the uniform bound on $\|A_2^*\|$:

$$\|A_2^* A_2 f\|_2 \ll \|A_2 f\|_2.$$

If I denotes the identity map on $\mathbb{C}^t$, then we may likewise assert that

$$\|A_2^*(I - E^{-1})A_2 f\|_2 \ll \|(I - E^{-1})A_2 f\|_2. \tag{23.1}$$

Here the dominating norm is

$$\left([x]^{-1} \sum_{2 \le n \le x} |f(n) - f(n-1)|^2 + [x]^{-1}|\mu|^2 \right)^{1/2},$$

where

$$\mu = \sum_{q \le x} \frac{1}{q}\left(1 - \frac{1}{q_0}\right) f(q).$$

To study the implications of an average bound on the difference $f(n) - f(n-1)$ we may consider the operator

$$A_2^*(I - E^{-1})A_2 : L^2(\mathbb{C}^s) \to L^2(\mathbb{C}^s).$$

In general this operator is not self-adjoint. For otherwise $A_2^* E^{-1} A_2 = A_2^* E A_2$. Then for all h, f in $\mathbb{C}^s$

$$0 = (h, A_2^*(E - E^{-1})A_2 f) = (A_2 h, (E - E^{-1})A_2 f),$$

the first inner product that of $L^2(\mathbb{C}^s)$, the second that of $M^2(\mathbb{C}^t)$. We choose h so that $h(t) = 0$ and

$$\sum_{q \le x} \frac{h(q)}{q}\left(1 - \frac{1}{q_0}\right) = 0.$$

The second inner product then has the representation

$$[x]^{-1} \sum_{2 \le n \le x-1} h(n)\{f(n+1) - f(n-1)\}. \tag{23.2}$$

Let p_k, p_{k+1} be successive prime numbers sufficiently large that $2p_k > p_{k+1}$. With $x = p_{k+1}$ we set $h(p_{k+1}) = 0$, $h(q) = \log q$ if $q \le x$ and $(q, p_k) = 1$, and

$$\frac{h(p_k)}{p_k}\left(1 - \frac{1}{p_k}\right) = -\sum_{\substack{q \le x \\ (q, p_k) = 1}} \frac{h(q)}{q}\left(1 - \frac{1}{q_0}\right).$$

The above two conditions on h are satisfied. With $f = \log$ those terms in the sum (23.2) with $(q, p_k) = 1$ contribute

$$[x]^{-1} \sum_{\substack{2 \le n \le x-1 \\ (n, p_k) = 1}} \log n \log\left(\frac{n+1}{n-1}\right) > \frac{3}{2x} \sum_{n \le x/3} \frac{\log n}{n} > \frac{(\log x)^2}{2x}$$

for all sufficiently large x. The only term with $(n, p_k) > 1$ occurs for $n = p_k$, and contributes

$$[x]^{-1} h(p_k) \log\left(\frac{p_k + 1}{p_k - 1}\right) \ll x^{-1} \sum_{q \le x} \frac{\log q}{q} \ll \frac{\log x}{x},$$

which is negligible in comparison with the lower bound. A contradiction is obtained.

Supposing the operator $A_2^* E^{-1} A_2$ to have a small norm we might still treat the left hand term in the inequality (23.1) by the method of Chapter 20. In fact $\|A_2^* E^{-1} A_2 f\|_2^2$ lies between positive absolute constant multiples of

$$x^{-2} \sum_{q \le x} q \left| \sum_{\substack{2 \le n \le x \\ n \cong 0 (\mathrm{mod} q)}} \{f(n-1) - \mu\} - \frac{1}{q}\left(1 - \frac{1}{q_0}\right) \sum_{2 \le n \le x} \{f(n-1) - \mu\} \right|^2$$

and it would be desirable for this to be $\ll (\log x)^{-c} \|f\|_2$ for some $c > 0$, uniform in f and $x \ge 2$. Since

$$\sum_{\substack{2 \le n \le x \\ n \cong 0 (\mathrm{mod} q)}} \mu - \frac{1}{q}\left(1 - \frac{1}{q_0}\right) \sum_{2 \le n \le x} \mu \ll |\mu| \ll \|f\|_2 (\log \log x)^{1/2}$$

uniformly in q, and

$$\left(x^{-2} \sum_{q \le x} q \right)^{1/2} \ll (\log x)^{-1/2},$$

it would suffice if

(23.3)
$$\sum_{q \le x} q \left| \sum_{\substack{m \le x-1 \\ m \equiv -1 (\mathrm{mod} q) \\ m \not\equiv -1 (\mathrm{mod} q q_0)}} f(m) - \frac{1}{q}\left(1 - \frac{1}{q_0}\right) \sum_{m \le x-1} f(m) \right|^2 \ll \frac{x^2}{(\log x)^{2c}} \|f\|_2^2.$$

Is it reasonable to expect such a result? The short answer to this must be 'No'.

The Bombieri version of the Bombieri–Vinogradov theorem asserts that given $A > 0$, there is a B such that

$$\sum_{D \le x^{1/2} (\log x)^{-B}} \max_{(r,D)=1} \max_{y \le x} \left| \sum_{\substack{p \le y \\ p \equiv r (\mathrm{mod} D)}} 1 - \frac{1}{\phi(D)} \sum_{p \le y} 1 \right| \ll x (\log x)^{-A}.$$

Bombieri's original account may be found in Bombieri [6]. A generalisation to multiplicative functions g, by Wolke [173], [174], Siebert and Wolke [161], has the similar form

$$\sum_{D\le x^{1/2}L^{-B}} \max_{(r,D)=1} \max_{y\le x} \left| \sum_{\substack{n\le y\\ n\equiv r(\operatorname{mod} D)}} g(n) - \frac{1}{\phi(D)} \sum_{\substack{n\le y\\ (n,D)=1}} g(n) \right| \ll xL^{-A},$$

where L denotes $\log x$. The proofs hinge upon a (mean-square) version of the Large Sieve for Dirichlet characters. Because of their generality, mean square inequalities remain available to us. However, prime numbers have special properties, and appropriate generalisations of these properties are also required by Wolke. For example, the Siegel–Walfisz theorem asserts that a typical summand in the Bombieri–Vinogradov theorem, with modulus D not exceeding a fixed power of $\log x$, is $\ll x\exp(-c\sqrt{\log x})$ for a certain $c > 0$. Such special properties are not available in the present situation. The inequality (23.3) is required to hold not only for functions essentially constant on the primes, but for all complex valued additive functions f. It is an abstract norm inequality.

There are other differences. The residue class representative -1 is fixed at the same value for all moduli q, and not required to vary over all the reduced residue classes (mod q), as is the case in the theorems of Bombieri, Siebert and Wolke. Moreover, the moduli need only be prime powers. In seeking a proof of (23.3) these might be advantages. The moduli q are required to run all the way up to x rather than to $x^{1/2}$. This might be a disadvantage.

Another difference lies deeper. The theorems of Bombieri and Wolke effect L^1 estimates of convolutions by means of L^2 estimates of the functions convoluted. An L^2 version of the Bombieri–Vinogradov theorem may be recovered by appealing to the uniform bound

$$\sum_{\substack{p\le x\\ p\equiv r(\operatorname{mod} D)}} 1 \ll \frac{x}{\phi(D)\log x}$$

of Brun and Titchmarsh, valid for D up to any fixed power x^{β} with $\beta < 1$. An analogue of the Brun–Titchmarsh theorem appropriate to the present circumstances would assert that

$$\left| \sum_{\substack{n\le x\\ n\equiv a(\operatorname{mod} D)}} f(n) \right|^2 \le \lambda \sum_{q\le x} \frac{|f(q)|^2}{q},$$

with λ about x/D, held uniformly for additive $f, (a, D) = 1, D \le x^\beta$. Such a result would be false. A theorem of Linnik (cf. Prachar, [134], Chapter X) asserts that for some $c > 0$ and all sufficiently large D, the interval $(1, D^c]$ contains primes in every reduced residue class (mod D). Choosing x to be such a prime exceeding $D^{1/2}$, and setting $a = x$, $f(x) = 1$, $f(q) = 0$ otherwise, would force $\lambda \ge x \ge (x/D)x^{1/c}$, a power of x too large.

It would perhaps not matter that an L^2 version of the theorem of Bombieri and Vinogradov might involve L^4 estimates and a damage factor of a few powers of $\log x$ in the final upper bound; for a saving over the trivial of an arbitrary power of $\log x$ is expected. In the case of (23.3) such a factor would be fatal, since (23.3) is false if $2c > 1$, [56]. At most a single power of $\log x$ may be saved within the generality required. A proof of (23.3) would seem to require an L^2 argument throughout.

Much militates against (23.3). However, each of the operators A_2^*, E, A_2 is bounded uniformly in $x \ge 2$; so with x^2 in place of $x^2(\log x)^{-2c}$ the inequality is valid; and as Professor Davenport once said to me: "When a proof is efficient, you think that with an effort you might get a little more."

Besides, the validity of (23.3) would be both pretty and useful.

Since an additive function is naturally a convolution, for each reduced residue class $r(\text{mod } q)$ there is a representation

$$\sum_{\substack{n \le x \\ n \equiv r(\text{mod} q)}} f(n) = \sum_{m \le x} \sum_{\substack{\ell \le x/m, (\ell, m)=1 \\ \ell m \equiv r(\text{mod} q)}} f(\ell).$$

Here ℓ denotes a prime power. We may rewrite the congruence condition on ℓ in the form $\ell \equiv \bar{m}r(\text{mod } q)$, where $m\bar{m} \equiv 1(\text{mod } q)$. As m traverses the reduced residue classes (mod q) so does $\bar{m}r$. With a view to applying the Cauchy–Schwarz inequality we might consider the sum

$$\phi(q) \sum_{\substack{t=1 \\ (t,q)=1}}^{q} \left| \sum_{\substack{\ell \le y \\ \ell \equiv t(\text{mod} q)}} f(\ell) - \frac{1}{\phi(q)} \sum_{\substack{\ell \le y \\ (\ell, q)=1}} f(\ell) \right|^2 .$$

This sum has the alternative representation

$$\sum_{\chi(\text{mod} q)}{}' \left| \sum_{\ell \le y} f(\ell)\chi(\ell) \right|^2, \tag{23.4}$$

where $'$ indicates that χ runs through the non-principal Dirichlet characters (mod q). Averaged over prime moduli, sums of these last two types are considered in the exercises of Chapter 4. On the group or on its dual; which form to adopt?

The success of his view in the Bombieri–Vinogradov theorem prompts that I continue to follow the dictum of Hecke: If you can expand a function as a Fourier series, then you should do so. To better prepare for generalisations, I argue from first principles.

When considering the sum (23.4) for q varying up to Q, a typical inner sum, which may be regarded as involving a primitive character χ_1 in place of χ, will occur once for every q which is a multiple of the modulus q_0^k to which χ_1 is defined; that is with a multiplicity of $[\log(Qq_0^{-k})/\log q_0]$. Again a factor of $\log x$ may intrude.

Disregarding objections so far, we reach a sum

$$\sum_j \left| \sum_{\ell \le y} f(\ell)\chi_j(\ell) \right|^2$$

which we wish to bound from above in function analytic terms, and where the χ_j run over the J distinct primitive characters to (prime power) moduli q not exceeding Q. An appropriate upper bound would have the form $\lambda \sum |f(\ell)|^2$, say.

It is easier to approach the dual norm inequality

$$\sum_{\ell \le y} \left| \sum_j c_j \chi_j(\ell) \right|^2 \le \lambda \sum |c_j|^2, \tag{23.5}$$

which is to be valid for all complex c_j. The mean square sum over ℓ (and j) represents an Hermitian form

$$\sum_{j,k} c_j \bar{c}_k \sum_{\ell \le y} \chi_j(\ell)\bar{\chi}_k(\ell),$$

and the best value for λ will be a maximal eigenvalue of the Hermitian matrix attached to this form. In other words, $\lambda^{1/2}$ is the spectral radius of the appropriate self-adjoint operator underlying the inequality (23.5).

We may assume λ to lie in one of the Gershgorin discs

$$\left| z - \sum_{\ell \le y} |\chi_j(\ell)|^2 \right| \le \sum_{k \ne j} \left| \sum_{\ell \le y} \chi_j(\ell)\bar{\chi}_k(\ell) \right|, \quad z \text{ in } \mathbb{C}.$$

The centres of these discs are the diagonal elements of the representing matrix, and are in size $\ll y(\log y)^{-1}$. Further progress depends upon information concerning character sums over prime powers. It is not straightforward to estimate such sums. Validity of the Riemann–Piltz conjecture (Prachar [134], Chapter VII, §5) would yield $\ll x^{1/2}\log(qx)$ for the sum of the values of a non-principal Dirichlet character (mod q) on the primes not exceeding x; in particular

$$\sum_{k\neq j}\left|\sum_{\ell\leq x}\chi_j(\ell)\bar{\chi}_j(\ell)\right| \ll Q^2x^{1/2}\log(Qx).$$

Whilst application of the Large Sieve will not currently yield a bound of this precision, it will give a bound of similar type with an added term $\ll x/\log x$, [55]. In the present circumstances that directly guarantees a bound

$$\sum_j\left|\sum_{\ell\leq x}f(\ell)\chi_j(\ell)\right|^2 \ll \left(\frac{x}{\log x}+D(Q)\right)\sum_{\ell\leq x}|f(\ell)|^2, \tag{23.6}$$

where $D(Q)$ does not increase more rapidly than a fixed power of Q. For the non-localised study of $f(n+1)-f(n)$ in mean square, in fact the existence of $D(Q)$ alone suffices, [39].

The Hermitian matrix attached to the inequality (23.6) is of size s-by-s, where (again) s denotes the number of prime powers not exceeding x. Since the sum of its eigenvalues is equal to its trace, there is an eigenvalue at least as large as

$$s^{-1}\sum_{\ell\leq x}\sum_j|\chi_j(\ell)|^2.$$

We may argue likewise with the matrix attached to inequality (23.5). Apart from extra eigenvalues zero these matrices have the same spectra, s^{-1} in the above lower bound may be replaced by J^{-1}. For x, Q exceeding a certain absolute constant,

$$\lambda \geq \frac{1}{2}\max(s,J) \geq \frac{1}{7}\max\left(\frac{x}{\log x},\frac{Q^2}{\log Q}\right)$$

must hold. The upper bound $\lambda \ll x/\log x + Q^{2+\varepsilon}$, vouchsafed indirectly in Chapter 4, exercise 54, is reasonably sharp. In particular, the upper bound factor in the operator bound (23.6) cannot be better than is allowed by the support of the appropriate dual operator, exemplifying a useful heuristic.

Altogether, although the function $f(n)$ is supported on the integers up to x, which are about x in number, we may view f as the convolution of the function **1** which is identically one on the integers, and a function $f(\ell)$ supported on the (smaller) set of prime powers ℓ up to x, about $x(\log x)^{-1}$ in number. We might hope to use Fourier analysis to strip the function **1** from f, and obtain (23.3) with $2c = 1$; or at least a version of (23.3) with the moduli q restricted not to exceed a fixed power of x, x^{β} with $\beta < 1/2$. There is little room to manoeuvre. We must tread carefully.

To develop a corresponding study of functions f with difference $f(n+1) - f(n)$ in $\mathcal{L}^{\alpha}$, $\alpha \neq 2$, would involve operators on the spaces $(L^2 \cap L^{\alpha'}(\mathbb{C}^s))'$ and $L^2 \cap L^{\alpha}(\mathbb{C}^s)$ introduced in Chapter 3. At the expense of a power of $\log\log x$ we might restrict our attention to the simpler spaces $L^{\alpha}(\mathbb{C}^s)$, and look for a bound of the form:

(23.7)

$$\sum_{q \le x^{\delta}} \phi(q)^{\alpha-1} \max_{(r,q)=1} \left| \sum_{\substack{n \le x \\ n \equiv r (\operatorname{mod} q)}} f(n) - \frac{1}{\phi(q)} \sum_{\substack{n \le x \\ (n,q)=1}} f(n) \right|^{\alpha} \ll \frac{x^{\alpha}}{(\log x)^{\theta}} \|f\|_{\alpha}^{\alpha},$$

valid for some $\delta > 0$, $\theta > 0$ and all complex additive f. Again to employ the convolution nature of an additive function we might hope to apply Hölder's inequality, rather than that of Cauchy, and reduce ourselves to the estimation of

$$\sum_{q \le x^{\delta}} \phi(q)^{\alpha-1} \sum_{\substack{r=1 \\ (r,q)=1}}^{q} \left| \sum_{\substack{\ell \le x \\ \ell \equiv r (\operatorname{mod} q)}} f(\ell) - \frac{1}{\phi(q)} \sum_{\substack{\ell \le x \\ (\ell,q)=1}} f(\ell) \right|^{\alpha}.$$

The efficacy of character sums in this more general setting is problematic. In particular, what can play the rôle of Parseval's identity?

Young–Hausdorff inequalities on a finite group. Riesz–Thorin theorem

Let G be a finite abelian group of order $|G|$. For $\alpha \ge 1$ let $L^{\alpha}(G)$ denote the linear space of complex valued functions h on G, topologised with norm

$$\left(|G|^{-1} \sum_{g \in G} |h(g)|^{\alpha} \right)^{1/\alpha}.$$

As usual $L^2(G)$ may be viewed as a Hilbert space, with inner product

$$(h, k) = |G|^{-1} \sum_{g \in G} h(g)\overline{k(g)}.$$

We likewise define $L^\alpha(\widehat{G})$, noting that $|\widehat{G}| = |G|$.

The Fourier transform $h \mapsto \hat{h}$ defined by

$$\hat{h}(\chi) = |G|^{-1/2} \sum_{g \in G} h(g)\overline{\chi(g)}$$

gives an isometry between $L^2(G)$ and $L^2(\widehat{G})$, (cf. Chapter 4, exercise 14). Denote the associated linear map by F. Then $F^* : L^2(\widehat{G}) \to L^2(G)$, adjoint to F, is given by

$$(F^*y)(g) = |G|^{-1/2} \sum_{\chi \in \widehat{G}} \chi(g)y(\chi).$$

Thus $F^*\hat{h} = h$, F^* is the inverse Fourier transform. In particular, $I - F^*F$, the operator corresponding to T in Chapter 20, is trivial.

If we identify $\widehat{\widehat{G}}$ and G (exemplifying the Pontryagin duality theorem), then the Fourier transform on $\widehat{G}$ coincides with the dual, F' of F. The function $F'\hat{h}$ at w coincides with h evaluated at w^{-1}.

Like $L^\alpha(\mathbb{C}^s)$, $M^\beta(\mathbb{C}^t)$ in Chapter 3, the spaces $L^\alpha(G)$ introduced here may be viewed as spaces of measurable functions, endowed with a mean-power norm. Let B be a linear map between the spaces $L(S), L(U)$ of appropriately measurable complex valued functions on S, U respectively. For $\beta \geq 1$, $\alpha \geq 1$ let $\|B\|_{\beta,\alpha}$ denote the norm of B viewed between $L^\beta(S)$, $L^\alpha(U)$, where these spaces are given β and α mean-power norms. The Riesz–Thorin theorem asserts that $\log \|B\|_{\beta,\alpha}$ is convex in $1/\alpha$, $1/\beta$, $\alpha \geq 1$, $\beta \geq 1$, including, suitably interpreted, $\alpha = \infty$, $\beta = \infty$, [5].

In the present circumstances each underlying measure is a frequency measure, on G or $\widehat{G}$ as the case may be, and $\|F\|_{2,2} = 1$. It is clear that $\|F\|_{1,\infty} \leq |G|^{1/2}$. The Young–Hausdorff inequalities, directly from the Riesz–Thorin theorem, assert that

$$\left(\sum_{\chi \in \widehat{G}} \left| \sum_{g \in G} b_g \overline{\chi(g)} \right|^{\alpha'} \right)^{1/\alpha'} \leq |G|^{1/\alpha'} \left(\sum_{g \in G} |b_g|^\alpha \right)^{1/\alpha}, \quad \alpha^{-1} + (\alpha')^{-1} = 1,$$

uniformly for $1 \leq \alpha \leq 2$, complex b_g. For our purposes this is an inequality in the wrong direction.

Interchanging the rôles of $G, \widehat{G}$, and considering the inverse Fourier transform (dualising), reverses the direction of the inequalities:

$$\left(\sum_{g \in G} \left| \sum_{\chi \in \widehat{G}} c_\chi \chi(g) \right|^{\alpha'} \right)^{1/\alpha'} \leq |G|^{1/\alpha'} \left(\sum_{\chi \in \widehat{G}} |c_\chi|^\alpha \right)^{1/\alpha}.$$

If we set

$$c_\chi = \sum_{w \in G} \overline{\chi(w)} d_w,$$

then by the orthogonality of characters

$$\left(\sum_{g \in G} |d_g|^{\alpha'} \right)^{1/\alpha'} \le |G|^{-1/\alpha} \left(\sum_{\chi \in \hat{G}} \left| \sum_{w \in G} \overline{\chi(w)} d_w \right|^{\alpha} \right)^{1/\alpha},$$

uniformly for $1 \le \alpha \le 2$, complex d_g, and where the conjugation of $\chi(w)$ may be deleted. With G the group of reduced residue classes $r (\operatorname{mod} q)$, and

$$d_r = \sum_{\substack{n \le x \\ n \equiv r (\operatorname{mod} q)}} a_n - \frac{1}{\phi(q)} \sum_{\substack{n \le x \\ (n,q)=1}} a_n,$$

we obtain

$$\left(\phi(q)^{\alpha'-1} \sum_{\substack{r=1 \\ (r,q)=1}}^{q} \left| \sum_{\substack{n \le x \\ n \equiv r (\operatorname{mod} q)}} a_n - \frac{1}{\phi(q)} \sum_{\substack{n \le x \\ (n,q)=1}} a_n \right|^{\alpha'} \right)^{1/\alpha'}$$

$$\le \left(\sum_{\chi \ne \chi_0 (\operatorname{mod} q)} \left| \sum_{n \le x} a_n \chi(n) \right|^{\alpha} \right)^{1/\alpha},$$

uniformly for $1 \le \alpha \le 2$, $x \ge 1$, complex a_n. Now in the right direction, this inequality is available only for $\alpha' \ge 2$, not in the more interesting range $1 < \alpha' < 2$. Even if we feel the Young–Hausdorff inequalities to be essentially equalities, our troubles are not over, as the following example with q a prime, $\alpha' = 4$, $\alpha = 4/3$, shows.

Let χ_j run through the $p-1$ Dirichlet characters defined to a prime modulus p. In order to estimate the norm

$$\left(\sum_j \left| \sum_{n \le x} a_n \chi_j(n) \right|^{4/3} \right)^{3/4} \tag{23.8}$$

it is natural to consider the dual norm

$$\left(\sum_{n \le x} \left| \sum_j c_j \chi_j(n) \right|^{4} \right)^{1/4} = S^{1/4},$$

say. The sum S may be treated directly. Expanding the summand and inverting the order of summation we have

$$\sum_{j_1} \cdots \sum_{j_4} c_{j_1} c_{j_2} \bar{c}_{j_3} \bar{c}_{j_4} \sum_{n \leq x} \chi(n)$$

where $\chi = \chi_{j_1} \chi_{j_2} \bar{\chi}_{j_3} \bar{\chi}_{j_4}$. If χ is non-principal then by the Pólya–Vinogradov inequality (Chapter 4, exercise 48) the inner sum is $\ll p^{1/2} \log p$. Otherwise the inner sum is $x(1-p^{-1}) + O(1)$, with an absolutely bounded error. Therefore

$$S = x(1-p^{-1}) \sum \cdots \sum c_{j_1} c_{j_2} \bar{c}_{j_3} \bar{c}_{j_4} + O\Big(p^{1/2} \log p \Big(\sum |c_j|\Big)^4\Big) \tag{23.9}$$

with leading summation confined to the four-tuples $(j_1, \ldots, j_4)$ for which $\chi_{j_1} \cdots \bar{\chi}_{j_4}$ is principal.

At this stage of the study, the objective has to be clarified. What (norm form) would be acceptable in an upper bound?

Consider the sum which is the coefficient of $x(1-p^{-1})$ in the representation of S at (23.9). Fix j_1, j_2. For each j_3 there will be exactly one value of j_4 so that $\chi_{j_1} \cdots \bar{\chi}_{j_4}$ is principal. Indeed, the (multiplicative) group of Dirichlet characters $(\operatorname{mod} p)$ is isomorphic to the additive group of residue classes mod $(p-1)$. Without loss of generality we may assume a correspondence $\chi_j \leftrightarrow j \operatorname{mod}(p-1)$, $0 \leq j \leq p-2$, with χ_0 principal. Given j_i, $i = 1, 2, 3$, we then ensure the product $\chi_j \prod_{i=1}^{3} \chi_{j_i}$ to be principal by choosing $j + j_1 + j_2 + j_3 \equiv 0(\operatorname{mod}(p-1))$.

The sum

$$N(\mathbf{c}) = \sum_{j+j_1+j_2+j_3 \equiv 0(\operatorname{mod}(p-1))} \cdots \sum c_{j_1} c_{j_2} \bar{c}_{j_3} \bar{c}_{j_4} = \sum_{t=0}^{p-2} \Bigg| \sum_{j+k \equiv t(\operatorname{mod}(p-1))} c_j c_k \Bigg|^2$$

is interesting in its own right. Here $\mathbf{c}$ denotes the vector in $\mathbb{C}^{p-1}$ with coordinates c_j, $0 \leq j \leq p-2$. $N(\mathbf{c})^{1/4}$ is a norm on $\mathbb{C}^{p-1}$. For any complex λ, $N(\lambda \mathbf{c}) = |\lambda|^4 N(\mathbf{c})$ is clearly satisfied. The necessary triangle property may be obtained using the representation

$$N(\mathbf{c}) = \lim_{x \to \infty} (x(1-p^{-1}))^{-1} S,$$

and the fact that for each $x \geq p$, $S^{1/4}$ is a norm on $\mathbb{C}^{p-1}$. Moreover, if $N(\mathbf{c}) = 0$, then for each $t(\operatorname{mod}(p-1))$, the sum $\sum c_j c_k$ taken over the j, k

with $j+k \equiv t(\bmod (p-1))$, vanishes. If ζ denotes a primitive $(p-1)^{\text{th}}$ root of unity, then this last condition implies

$$\left(\sum_{j=0}^{p-2} c_j \zeta^{jw}\right)^2 = \sum_{t=0}^{p-2} \zeta^{tw} \sum_{j+k \equiv t(\bmod (p-1))} c_j c_k = 0.$$

The sum $\sum c_j \zeta^{jw}$ vanishes for every $w(\bmod (p-1))$. Equivalently, for every $r(\bmod (p-1))$

$$c_r = (p-1)^{-1} \sum_{w=0}^{p-2} \zeta^{-rw} \sum_{j=0}^{p-2} c_j \zeta^{jw} = 0.$$

In order to gain advantage from the inequality

$$S^{1/4} \ll x^{1/4} N(\mathbf{c})^{1/4} + (p^{1/2} \log p)^{1/4} \sum_{j=1}^{p-2} |c_j|, \tag{23.10}$$

we need to identify reasonably simply the dual of the space given the upper bound at (23.10) as norm.

In pursuit of a more simply presented inequality we apply the Cauchy–Schwarz inequality to show that $N(\mathbf{c})$ does not exceed

$$(p-1) \sum_{t=0}^{p-2} \sum_{j+k \equiv t(\bmod (p-1))} |c_j c_k|^2 = (p-1) \left(\sum_{j=0}^{p-2} |c_j|^2\right)^2.$$

As a consequence

$$S^{1/4} \ll (xp)^{1/4} \left(\sum |c_j|^2\right)^{1/2} + (p^{1/2} \log p)^{1/4} \sum |c_j|.$$

The factor $p^{1/4}$ which appears alongside $x^{1/4}$ indicates that some ground has been lost, but the upper bound is now a norm of the $L^2 \cap L^1$ type, and its dual norm may be reasonably computed.

Since this is only an example, I apply the Cauchy–Schwarz inequality again, this time to $\sum |c_j|$. We reach

$$S^{1/4} \ll \{(xp)^{1/4} + p^{5/8} (\log p)^{1/4}\} \left(\sum |c_j|^2\right)^{1/2},$$

and an upper bound of L^2 type.

If we regard the operator

$$D : \mathbf{c} \mapsto \sum_J c_j \chi_j(n), \qquad 1 \le n \le x,$$

as between $\mathbb{C}^{p-1}$ with standard square norm, and $\mathbb{C}^{[x]}$ with standard fourth power norm, then $\|D\| \ll (xp)^{1/4} + p^{5/8}(\log p)^{1/4}$. The dual map will then be between $\mathbb{C}^{[x]}$ with 4/3-power norm, and $\mathbb{C}^{p-1}$ again with mean square:

$$\left(\sum_j \left| \sum_{n \le x} a_n \chi_j(n) \right|^2 \right)^{1/2} \ll \{(xp)^{1/4} + p^{5/8}(\log p)^{1/4}\} \left(\sum_{n \le x} |a_n|^{4/3} \right)^{3/4}.$$

This does not directly estimate the sum (23.8) with which we set out, but we could reach that sum from the mean-square estimate given here by applying Hölder's inequality. We might otherwise force the issue, and this step is effectively equivalent, by replacing $\left(\sum |c_j|^2 \right)^{1/2}$ with $(p-1)^{1/4}\left(\sum |c_j|^4 \right)^{1/4}$, and regarding D as between two fourth-power (norm) spaces. In this way we reach

$$\left(\sum_j \left| \sum_{n \le x} a_n \chi_j(n) \right|^{4/3} \right)^{3/4} \ll \{x^{1/4}p^{1/2} + p^{7/8}(\log p)^{1/4}\} \left(\sum_{n \le x} |a_n|^{4/3} \right)^{3/4}.$$

In order to derive this upper bound for our initial sum (23.8) we degraded the natural upper bound (23.10) three times. We might have stopped at each of four stages, with differing advantages. When not studying character sums for their own sake an aesthetic may be derived from the application in view, and the above analysis terminated appropriately. An example, although not expressed in these terms, may be found in Vaughan, [166] Lemma 8.2, applied in the context of the Hardy–Littlewood circle method. Note the loss of logarithms there, with no detriment to the application. See also Forti and Viola, [87].

We seek for $\alpha \ne 2$, inequalities as canonical as those for $\alpha = 2$. To argue with characters in our present situation can be to sink into a morass, gaining non-trivial estimates all the way. Accordingly, I eschew Hecke's dictum and devise a fractional-power version of the Large Sieve within the aesthetic of the study of additive functions in functional analytic terms, constructing sieve and dual sieve together.

In Chapters 25, 27 and 28 I show that a result of the type (23.7) is available for each α-power with $\alpha > 1$. I was driven to the mean square result in stages, [54], [56], [60]. An extra uniformity over the innermost summation in (23.7), to fully mimic the Bombieri–Vinogradov theorem, is not needed for my immediate application, and I defer its consideration to Chapter 29.

I separate convoluted variables by a Mellin transform in both Chapters 24 and 28. I apply a Mellin transform to obtain maximal inequalities in Chapter 29.

Bolstered by these results I reconsider (23.3) and analogues, in Chapter 30, introducing a conjecture of function analytic type.

If we view $L^2(\mathbb{C}^s)$ as a space of functions on a set of s points, then the total measure assigned to these points is

$$\sum_{q \le x} \frac{1}{q}\left(1 - \frac{1}{q_0}\right) = \log\log x + O(1), \quad x \ge 2.$$

For $0 < \beta \le 1$, the (coordinate) points with q not exceeding x^β also have measure $\log\log x + O(1)$, almost all of the total, so-to-speak.

With the versions (23.7) currently available, a treatment of the differences of additive functions $f(n) - f(n-1)$ in the exact manner of Chapters 20 and 21 does not seem possible. However, the method of the stable dual proceeds. Essentially we study the additive function on a patch of spaces $\mathbb{C}^{s_1}$, where s_1 denotes the number of prime powers up to x_1, and x_1 is allowed to vary over an interval $x \le x_1 \le x^\delta$, $\delta > 1$. We may view the interval $[x, x^\delta]$ in the multiplicative group of the positive reals. Its Haar measure, renormalised as in the historical remarks closing Chapter 6, is then $\delta - 1$, and independent of x. In this sense the patch is not large.

The diffusion through analytic number theory of operators on finite dimensional spaces, studied through their duals, was in part provoked by the challenge of understanding the particular Large Sieve inequality of Linnik.

24

Exercises: Multiplicative functions on arithmetic progressions; Wiener phenomenon

The pursuit of approximate functional equations within the method of the stable dual leads naturally to the study of arithmetic functions on residue classes to prime moduli. The following exercises begin such a study for multiplicative functions, in particular for the characters on Q^*.

For an arithmetic function h define

$$\Delta(h) = \left(\sum_{\omega < p \le Q} \phi(p) \sum_{r=1}^{p-1} \left| \sum_{\substack{n \le x \\ n \equiv r (\operatorname{mod} p)}} h(n) - \frac{1}{\phi(p)} \sum_{\substack{n \le x \\ (n,p)=1}} h(n) \right|^2 \right)^{1/2}.$$

There are several ways to step from a multiplicative function to a Dirichlet convolution. One way is exposed in exercise 12 of Chapter 12. Another is given in exercise 13 of Chapter 15. The next exercises exploit a third way, indicated in the remarks following exercise 21 of that same chapter.

1. Prove that for a completely multiplicative function g with values in the complex unit disc

$$\Delta(g) = \frac{1}{\log x} \Delta(g \log) + O\left(\frac{x}{\log x} \right),$$

uniformly in $\omega \le Q \le \sqrt{x}$, $x \ge 2$. Here $g(n) \log n = \sum_{dm=n} g(d)\Lambda(d)g(m)$, with Λ the von Mangoldt function; $g \log = g\Lambda * g$.

2. Prove that for any positive integer D

$$\phi(D) \sum_{\substack{r=1 \\ (r,D)=1}}^{D} \left| \sum_{\substack{n \le x \\ n \equiv r (\operatorname{mod} D)}} h(n) - \frac{1}{\phi(D)} \sum_{\substack{n \le x \\ (n,D)=1}} h(n) \right|^2 = \sum_{\chi} \left| \sum_{n \le x} h(n)\chi(n) \right|^2,$$

χ running over all non-principal Dirichlet characters $(\operatorname{mod} D)$.

In terms of Dirichlet characters χ, the study of Δ reduces to that of convolution bilinear forms $B = \sum_{dm \le x} u_d v_m$ with $u_d = g(d)\Lambda(d)\chi(d)$, $v_m = g(m)\chi(m)$. For x not an integer, we can convert the form to a product using Fourier analysis on the multiplicative group of positive reals:

$$B = \frac{1}{2\pi i}\int \sum_{d\le x} u_d d^{-s} \sum_{m \le x} v_m m^{-s} s^{-1} x^s ds, \quad s = \sigma + i\tau,$$

the integral taken over any line $\mathrm{Re}\,(s) = \sigma > 0$ in the complex plane and interpreted according to the practical remarks made following section **14** of Chapter 0. It is tempting to estimate $|B|^2$ through the agency of the Cauchy–Schwarz inequality, but the kernel $s^{-1}x^s$ is not Lebesgue integrable over the whole (real τ) line; cf. exercise 21 of Chapter 7. We encounter in clear the phenomenon of Wiener; the Mellin transform of a single step is not very small at infinity.

For $y \ge 1/2$, $s \ne 0, -1$, define

$$K(y,s) = \frac{4((y+\frac{1}{4})^{s+1} - y^{s+1})}{s(s+1)}.$$

3. Prove that if y is half an odd positive integer, then

$$\sum_{d\le y} c_d = \frac{1}{2\pi i}\int \sum_{d \le x} c_d d^{-s} K(y,s) ds,$$

uniformly for $1/2 \le y \le x$, the integral taken over $\mathrm{Re}\,(s) = \sigma > 0$.

4. Prove that over the line $\mathrm{Re}\,(s) = \sigma (>0)$,

$$\int \sup_{1/2 \le y \le x} |K(y,s)| d\tau \ll x^\sigma \log \frac{x}{\sigma},$$

The implied constant absolute. A proof is included in Lemma 28.2.

5. Prove that if $x \ge 2$, then

$$\max_{y \le x}\left|\sum_{d\le y} c_d\right| \ll \max_{\tau\in\mathbb{R}}\left|\sum_{d \le x} c_d d^{-\sigma_0 - i\tau}\right| \log x,$$

with $\sigma_0 = (\log x)^{-1}$.

6. Prove that

$$\max_{\tau \in \mathbb{R}}\left|\sum_{p\le x} c_p p^{-i\tau}\right| = \max_{|z_p|=1}\left|\sum_{p \le x} c_p z_p\right|.$$

Hint. Kronecker.

At the expense of a factor of $\log x$, exercise 5 dominates a maximum on the real line with another over rotations in the complex plane.

7. Let $T : \mathbb{C}^m \to \mathbb{C}^n$ be a linear operator, given by

$$(Ta)_k = \sum_{j=1}^{m} c_{kj} a_j, \qquad a \text{ in } \mathbb{C}^m, \ 1 \le k \le r.$$

For each k choose a positive integer t_k, not exceeding m, and define a (maximal) operator $\widetilde{T}$ by

$$(\widetilde{T}a)_k = \sum_{j=1}^{t_k} c_{kj} a_j, \qquad a \text{ in } \mathbb{C}^m, \ 1 \le k \le r.$$

Prove that with respect to standard mean-power norms $\|\widetilde{T}\| \ll \|T\| \log 2m$, the implied constant absolute.

In the notation following exercise 2, and for $A > 0$, define h_j, $1 \le j \le 3$, by

$$h_1(n) = \sum_{\substack{dm=n \\ \max(d,m) \le x(\log x)^{-A}}} u_d v_m,$$

and h_2, h_3 similarly, the maximum-condition replaced by $m > x(\log x)^{-A}$, $d > x(\log x)^{-A}$, respectively. Assume x to be sufficiently large that $(\log x)^A \le x(\log x)^{-A}$.

We treat $\Delta(h_1)$ with the Large Sieve.

8. Let M, N be reals exceeding $2/3$. Prove that

$$\sum_{D \le t} \sideset{}{^*}\sum_{\chi (\operatorname{mod} D)} \max_y \left| \sum_{\substack{M < m \le 2M \\ N < n \le 2N \\ mn \le y}} \alpha_m \beta_n \chi(mn) \right|$$
$$\ll \|\alpha\| \|\beta\| (M^{1/2} + t)(N^{1/2} + t) \log 4MN$$

with $\|\alpha\| = (\sum |\alpha_m|^2)^{1/2}$, $\|\beta\| = (\sum |\beta_n|^2)^{1/2}$, the implied constant absolute. Here * again denotes summation over primitive characters.

9. Prove that for $\omega \ge 1$, $x \ge 2$

$$\sum_{\omega < p \le Q} \max_{y \le x} \max_{(r,p)=1} \left| \sum_{\substack{n \le y \\ n \equiv r (\operatorname{mod} p)}} h_1(n) - \frac{1}{\phi(p)} \sum_{\substack{n \le y \\ (n,p)=1}} h_1(n) \right|$$
$$\ll \omega^{-1} x \log x + Q x^{1/2} \log x + x(\log x)^{2-A/2},$$

the implied constant absolute.

Hint. Represent the condition $n \equiv r(\operatorname{mod} p)$ using Dirichlet characters. Cover the region $dm \le x$ by rectangles of the form $N < d \le 2N$, $M < m \le 2M$.

10. Prove that

$$\sum_{w<p\le Q} \max_{y\le x} \max_{(r,p)=1} \phi(p) \left| \sum_{\substack{n\le y \\ n\equiv r(\operatorname{mod} p)}} h_1(n) - \frac{1}{\phi(p)} \sum_{\substack{n\le y \\ (n,p)=1}} h_1(n) \right|^2$$
$$\ll \omega^{-1}x^2(\log x)^2 + Qx^{3/2}(\log x)^2 + x^2(\log x)^{3-A/2},$$

the implied constant absolute.

Note that exercises 8–10 apply inequalities of type L^2 to gain those of type L and from these others of type L^2.

11. The function h_2 largely coincides with g. Prove that

$$\Delta(h_2) \ll x \log\log x + Qx^{1/2}(\log x)^{A/2}.$$

12. Let $\varepsilon > 0$. Prove that

$$\Delta(h_3) \ll x \log\log x + Q^{1+\varepsilon}x^{1/2}(\log x)^{A+1}(\log\log x)^{1/2}.$$

13. Let $\varepsilon > 0$. Prove that for a completely multiplicative function g with values in the complex unit disc

$$\sum_{(\log x)^2<p\le x^{\frac{1}{2}-\varepsilon}} \phi(p) \max_{(r,p)=1} \left| \sum_{\substack{n\le x \\ n\equiv r(\operatorname{mod} p)}} g(n) - \frac{1}{\phi(p)} \sum_{\substack{n\le x \\ (n,p)=1}} g(n) \right|^2$$
$$\ll \left(\frac{x\log\log x}{\log x}\right)^2,$$

the implied constant depending at most upon ε.

14. The method of exercises 8–13 iterates. Prove that for a suitable value of the implicit parameter A,

$$\sum_{(\log x)^4<p\le x^{\frac{1}{2}-\varepsilon}} \phi(p) \max_{(r,p)=1} \left| \sum_{\substack{n\le x \\ n\equiv r(\operatorname{mod} p)}} g(n)\log n - \frac{1}{\phi(p)} \sum_{\substack{n\le x \\ (n,p)=1}} g(n)\log n \right|^2$$

differs from

$$\sum_{(\log x)^4 < p \le x^{\frac{1}{2}-\varepsilon}} \phi(p) \max_{(r,p)=1} \left| \sum_{\substack{n \le x \\ n \equiv r (\operatorname{mod} p)}} h_3(n) - \frac{1}{\phi(p)} \sum_{\substack{n \le x \\ (n,p)=1}} h_3(n) \right|^2$$

by $O(x(\log\log x)^4(\log x)^{-2})$ uniformly in $g, x \ge 2$. The function h_3 controls the distribution of the (completely) multiplicative function g in residue classes. When g is a particular function, such as that of Liouville, $\Delta(h_3)$ is small. In general we are driven to apply inequalities of the type given in exercise 55 of Chapter 4.

15. Let g be a multiplicative function with values in the complex unit disc. Define a completely multiplicative function g_1 by $g_1(p) = g(p)$, and the (multiplicative) function w by Dirichlet convolution, $g = w * g_1$. Prove that

$$\sum_{r \le x} r^{-1/2}|w(r)| \ll (\log x)^2,$$

uniformly for $x \ge 2$ and g.

16. Prove that the inequality of exercise 13 remains valid for standard multiplicative functions g.

Initially, the method of the stable dual is concerned with prime moduli sufficiently large that an analogue of the classical Siegel–Walfisz theorem is not required. This is fortunate, since for multiplicative functions near to a Dirichlet character, the natural such analogue is false. For multiplicative functions on progressions to composite moduli this difficulty must be faced. It is particularly advantageous in the study of harmonic analysis on Q^* that it can be somewhat overcome, [75], [78]. I do not pursue such matters here.

The Bombieri(–Vinogradov) theorem possesses a uniformity over the range of the inner sum(mands). To effect a corresponding uniformity in the inequality of exercise 13 an appeal to exercise 5 will not suffice, the factor $(\log x)^2$ so introduced would vitiate everything gained by the original argument. Two methods suggest themselves. We can modify the kernel $K(y, s)$. This procedure is begun in Chapter 25; it leads to certain complications. Otherwise, we may employ the duality principle to seek maximal inequalities from the outset. Examples are furnished in Chapter 29.

Is the Wiener phenomenon a manifestation of the size of the group $\mathbb{R}$? Could we avoid it by filtering through a smaller group?

Wiener phenomenon on finite groups

17. Let G be a finite abelian group. Prove that $\det(\chi(g))$, g in G, χ in $\widehat{G}$, is non-zero.

18. If the function $f : G \to \mathbb{C}$ has the representation $\sum_{j=1}^{|G|} c_j \chi_j$, where the χ_j belong to $\widehat{G}$, what is the value of $\hat{f}$ at χ_j? Prove that given χ_j, $1 \le j \le b$, in $\widehat{G}$, with $b < |G|$, there is a non-trivial function $f : G \to \mathbb{C}$ such that $\hat{f}$ vanishes at each χ_j.

19. Prove that if the sum of the positive integers a, b is less than $|G|$, then the following statement is true: Given any members g_i, $1 \le i \le a$, of G, and characters χ_j, $1 \le j \le b$, of $\widehat{G}$, there is a non-trivial function $f : G \to \mathbb{C}$, which vanishes on the g_i, and is such that $\hat{f}$ vanishes on the χ_j.

20. Evaluate as a product the van der Monde determinant with entries z_i^j, $z_i^0 = 1$, $1 \le i \le m$, $0 \le j \le m-1$.

21. Let G be a finite cyclic group generated by g. Let $f : G \to \mathbb{C}$ vanish on g^v for v belonging to a consecutive integer values, and let $\hat{f}$ vanish on b characters in $\widehat{G}$. Prove that if $a + b \ge |G|$, then f vanishes identically.

This result is applicable to every additive group $\mathbb{Z}/m\mathbb{Z}$, $m \ge 1$, to the multiplicative group of each finite field, in particular to the group of reduced residue classes $(\bmod\, p)$. In applications of Fourier analysis, Wiener's phenomenon is to be expected.

25

Fractional power Large Sieves. Operators involving primes

In this chapter I continue with the notation that if q is a prime power, then q_0 is the prime of which it is a power.

Let k be a positive integer, Q a real, $Q \geq 2$. Let $\mathbb{C}^u$ be the space of vectors h with one coordinate $h(q,r)$ for each pair (q,r) with $q = q_0^k$ a prime power not exceeding Q, and r one of a complete set of residue class representatives $(\operatorname{mod} q)$ chosen from the interval $[1, qq_0]$. It is convenient to allow r the freedom to belong to this larger interval, rather than the interval $[1,q]$. I topologise $\mathbb{C}^u$ with the norm

$$\|h\|_\beta = \left(\sum_{\substack{q \leq Q \\ q = q_0^k}} \sum_r \frac{1}{q} \left(1 - \frac{1}{q_0}\right) |h(q,r)|^\beta \right)^{1/\beta},$$

and denote the corresponding measure space by $J^\beta(\mathbb{C}^u)$. The dimension u of this space is

$$\sum_{q_0^k \leq Q} q = (1 + o(1)) \frac{kQ^{1+1/k}}{(k+1)\log Q}, \qquad Q \to \infty.$$

The measure assigned to the support set of u points is

$$\sum_{q_0^k \leq Q} \left(1 - \frac{1}{q_0}\right) = (1 + o(1)) \frac{kQ^{1/k}}{\log Q}, \qquad Q \to \infty.$$

I shall suppress the notation that the q are all k^{th} powers of their primes.

Let $2 \leq \rho \leq x$. With $v = [x] - [x - \rho]$ define a space $K^\beta(\mathbb{C}^v)$ of vectors a, one coordinate to each integer n in the interval $x - \rho < n \leq x$, topologised by

$$\|a\|_\beta = \left(\frac{1}{[x] - [x-\rho]} \sum_{x - \rho < n \leq x} |a_n|^\beta \right)^{1/\beta}.$$

Associated with $K^\beta(\mathbb{C}^v)$ is the frequency measure on a set of v points.

For a given positive Q define the map $W:\mathbb{C}^u\to\mathbb{C}^v$ by

$$(Wh)(n)=\sum_{\substack{r,q\\ q\|(n-r)}} h(q,r)-\sum_{r,q}\frac{1}{q}\left(1-\frac{1}{q_0}\right)h(q,r).$$

It is convenient to denote by W_α the map W viewed between the spaces $(J^2\cap J^{\alpha'}(\mathbb{C}^u))'$, $K^\alpha(\mathbb{C}^v)$ if $1<\alpha\le 2$; between the spaces $J^2\cap J^\alpha(\mathbb{C}^u)$, $K^\alpha(\mathbb{C}^v)$ if $\alpha\ge 2$. When no confusion so arises I omit the suffix.

Theorem 25.1. *For $1<\alpha\le 2$ the operator*

$$W'_{\alpha'}:(K^{\alpha'}(\mathbb{C}^v))'\to(J^2\cap J^{\alpha'}(\mathbb{C}^u))'$$

has norm $\ll(1+\rho^{-1}Q^{2\alpha'+3})^{1/\alpha'}$. For $\alpha\ge 2$ the operator

$$W'_{\alpha'}:(K^{\alpha'}(\mathbb{C}^v))'\to(J^2\cap J^{\alpha}(\mathbb{C}^u))''\cong J^2\cap J^\alpha(\mathbb{C}^u)$$

has norm $\ll(1+\rho^{-1}Q^7)^{1/\alpha'}$. As usual, $\alpha^{-1}+(\alpha')^{-1}=1$.

The operator dual to W is given by $(W'a)(q,r)=$

$$\frac{q}{[x]-[x-\rho]}\left(1-\frac{1}{q_0}\right)^{-1}\left(\sum_{\substack{x-\rho<n\le x\\ n\cong r(\mathrm{mod}\, q)}} a_n-\frac{1}{q}\left(1-\frac{1}{q}\right)\sum_{x-\rho<n\le x} a_n\right).$$

Theorem 25.1 is most simply illustrated when $\alpha=2$. In particular it asserts that

$$\sum_{q\le Q} q\sum_{r=0}^{q-1}\left|\sum_{\substack{x-\rho<n\le x\\ n\cong r(\mathrm{mod}\, q)}} a_n-\frac{1}{q}\left(1-\frac{1}{q_0}\right)\sum_{x-\rho<n\le x} a_n\right|^2\ll(\rho+Q^7)\sum_{x-\rho<n\le x}|a_n|^2$$

holds for all complex a_n, real $x\ge\rho\ge 2$, $Q\ge 2$. This is an arithmetical version of the Large Sieve inequality, save that we have prime power moduli in place of the usual primes, Q^7 in place of the usual Q^2. It may be compared with the inequalities in exercises 18, 25 and 26 of Chapter 4. As will be seen in the following argument, I make no attempt to gain a small power of Q, and for various values of α improvements can certainly be made.

For $1 < \alpha < 2$ the result of Theorem 25.1 is more complicated to present. We may assert that

$$\sum_{q\le Q} q^{\alpha-1} \sum_{r=0}^{q-1} \left| \sum_{\substack{x-\rho<n\le x \\ n\cong r(\bmod q)}} a_n - \frac{1}{q}\left(1-\frac{1}{q_0}\right) \sum_{x-\rho<n\le x} a_n \right|^{\alpha}$$

$$\ll (\rho^{\alpha-1} + Q^{5\alpha-3}) \sum_{x-\rho<n\le x} |a_n|^{\alpha}$$

provided the pairs (q, r) are restricted to those for which

$$\frac{q}{\rho} \left| \sum_{\substack{x-\rho<n\le x \\ n\cong r(\bmod q)}} a_n - \frac{1}{q}\left(1-\frac{1}{q_0}\right) \sum_{x-\rho<n\le x} a_n \right| > \|W'\| \left(\frac{1}{\rho} \sum_{x-\rho<n\le x} |a_n|^{\alpha} \right)^{1/\alpha}.$$

Those pairs for which this last inequality fails give rise to the bound

$$\left(\sum_{q\le Q} q \sum_{r=0}^{q-1} \left| \sum_{\substack{x-\rho<n\le x \\ n\cong r(\bmod q)}} a_n - \frac{1}{q}\left(1-\frac{1}{q_0}\right) \sum_{x-\rho<n\le x} a_n \right|^2 \right)^{1/2}$$

$$\ll \rho(1+\rho^{-1}Q^{2\alpha'+3})^{1/\alpha'} \left(\rho^{-1} \sum_{x-\rho<n\le x} |a_n|^{\alpha} \right)^{1/\alpha}.$$

According to the remark following Lemma 2.7, when $Q^{2\alpha'+3} \le \rho$ these assertions remain valid with $\|W'\|$ replaced by 1.

Replacing the norm on $(L^2 \cap L^{\alpha'}(\mathbb{C}^s))'$ by that on the space $L^{\alpha}(\mathbb{C}^s)$ introduces a scaling factor of $\ll (\log\log x)^{1/\alpha-1/2}$. When endeavouring to establish a norm estimate which gains a power of $\log x$ over the trivial bound, this factor is not a serious loss. A similar replacement between the spaces $(J^2 \cap J^{\alpha'}(\mathbb{C}^u))'$ and $J^{\alpha}(\mathbb{C}^u)$ would introduce a scaling factor on the order of $(Q^{1+1/k}(\log Q)^{-1})^{1/\alpha-1/2}$, which would be seriously large. Indeed, when Q is a power of x, which I shall shortly require, such a factor would swamp any saving of a power of $\log x$.

In the study of generalisations of the mean-square Large Sieve, the formulation of appropriate inequalities is part of the difficulty. The interpretation of Large Sieve inequalities in functional analytic terms is then most helpful.

I began with the operator W'. That it is appropriate to the study of fractional power Large Sieves is clear from the discussion in Chapter 23, and the results of Chapters 3 and 4. The operator W, dual to W', is formally acquired, and since to compute moments of $(Wh)(n)$ is to estimate integers in residue classes, progress seems possible. The reciprocating action of dualisation comes into play.

It is a little awkward to visualise a typical function Wh. Suppose for the moment that q is confined to the interval $(Q/2, Q]$ for an integer Q; and r is constrained to run over the positive integers not exceeding Q, a range independent of the q. There will be a representation

$$(Wh)(n) = \sum_{r=1}^{Q} f_r(n-r)$$

with (differing) additive function f_j, $1 \le j \le Q$. The form of the results in Chapters 2, 3 suggests consideration of W between the spaces $J^2 \cap J^\alpha(\mathbb{C}^u)$ and $K^\alpha(\mathbb{C}^v)$, $\alpha \ge 2$.

Setting out to estimate powers of $(Wh)(n)$ by modifying the argument of Chapter 2, we encounter the correlation

$$x^{-1} \sum_{n \le x} g_1(n-1) \cdots g_Q(n-Q), \qquad x \ge Q+1,$$

with the g_j multiplicative functions uniformly bounded on prime powers. Standard procedures, such as representing each g_j by a Dirichlet convolution $\mathbf{1} * k_j$, are feasible only if $Q^Q \ll x$, restricting Q to a value $\ll \log x / \log\log x$. For Q allowed as large as a fixed power of x, our present intent, neither a method to produce a non-trivial upper bound for this correlation, nor an aesthetic to lend one form, suggests itself. We keep the operator W in view, abandon the use of multiplicative functions in the style of Chapter 2, and revert to the generality of Theorem 25.1.

In mean-square spaces the inequality of Theorem 25.1 may be obtained by representing W' in terms of additive characters, and appealing to Parseval's relation. For spaces of L^α type, $1 < \alpha < 2$, this relation is not available. Instead I argue indirectly, using duality and interpolation. The interpolation weakens the uniformity in Q. The following result is an essential step.

Lemma 25.2. *For each positive integer m, the map*

$$W : J^2 \cap J^{2m}(\mathbb{C}^u) \to K^{2m}(\mathbb{C}^v)$$

has norm $\ll (1 + \rho^{-1} Q^{4m-1})^{1/(2m)}$.

It will be convenient to appeal to an inequality of Rosenthal [144], from the theory of probability.

Lemma 25.3. *Let m be a positive integer. For independent random variables H_i of mean zero, $i = 1, \ldots, s$,*

$$E\left(\sum_{i=1}^{s} H_i\right)^{2m} \ll \sum_{i=1}^{s} E(H_i^{2m}) + \left(\sum_{i=1}^{s} E(H_i^2)\right)^{m}.$$

where E denotes expectation with respect to the underlying probability measure.

The form of this inequality may be compared with that of Theorem 2.1. I indicate a short proof of Lemma 25.3 in Chapter 26.

Proof of Lemma 25.2. We estimate the norm of W, in this context, directly. For pairs q, r subject to the restrictions in Theorem 25.1, and integers n, define

$$\theta(n, q, r) = \begin{cases} 1 & \text{if} \quad q \,\|\, (n-r), \\ 0 & \text{otherwise.} \end{cases}$$

Then

$$\Lambda = v\|Wh\|_{2m}^{2m} = \sum_{x-\rho<n\leq x} \left| \sum_{r,q} h(q,r) \left(\theta(n,q,r) - \frac{1}{q}\left(1 - \frac{1}{q_0}\right)\right)\right|^{2m}.$$

Without loss of generality we may assume the $h(q,r)$ to be real valued. Expanding the innermost power and changing the order of summation we obtain for Λ the representation

$$\sum_{\substack{r_j, q_j \\ j=1,\ldots,2m}} \prod_{j=1}^{2m} h(q_j, r_j) \sum_{x-\rho<n\leq x} \prod_{j=1}^{2m} \left\{ \theta(n, q_j, r_j) - \frac{\phi(q_j)}{q_j^2} \right\}. \tag{25.1}$$

We are reduced to considering sums of the form

$$\sum_{x-\rho<n\leq x} \prod_{t=1}^{d} \theta(n, q_t, r_t).$$

We need an estimate for the number of integers n, in the interval $x - \rho < n \leq x$, which satisfy $n \cong r_t (\operatorname{mod} q_t)$ for $1 \leq t \leq d$, say. Bearing in mind that the q_t are all k^{th} powers of their primes, these congruences will be consistent if and only if the q_t are distinct, or when two q_t coincide so do the corresponding r_t. The Chinese remainder theorem together with an elementary argument shows the number of integers n to have the form

$\rho\Delta(q_1, \dots, q_d, r_1, \dots, r_d) + O(1)$, the error term absolutely bounded. From this we readily obtain for the inner sum at (25.1) an estimate of the form

$$\rho\psi(q_1, \dots, q_{2m}, r_1, \dots, r_{2m}) + O\left(\prod_{j=1}^{2m} \frac{q_j}{\phi(q_j)}\right).$$

The contribution of error terms to the whole sum at (25.1) is

$$\begin{aligned}
&\ll \left(\sum_{r,q} \frac{q}{\phi(q)} |h(q,r)|\right)^{2m} \\
&\ll \|h\|_{2m}^{2m} \left(\sum_{r,q} \left(q^{\frac{1}{2m}} \frac{q}{\phi(q)}\right)^{\frac{2m}{2m-1}}\right)^{2m-1} \\
&\ll \frac{Q^{4m-1}}{(\log Q)^{2m-1}} \|h\|_{2m}^{2m},
\end{aligned}$$

the penultimate step by an application of Hölder's inequality. When $q = q^k$ and $k > 1$, the last of these steps is wasteful. The constant implied in the last bound is not greater than c^m for a computable absolute $c > 1$.

Apart from the factor ρ, what remains to Λ is the sum

$$\sum_{\substack{r_j, q_j \\ j=1,\dots,2m}} \prod_{j=1}^{2m} h(q_j, r_j)\psi(q_1, \dots, q_{2m}, r_1, \dots, r_{2m}). \tag{25.2}$$

This we compute with help from the theory of probability.

Let D be the product of the prime powers $q_0^{k_0+1}$ with $k_0 = [\log Q / \log q_0]$. Consider the probability space consisting of the residue classes of the integers $(\operatorname{mod} D)$, with the natural frequency measure which assigns D^{-1} as the measure of each class. On this space the functions $\theta(n, q, r)$ naturally induce random variables. For distinct q the random variables

$$U(q) = \sum_r \theta(n, q, r)h(q, r)$$

are independent. This is justified by the Chinese remainder theorem. If, as above, $E(\ )$ denotes expectation with respect to the underlying probability measure, then for the random variable

$$X = \sum_{q \le Q} U(q)$$

we have

$$E(X) = \sum_{r,q} \frac{1}{q}\left(1 - \frac{1}{q_0}\right) h(q,r).$$

The sum (25.2) has the alternative representation

$$E((X - E(X))^{2m}).$$

We now appeal to Lemma 25.3, with the various $U(q) - E(U(q))$ taken for the H_i. The sum (25.2) is therefore

$$\ll \sum_{q \le Q} E(\{U(q) - E(U(q))\}^{2m}) + \sum_{q \le Q} (E(\{U(q) - E(U(q))\}^2))^m.$$

The second of these terms is

$$\sum_{q \le Q}\left(\sum_r \frac{1}{q}\left(1 - \frac{1}{q_0}\right)|h(q,r)|^2 - \left\{\sum_r \frac{1}{q}\left(1 - \frac{1}{q_0}\right)h(q,r)\right\}^2\right)^m,$$

which does not exceed $\|h\|_2^{2m}$. The first is likewise $\ll \|h\|_{2m}^{2m}$. Lemma 25.2 is established.

Lemma 25.4. *For $\beta \ge 2$ the operator*

$$W_\beta : J^2 \cap J^\beta(\mathbb{C}^u) \to K^\beta(\mathbb{C}^v)$$

has norm $\ll (1 + \rho^{-1}Q^{2\beta+3})^{1/\beta}$. For $1 < \beta \le 2$ the operator

$$W_\beta : (J^2 \cap J^{\beta'}(\mathbb{C}^u))' \to K^\beta(\mathbb{C}^v)$$

has norm $\ll (1 + \rho^{-1}Q^7)^{1/\beta}$.

Proof when $\beta \ge 2$. Let $2m \le \beta < 2(m+1)$ for the positive integer m. We would like to interpolate between the cases $\beta = 2m,\ 2(m+1)$ of Lemma 25.2, but the Riesz–Thorin theorem is not immediately applicable.

To overcome this let S be a finite collection of pairs (q,r), and let the projection P leave the (q,r)-coordinate of a vector in $\mathbb{C}^u$ invariant if (q,r) belongs to S, and otherwise take it to zero. For $\delta \ge 1$ we may view WP as a map between $J^\delta(P\mathbb{C}^u)$ and $K^\delta(\mathbb{C}^v)$. Let $\gamma(\delta)$ denote its norm.

By Hölder's inequality

$$\|Ph\|_2 \le \|Ph\|_\delta \|P\nu\|_1^{\frac{1}{2} - \frac{1}{\delta}}$$

uniformly for $\delta \geq 2$, where the function ν is 1 on all pairs (q, r). We may apply Lemma 25.2 to obtain

$$\gamma(2m) \ll (1 + \rho^{-1}Q^{4m-1})^{\frac{1}{2m}}(1 + \|P\nu\|_1^{\frac{1}{2}-\frac{1}{2m}}).$$

A similar bound may be effected for $\gamma(2m+2)$.

Defining θ by

$$\frac{1}{\beta} = \frac{\theta}{2m} + \frac{1-\theta}{2m+2},$$

we apply the Riesz–Thorin interpolation theorem to obtain

$$\gamma(\beta) \leq \gamma(2m)^{\theta}\gamma(2m+2)^{1-\theta}.$$

This in turn yields

$$\gamma(\beta) \ll (1 + \rho^{-1}Q^{2\beta+3})^{1/\beta}(1 + \|P\nu\|_1^c)$$

for some positive constant c. This is a wasteful step.

Let h on $J^2 \cap J^\beta(\mathbb{C}^u)$ be given. We take S to be the set of pairs for which $|h(q,r)| > \|h\|_2$. Then $\|P\nu\|_1 \ll 1$, and for that particular h

$$\|WPh\|_\beta \ll (1 + \rho^{-1}Q^{2\beta+3})^{1/\beta}\|Ph\|_\beta.$$

For the remaining values of $h(q,r)$ we argue crudely. Let t denote the vector $h - Ph$. With another application of Lemma 25.2.

$$\|Wt\|_\beta \leq \|Wt\|_{2m+2} \ll (1 + \rho^{-1}Q^{4m+3})^{\frac{1}{2m+2}}(\|t\|_{2m+2} + \|t\|_2).$$

Here

$$\begin{aligned}\|t\|_{2m+2}^{2m+2} &= \sum_{q \leq Q}\sum_{r} \frac{1}{q}\left(1 - \frac{1}{q_0}\right)|t(q,r)|^{2m+2} \\ &\leq \|h\|_2^{2m}\sum_{q,r}\frac{1}{q}\left(1 - \frac{1}{q_0}\right)|t(q,r)|^2 \leq \|h\|_2^{2m+2}.\end{aligned}$$

Altogether

$$\|W\|_\beta \ll \max((1 + \rho^{-1}Q^{2\beta+3})^{1/\beta},\ (1 + \rho^{-1}Q^{4m+3})^{1/(2m+2)}),$$

which gives the asserted bound. In particular we may recover the cases $\alpha \geq 2$ of Theorem 2.1.

Proof when $1 < \beta < 2$. Let $\| \ \|$ denote the norm on $(J^2 \cap J^{\beta'}(\mathbb{C}^u))'$. We decompose h in this space into the sum $h_1 + h_2$, where $h_1(q,r)$ is defined to be $h(q,r)$ if $|h(q,r)| \le \|h\|$, and to be zero otherwise. From Hölder's inequality and the case $\beta = 2$ already established

$$\|Wh_1\|_\beta \le \|Wh_1\|_2 \ll (1+\rho^{-1}Q^7)^{1/2}\|h_1\|_2 \ll (1+\rho^{-1}Q^7)^{1/2}\|h\|.$$

Here I have employed the characterisation of $(J^2 \cap J^{\beta'})'$ implied by Lemma 3.2.

We may continue along the lines of the treatment of additive functions given in Chapter 2. By Hölder's inequality, for each positive integer n

$$|(Wh_2)(n)|^\beta \le T(n)^{\beta-1} \sum_{q\|(n-r)} |h_2(q,r)|^\beta$$

where, in the notation of Lemma 25.2,

$$T(n) = {\sum_{\ell,s}}' \theta(n,\ell,s),$$

and $'$ denotes that summation is confined to the pairs ℓ, s for which $h_2(\ell,s)$ is non-zero. Preparing to invert an order of summation we note that

$$\begin{aligned}\sum_{n\cong r(\operatorname{mod} q)} T(n) &= {\sum_{\ell,s}}' \sum_{n\cong r(\operatorname{mod} q)} \theta(n,\ell,s)\\ &\le {\sum_{\ell,s}}' \left(\frac{\rho}{\ell q}+1\right) \le 2^{1+\beta}(\rho+Q^2)q^{-1},\end{aligned}$$

since

$${\sum_{\ell,s}}' \frac{1}{\ell} \le \|h\|^{-\beta} \sum_{\ell,s} |h_2(\ell,s)|^\beta \frac{2}{\ell}\left(1-\frac{1}{\ell_0}\right) \le 2^{1+\beta},$$

again by appeal to Lemma 3.2. Bearing in mind that $\beta - 1 \le 1$,

$$\|Wh_2\|_\beta \le \left(\frac{8(\rho+Q^2)}{[x]-[x-\rho]} \sum_{q,r} |h_2(q,r)|^\beta \frac{1}{q}\right)^{1/\beta} \le (2^7(1+\rho^{-1}Q^2))^{1/\beta}\|h\|,$$

after a third appeal to Lemma 3.2.

The cases $1 < \beta \le 2$ are established.

Proof of Theorem 25.1. This follows at once from Lemma 25.4, since $W'_{\alpha'}$ and $W_{\alpha'}$ have the same norm.

As in Chapters 2 and 3, we have constructed families of spaces looped by an operator and its dual:

$$(J^2 \cap J^{\alpha'})' \xrightarrow{W_\alpha} K^\alpha \simeq (K^{\alpha'})' \xrightarrow{W'_{\alpha'}} (J^2 \cap J^{\alpha'})'$$

if $1 < \alpha \le 2$, and

$$J^2 \cap J^\alpha \xrightarrow{W_\alpha} K^\alpha \simeq (K^{\alpha'})' \xrightarrow{W'_{\alpha'}} (J^2 \cap J^\alpha)'' \simeq J^2 \cap J^\alpha$$

if $\alpha \ge 2$. To continue with the analogy, we seek a large eigenvalue of the self-adjoint $W^*W : J^2(\mathbb{C}^u) \to J^2(\mathbb{C}^u)$. In the terminology of Lemma 25.2, the trace of this operator is

$$\sum_n \sum_{q,r} \frac{q(1-q_0^{-1})^{-1}}{[x]-[x-\rho]} \left(\theta(n,q,r) - \frac{1}{q}\left(1 - \frac{1}{q_0}\right)\right)^2,$$

which is $\ge u/5$ provided $2 \le Q \le \rho/8$. Under this same condition

$$\|W'_{\alpha'}\| \|W_\alpha\| \ge \|W'_{\alpha'} W_\alpha\| \ge \|W^*W\| \ge 1/5.$$

Combined with Theorem 25.1 this shows that if $1 < \alpha \le 2$, $Q \ge 2$, $8Q^{2\alpha'+3} \le \rho \le x$, then $1 \ll \|W_\alpha\| \ll 1$. A similar assertion may be made for the cases $\alpha > 2$ when $Q \ge 2$, $8Q^{2\alpha+3} \le \rho \le x$. To this extent Theorem 25.1 is best possible.

One may view the standard L^2 form of the Large Sieve as depending upon certain random variables being approximately uncorrelated, an idea of Rényi. The present argument uses the independence of random variables taken m at a time, where m increases unboundedly as α approaches 1.

It is useful to have a variant of Theorem 25.1 with a simpler congruential condition. The following example serves. Define the operator $\widetilde{W}'$ from $\mathbb{C}^v$ to $\mathbb{C}^u$ by

$$(\widetilde{W}'a)(q,r) = \frac{q(1-q_0^{-1})^{-1}}{[x]-[x-\rho]} \left(\sum_{\substack{x-\rho<n\le x \\ n\equiv r (\mathrm{mod}\, q)}} a_n - \frac{1}{q} \sum_{x-\rho<n\le x} a_n \right).$$

Theorem 25.5. *Let $1 < \alpha \le 2$. Considered between*

$$(K^{\alpha'}(\mathbb{C}^v))' \quad \textit{and} \quad (J^2 \cap J^{\alpha'}(\mathbb{C}^u))'$$

the operator $\widetilde{W}'$ has norm $\ll (1 + \rho^{-1}Q^{(2\alpha'+3)\alpha'})^{1/\alpha'}$.

Remark. This assertion continues to hold if in the definition of $\widetilde{W}'a$ the sum

$$\frac{1}{q} \sum_{x-\rho<n\le x} a_n \quad \text{is replaced by} \quad \frac{1}{\phi(q)} \sum_{\substack{x-\rho<n\le x \\ (n,q)=1}} a_n.$$

The difference between the variant operators is readily estimated using Hölder's inequality.

Proof of Theorem 25.5. A modification of Theorem 25.1 may be effected by employing the representation $(\widetilde{W}'a)(q,r) =$

$$\sum_{j=0}^{\infty} q_0^{-j} \frac{qq_0^j(1-q_0^{-1})^{-1}}{[x]-[x-\rho]} \left(\sum_{\substack{x-\rho<n\le x \\ n\cong r (\bmod qq_0^j)}} a_n - \frac{1}{qq_0^j}\left(1-\frac{1}{q_0}\right) \sum_{x-\rho<n\le x} a_n \right).$$

Let $t(\alpha-1)>1$. Those terms in the double sum with $j \ge t$ may be collected together, and in absolute value do not exceed

$$\frac{2q}{\rho} \sum_{\substack{x-\rho<n\le x \\ n\equiv r (\bmod qq_0^t)}} |a_n| + \frac{2}{\rho q_0^t} \sum_{x-\rho<n\le x} |a_n|.$$

Applications of Hölder's inequality show that the corresponding contribution to $\|\widetilde{W}'a\|$ is

$$\ll \left(\sum_{q,r} \frac{1}{q} \left(\frac{q}{\rho}\right)^{\alpha} \left(\sum_{\substack{x-\rho<n\le x \\ n\equiv r (\bmod qq_0^t)}} 1 \right)^{\alpha-1} \sum_{\substack{x-\rho<n\le x \\ n\equiv r (\bmod qq_0^t)}} |a_n|^{\alpha} \right)^{1/\alpha}$$

$$+ \left(\sum_{q,r} \frac{1}{q} \left(\frac{1}{\rho q_0^t} \right)^{\alpha} \rho^{\alpha-1} \sum_{x-\rho<n\le x} |a_n|^{\alpha} \right)^{1/\alpha}.$$

The first of these expressions is

$$\ll \|a\|_\alpha \left(\sum_q \frac{1}{q_0^{t(\alpha-1)}} + \frac{Q^\alpha}{\rho^{\alpha-1}\log Q} \right)^{1/\alpha},$$

and the second

$$\ll \|a\|_\alpha \left(\sum_q \frac{1}{q_0^{t\alpha}} \right)^{1/\alpha}.$$

Bearing in mind that the q are fixed powers of their prime q_0, the multiples of $\|a\|_\alpha$ in these bounds clearly do not exceed $(1+\rho^{-1/\alpha'}Q)$.

The contribution to $\widetilde{W}'a$ arising from the pairs (qq_0^j, r) with a fixed value of j, $0 \le j \le t-1$, we may estimate by considering the appropriate projection of W'. In this projection the moduli qq_0^j will not exceed q^t, and so Q^t. Bearing in mind the second remark following Lemma 3.2, we see that

$$\|\widetilde{W}'a\| \ll \|a\|_\alpha (1 + \rho^{-1} Q^{(2\alpha'+3)t})^{1/\alpha'} \sum_{j=0}^{t} \max_q q_0^{-j}.$$

The choice $t = 1 + [(\alpha - 1)^{-1}]$ gives the asserted bound.

Operators involving primes

I shall apply a version of Theorem 25.1 with the coordinate variable in $K^\beta(\mathbb{C}^v)$ running over prime powers rather than integers.

Let k be a positive integer, Q a real exceeding 2. Let F be the space of vectors h with one coordinate for each pair (q, r) where $q = q_0^k$ is a prime power in the interval $(Q^{1/2}, Q]$, and for each q, r runs through a complete set of reduced residue class representatives $(\bmod\, q)$, chosen from the interval $[1, qq_0]$. For $\beta \ge 1$, I give F the norm

$$\left(\sum_{Q^{1/2} < q \le Q} \sum_r \frac{1}{q} \, |h(q, r)|^\beta \right)^{1/\beta}$$

to define a (measure) space $\lambda^\beta(F)$.

For $2 \le \rho \le x$, define a space $\mu^\beta(G)$, with one coordinate for each prime power ℓ in the interval $x - \rho < \ell \le x$, and topologised by

$$\|b\|_\beta = \left(\frac{\log x}{\rho} \sum_{x-\rho<\ell\le x} |b_\ell|^\beta \right)^{1/\beta}.$$

Here the prime powers ℓ are not restricted to a particular power of their prime ℓ_0. A natural scaling factor for this β-power norm would be the number of prime powers ℓ in $(x - \rho, x]$, but this may be zero. On the Riemann hypothesis we would have

$$\sum_{p \le y} 1 = \int_2^y \frac{du}{\log u} + O(y^{\frac{1}{2}+\varepsilon}),$$

for each fixed $\varepsilon > 0$. Then if $x^{\frac{1}{2}+2\varepsilon} < \rho \le x$,

$$\sum_{x-\rho<p\le x} 1 \ge \int_{x-\rho}^{x} \frac{du}{\log u} + O(x^{\frac{1}{2}+\varepsilon}) > \frac{\rho}{\log x}(1 + O(x^{-\varepsilon})).$$

The factor $\rho/\log x$ is therefore not inappropriate.

Define a map $\omega : F \to G$ by

$$(\omega h)(\ell) = \sum_{\substack{q,r \\ q \| (\ell - r)}} h(q,r) - \sum_{q,r} q^{-1} h(q,r).$$

As with the specialisations of W, ω_α will denote the operator ω between $(\lambda^2 \cap \lambda^{\alpha'}(F))'$ and $\mu^\alpha(G)$ if $1 < \alpha \le 2$; between $\lambda^2 \cap \lambda^\alpha(F)$ and $\mu^\alpha(G)$ if $\alpha \ge 2$.

Theorem 25.6. *Let $\varepsilon > 0$. For $1 < \alpha \le 2$ the operator*

$$\omega'_{\alpha'} : (\mu^{\alpha'}(G))' \longrightarrow (\lambda^2 \cap \lambda^{\alpha'}(F))'$$

has norm $\ll (1 + \rho^{-1} x^\varepsilon Q^{2\alpha'+3})^{1/\alpha'}$. For $\alpha \ge 2$ the operator

$$\omega'_{\alpha'} : (\mu^{\alpha'}(G))' \longrightarrow (\lambda^2 \cap \lambda^\alpha(F))'' \simeq \lambda^2 \cap \lambda^\alpha(F)$$

has norm $\ll (1 + \rho^{-1} x^\varepsilon Q^7)^{1/\alpha'}$.

The operator dual to ω is given by

$$(\omega' b)(q,r) = \frac{q \log x}{\rho} \left(\sum_{\substack{x-\rho<\ell\le x \\ \ell \cong r (\mathrm{mod}\, q)}} b_\ell - \frac{1}{q} \sum_{x-\rho<\ell\le x} b_\ell \right).$$

Theorem 25.6, too, is most simply illustrated by the case $\alpha = 2$. In particular it then gives

$$\sum_{Q^{1/2}<q\le Q} q \sum_{\substack{r=1 \\ (r,q)=1}}^{q} \left| \sum_{\substack{x-\rho<\ell\le x \\ \ell \cong r (\mathrm{mod}\, q)}} b_\ell - \frac{1}{q} \sum_{x-\rho<\ell\le x} b_\ell \right|^2$$
$$\ll \left(\frac{\rho}{\log x} + x^\varepsilon Q^7 \right) \sum_{x-\rho<\ell\le x} |b_\ell|^2$$

uniformly for all complex b_ℓ, real $x \ge \rho \ge 2$. Exercise 54 of Chapter 4 shows that for prime moduli the factor Q^7 is larger than need be, moreover the restriction $Q^{1/2} < q$ in the outermost summation may be removed.

Even in the case $\alpha = 2$ the estimate of Theorem 25.6 is not straightforward to obtain. As α approaches 1 in value the assertion of Theorem 25.6 becomes deeper.

Given sufficiently accurate information concerning the distribution of primes in residue classes to large moduli, Theorem 25.6 could be established by straightforward modification of the method for Theorem 25.1. A sufficient knowledge would follow from the validity of the analogue of the Riemann hypothesis for Dirichlet L-series. To avoid appeal to such an hypothesis, I adopt a route which may appear artificial. I employ a result which is extracted from the mechanics of the Selberg sieve method.

Lemma 25.7. *Given $z \geq 2$ and a product P of distinct primes not exceeding z, there are real λ_d, one for each divisor d of P, which satisfy $|\lambda_d| \leq 1$ and*

$$\sum_{\substack{d_1 \leq z \\ d_1 | P}} \sum_{\substack{d_2 \leq z \\ d_2 | P}} \frac{\lambda_{d_1} \lambda_{d_2}}{[d_1, d_2]} \leq c \prod_{p|P} \left(1 - \frac{1}{p}\right),$$

where c is a positive absolute constant.

The following treatment of Selberg's sieve may be compared with that in the exercises of Chapter 4. I begin in almost the same way.

The Selberg sieve method estimates the number of primes in the interval $(z, x]$ not to exceed

$$S = \sum_{n \leq x} \left(\sum_{d|(n,P_1)} \lambda_d \right)^2,$$

where P_1 is the product of the primes $p \leq z$, and the real λ_d are constrained by $\lambda_1 = 1$. Interchanging the order of summation and appealing to the estimate $xD^{-1} + O(1)$ for the number of integers $n(\leq x)$ which satisfy $n \equiv 0(\operatorname{mod} D)$, we see that

$$S \leq x \sum_{d_j | P_1} \frac{\lambda_{d_1} \lambda_{d_2}}{[d_1, d_2]} + O\left(\sum_{d_j | P_1} |\lambda_{d_1} \lambda_{d_2}| \right).$$

Typically we set $\lambda_d = 0$ if $d > z$, and choose the λ_d to minimise the quadratic form appearing as the coefficient of x. This may be effected via exercise 37 of Chapter 4. Rather than appeal to the results of that chapter, I argue in the classical manner of Selberg.

A detailed account of such matters may be found in Halberstam and Richert, [93], and an account sufficiently general for the purposes of the present volume, in [41], Chapter 2. A simple modification of the initial argument shows that the quadratic form appearing in Lemma 25.7 is certainly non-negative definite.

Proof of Lemma 25.7. Until further notice in the proof of this lemma, integer variables are confined to divisors of P.

For $y > 0$ define

$$G(y) = \sum_{m \leq y} \frac{1}{\phi(m)}, \qquad G_d(y) = \sum_{\substack{m \leq y \\ (m,d)=1}} \frac{1}{\phi(m)}.$$

We define

$$\lambda_d = \frac{\mu(d)d}{\phi(d)} \frac{G_d(z/d)}{G(z)}, \qquad d \le z,$$

and demonstrate that these λ_d have the asserted properties.

Collecting together terms for which (m, d) has the particular value r,

$$G(z) = \sum_{r|d} \frac{1}{\phi(r)} \sum_{\substack{k \le z/r \\ (k,d)=1}} \frac{1}{\phi(k)} = \sum_{r|d} \frac{1}{\phi(r)} G_d\left(\frac{z}{r}\right).$$

Here $\phi(r)^{-1}$, taken over the divisors of d, sums to $d(\phi(d))^{-1}$. Hence

$$G_d\left(\frac{z}{d}\right) \le \frac{\phi(d)}{d} G(z) \le G_d(z). \tag{25.3}$$

In particular, $|\lambda_d| \le 1$. Note that $\lambda_1 = 1$.

The quadratic form in the statement of Lemma 25.7 has the representation

$$\frac{1}{G(z)^2} \sum_{\substack{d_j \le z \\ j=1,2}} \frac{1}{[d_1, d_2]} \frac{\mu(d_1)d_1\mu(d_2)d_2}{\phi(d_1)\phi(d_2)} G_{d_1}\left(\frac{z}{d_1}\right) G_{d_2}\left(\frac{z}{d_2}\right).$$

Since

$$\frac{d_1 d_2}{[d_1, d_2]} = (d_1, d_2) = \sum_{\substack{k|d_1 \\ k|d_2}} \phi(k),$$

we can further represent the form by

$$\frac{1}{G(z)^2} \sum_{k \le z} \phi(k) \left(\sum_{\substack{d \le z \\ d \equiv 0 \pmod{k}}} \frac{\mu(d)}{\phi(d)} G_d\left(\frac{z}{d}\right) \right)^2.$$

A typical sum to be squared has the alternative representations

$$\sum_{\substack{md \le z \\ d \equiv 0 \pmod{k} \\ (m,d)=1}}\sum \frac{\mu(d)}{\phi(d)\phi(m)} = \frac{\mu(k)}{\phi(k)} \sum_{\substack{tm \le z/k \\ (kt,m)=1 \\ (k,t)=1}}\sum \frac{\mu(t)}{\phi(t)\phi(m)} = \frac{\mu(k)}{\phi(k)} \sum_{\substack{v \le z/k \\ (v,k)=1}} \frac{1}{\phi(v)} \sum_{t|v} \mu(t).$$

The innermost sum is zero unless $v = 1$, when it is 1. The quadratic form has the value

$$G(z)^{-2} \sum_{k \le x} \frac{1}{\phi(k)} = G(z)^{-1}.$$

It remains to produce a lower bound for $G(z)$, and to this end we employ the upper bound at (25.3), replacing P by the product P_2 of all primes $p \le z$ which do not divide P. Integer variables are now freed of the constraint that they divide P. Then

$$\sum_{\substack{m \le z \\ (m,P_2)=1}} \frac{\mu^2(m)}{\phi(m)} \ge \frac{\phi(P_2)}{P_2} \sum_{m \le z} \frac{\mu^2(m)}{\phi(m)}.$$

Here

$$\frac{1}{\phi(m)} = \frac{1}{m} \prod_{p|m} \left(1 + \frac{1}{p} + \frac{1}{p^2} + \cdots\right).$$

Collecting up integers with the same distinct prime divisors ('squarefree kernel') shows that

$$\sum_{m \le z} \frac{\mu^2(m)}{\phi(m)} > \sum_{m \le z} \frac{1}{m} \ge \int_1^z \frac{du}{u} = \log z.$$

Moreover, from the elementary estimate

$$\sum_{p \le x} \frac{1}{p} = \log\log x + c_1 + O\left(\frac{1}{\log x}\right)$$

we derive

$$\prod_{p \le x} \left(1 - \frac{1}{p}\right) = \frac{c_2}{\log x}\left(1 + O\left(\frac{1}{\log x}\right)\right), \qquad c_2 > 0,$$

a theorem of Mertens. In fact c_2 has the form $e^{-\gamma}$ with Euler's constant γ. Hence

$$\sum_{\substack{m \le z \\ m|P}} \frac{1}{\phi(m)} \ge \prod_{\substack{p \le z \\ (p,P)=1}} \left(1 - \frac{1}{p}\right) c_2 \left(1 + O\left(\frac{1}{\log z}\right)\right) \prod_{p \le z} \left(1 - \frac{1}{p}\right)^{-1},$$

and the proof of the lemma is complete.

Remarks. The argument preceding the proof of Lemma 25.7 yields

$$\pi(x) \ll \frac{x}{\log z} + z^2.$$

With a suitable choice for z we recover a bound of Chebyshev type. With only a slight modification, the same argument gives $\pi(x) - \pi(x - \rho) \ll$

$\rho/\log\rho$ uniformly for $x \ge \rho \ge 2$. As a consequence the number of prime powers which belong to such an interval $(x-\rho, x]$ is also $\ll \rho/\log\rho$. Indeed, the asserted inequality is clear if $\rho > x/2$. In the contrary case we need an upper bound for the sum

$$\sum_{1\le s\le \frac{\log x}{\log 2}} \sum_{(x-\rho)^{1/s}<p\le x^{1/s}} 1.$$

For each s, an application of the mean value theorem of calculus shows that the interval $((x-\rho)^{1/s}, x^{1/s}]$ has length $\ll \rho x^{-1/2}$ uniformly in $x \ge 2\rho \ge 4$, $s \ge 2$. The double sum is therefore

$$\ll \frac{\rho}{\log\rho} + \frac{\rho\log x}{x^{1/2}} \ll \frac{\rho}{\log\rho},$$

as asserted.

In a similar way we obtain the following generalisation of the Brun–Titchmarsh inequality (cf. Halberstam and Richert, [93], pp. 105–110). Let $\varepsilon > 0$. Then the number of prime powers q in the interval $(x-\rho, x]$ which satisfy $q \equiv r(\mathrm{mod}\, D)$, for a given reduced residue class $r(\mathrm{mod}\, D)$, is

$$\ll \frac{\rho}{\phi(D)\log\rho} + \rho^{\varepsilon}$$

uniformly in $x \ge \rho \ge 2$, $D \ge 1, r$.

Proof of Theorem 25.6, beginning. The fundament is provided by an analogue of Lemma 25.2. This asserts that for $\varepsilon > 0$,

$$\omega : \lambda^2 \cap \lambda^{2m}(F) \to \mu^{2m}(G)$$

has norm $\ll (1+\rho^{-1}x^{\varepsilon}Q^{4m-1})^{1/(2m)}$.

We require an estimate for the sum

$$\sum_{x-\rho<\ell\le x} \left| \sum_{q,r} h(q,r)\left\{\theta(\ell,q,r) - \frac{1}{q}\right\}\right|^{2m}. \tag{25.4}$$

Without loss of generality we may assume x sufficiently large that $x^{\varepsilon/4} < (x/2)^{\varepsilon/2}$. The contribution to the sum (25.4) which arises from the terms $\ell = \ell_0^s$ with $s > 2/\varepsilon$ is by Hölder's inequality

$$\ll \frac{Q^{4m-1}}{(\log Q)^{2m-1}} \|h\|_{2m}^{2m} \sum_{\substack{\ell_0^s \le x \\ s>2/\varepsilon}} 1,$$

and the sum over the ℓ_0 is $\ll x^{\varepsilon/2}$. I shall therefore assume that the $\ell = \ell_0^s$ in (25.4) satisfy $s \le 2/\varepsilon$.

Suppose that $\rho \le x/2$. In particular we may then assume $\ell_0 > x^{\varepsilon/4}$. Let $z \le x^{\varepsilon/4}$, and let P be the product of the primes not exceeding z, the primes in the interval $(Q^{1/(2k)}, Q^{1/k}]$ excepted. As usual an empty product is given the value 1. No prime divisor of P will be an ℓ_0. In the notation of Lemma 25.7 the sum (25.4) does not exceed

$$\begin{aligned}\Gamma &= \sum_{x-\rho<n\le x}\Bigg(\sum_{d|(n,P)}\lambda_d\Bigg)^2\left|\sum_{q,r}h(q,r)\left\{\theta(n,q,r)-\frac{1}{q}\right\}\right|^{2m}\\ &= \sum_{\substack{d_j|P\\ d_j\le z}}\lambda_{d_1}\lambda_{d_2}T(d_1,d_2),\end{aligned}$$

say.

A typical sum $T(d_1, d_2)$ may be treated exactly as was the sum Λ during the proof of Lemma 25.2. The side conditions $(r, q) = 1$, and so on, do not cause any new difficulties. There is only one important change. We are reduced to considering sums of the form

$$\sum_{x-\rho<n\le x}\prod_{t=1}^{\delta}\theta(n,q_t,r_t)$$

with the additional constraint $n \equiv 0(\operatorname{mod}[d_1, d_2])$. Of course $(r_t, q_t) = 1$ is inherited from the outer summation conditions in the analogue of (25.1). Here the congruence conditions $s[d_1, d_2] \cong r_t(\operatorname{mod} q_t)$ in s, for $1 \le t \le \delta$, are compatible if and only if the conditions $s \cong r_t(\operatorname{mod} q_t)$ for $1 \le t \le \delta$ are compatible. It was to ensure the mutual primality of pairs d_i, q_t that the primes in $(Q^{1/(2k)}, Q^{1/k}]$ were excluded from P.

In an abuse of notation, the number of integers $n = s[d_1, d_2]$ in the interval $(x - \rho, x]$ which satisfy these congruence conditions is

$$[d_1,d_2]^{-1}\rho\Delta(q_1,\dots,q_\delta,\ r_1,\dots,r_\delta)+O(1),$$

the error term again absolute.

In a further abuse of notation, the sum $T(d_1, d_2)$ has the estimate

$$\frac{\rho}{[d_1,d_2]}\,E(\{X-Y\}^{2m})+O\left(\frac{Q^{4m-1}}{(\log Q)^{2m-1}}\,\|h\|_{2m}^{2m}\right),$$

where

$$Y=\sum_{q,r}\frac{1}{q}\,h(q,r).$$

Here it is understood that the modifications $(r_t, q_t) = 1$ and so on, in the side conditions, are observed in the various definitions and computations. It is important that the expectation does not depend upon the integers d_1 and d_2. In particular

$$\Gamma = \rho E(\{X-Y\}^{2m}) \sum_{\substack{d_j|P \\ d_j \le z}} \frac{\lambda_{d_1}\lambda_{d_2}}{[d_1,d_2]} + O\left(\left(\sum_{d\le z} |\lambda_d| \right)^2 \frac{Q^{4m-1}}{(\log Q)^{2m-1}} \|h\|_{2m}^{2m} \right).$$

The initial quadratic form in the λ_d is by Lemma 25.7

$$\ll \prod_{p\le z}\left(1-\frac{1}{p}\right) \prod_{Q^{1/(2k)}<p\le Q^{1/k}} \left(1-\frac{1}{p}\right)^{-1} \ll \frac{1}{\log z},$$

and the sum of the $|\lambda_d|$ is clearly at most z. We choose $z = x^{\epsilon/4}$.

Moreover,

$$E(\{X-Y\}^{2m}) \ll E(\{X-E(X)\}^{2m}) + (Y-E(X))^{2m}.$$

An application of Hölder's inequality shows that

$$(Y-E(X))^{2m} \le \|h\|_{2m}^{2m} \left(\sum_{q,r} \frac{1}{qq_0} \right)^{2m-1}.$$

The sum in the $(2m-1)^{\text{th}}$ power is

$$\le \sum_{Q^{1/(2k)}<q_0\le Q^{1/k}} \frac{1}{q_0} \ll 1,$$

the bound uniform in $Q \ge 2$, $k \ge 1$.

The appropriate analogue of Lemma 25.2 is thus established subject to the condition $\rho \le x/2$.

If $\rho > x/2$, we widen the outer summation in (25.4) to $\ell \le x$, and cover the interval of summation by subintervals $(x2^{-j} - x2^{-(j+1)}, x2^{-j}]$, $j = 0, 1, 2, \ldots$. To those intervals with $2^j \le x^{1-\varepsilon/2}$ we apply what we have already established. The analogue of the sum at (25.4) for $x^{\varepsilon/2} < \ell \le x$ is thus

$$\ll \left(\frac{x}{\varepsilon \log x} \sum_{j\ge 0} 2^{-j} + x^{\varepsilon/2} \log x Q^{4m-1} \right) \|h\|_{2m}^{2m}.$$

The remaining intervals are contained in $[1, x^{\varepsilon/2}]$ and an application of Hölder's inequality shows that the contribution arising from those ℓ not exceeding $x^{\varepsilon/2}$, like that from the ℓ_0^s with $s > 2/\varepsilon$, falls within the asserted bound.

An analogue of Lemma 25.4 follows rapidly.

Lemma 25.8. *Let $\varepsilon > 0$. For $\beta \geq 2$ the operator*

$$\omega_\beta : \lambda^2 \cap \lambda^\beta(F) \to \mu^\beta(G)$$

has norm $\ll (1 + \rho^{-1} x^\varepsilon Q^{2\beta+3})^{1/\beta}$. For $1 < \beta \leq 2$ the operator

$$\omega_\beta : (\lambda^2 \cap \lambda^{\beta'}(F))' \to \mu^\beta(G)$$

has norm $\ll (1 + \rho^{-1} x^\varepsilon Q^7)^{1/\beta}$.

Proof. With two minor modifications the proof follows that of Lemma 25.4. When $\beta \geq 2$ the inequality $\|Wt\|_\beta \leq \|Wt\|_{2m+2}$ of Lemma 25.4 is (here) replaced by

$$\|\omega t\|_\beta \leq \left(\frac{\log x}{\rho} \sum_{x-\rho<\ell\leq x} 1 \right)^{\frac{1}{\beta}-\frac{1}{2m+2}} \|\omega t\|_{2m+2}.$$

From the remarks following the proof of Lemma 25.7 this scaling factor is $\ll (\log x / \log \rho)^{\frac{1}{\beta}-\frac{1}{2m+2}}$, which for $\rho > x^\varepsilon$ is bounded above uniformly in $x \geq 2$. For $\rho \leq x^\varepsilon$ both Lemma 25.4 and its analogue for a prime power variable follow at once from an application of Hölder's inequality.

When $1 < \beta \leq 2$ we treat the analogue of $T(n)$ by appeal to the generalised Brun–Titchmarsh theorem established in the last of the same remarks.

Proof of Theorem 25.6. The operators $\omega'_{\alpha'}, \omega_{\alpha'}$ have the same norms.

For the third and last time in the present volume, we have constructed families of spaces looped by an operator and its dual:

$$(\lambda^2 \cap \lambda^{\alpha'})' \xrightarrow{\omega_\alpha} \mu^\alpha \simeq (\mu^{\alpha'})' \xrightarrow{\omega'_{\alpha'}} (\lambda^2 \cap \lambda^{\alpha'})'$$

if $1 < \alpha \leq 2$, and

$$\lambda^2 \cap \lambda^\alpha \xrightarrow{\omega_\alpha} \mu^\alpha \simeq (\mu^{\alpha'})' \xrightarrow{\omega'_{\alpha'}} (\lambda^2 \cap \lambda^\alpha)'' \simeq \lambda^2 \cap \lambda^\alpha$$

if $\alpha \geq 2$. Again, we seek a large eigenvalue of the self-adjoint $\omega^*\omega : \lambda^2(F) \to \lambda^2(F)$. The trace of this operator is

$$\sum_\ell \sum_{q,r} \frac{q \log x}{\rho} \left(\theta(\ell, q, , r) - \frac{1}{q} \right)^2,$$

and not quite as straightforward to estimate as that of W^*W. Note that independent of q,

$$\sum_r \sum_{\ell \equiv r (\operatorname{mod} q)} 1$$

counts the number of prime powers ℓ in the interval $(x-\rho, x]$. Let $0 < \delta < 1/8$. We see from Lemma 6.2 that for $\rho \geq \max(x^{\frac{5}{8}+\delta}, Q^{1+\delta})$ and $Q^{1/k}$ sufficiently large (in terms of δ), the trace exceeds half the dimension of F. Then

$$\|\omega'_{\alpha'}\| \|\omega_\alpha\| \geq \|\omega'_{\alpha'}\omega_\alpha\| \geq \|\omega^*\omega\| \geq \frac{1}{2}.$$

The further conditions $x^\varepsilon Q^{2\alpha'+3} \leq \rho$, $x^\varepsilon Q^{2\alpha+3} \leq \rho$ guarantee the bounds $1 \ll \|\omega_\alpha\| \ll 1$ for the ranges $1 < \alpha \leq 2$, $\alpha \geq 2$ respectively. Under these various terms Theorem 25.6 is best possible.

Again it is useful to consider a variant operator $\tilde{\omega}'$, defined by

$$(\tilde{\omega}'b)(q,r) = \frac{q \log x}{\rho} \left(\sum_{\substack{x-\rho<\ell\leq x \\ \ell \equiv r (\operatorname{mod} q)}} b_\ell - \frac{1}{\phi(q)} \sum_{x-\rho<\ell\leq x} b_\ell \right).$$

The analogue of Theorem 25.5 is then

Theorem 25.9. *Let $1 < \alpha \leq 2$, $\varepsilon > 0$. Viewed from*

$$(\mu^{\alpha'}(G))' \quad \text{to} \quad (\lambda^2 \cap \lambda^{\alpha'}(F))',$$

the operator $\tilde{\omega}'$ has norm $\ll (1 + \rho^{-1} x^\varepsilon Q^{(2\alpha'+3)\alpha'})^{1/\alpha'}$.

Proof. We may apply the argument given for Theorem 25.5. The only significant change is that in place of the bound

$$\sum_{\substack{x-\rho<n\leq x \\ n \equiv r (\operatorname{mod} qq_0^t)}} 1 \ll \frac{\rho}{qq_0^t} + 1$$

we apply

$$\sum_{\substack{x-\rho<\ell\leq x \\ \ell \equiv r (\operatorname{mod} qq_0^t)}} 1 \ll \frac{\rho}{\phi(qq_0^t) \log \rho} + \rho^\varepsilon,$$

as guaranteed in the remarks following the proof of Lemma 25.7. For $x^\varepsilon < \rho \leq x$ this bound amply fulfills the requirements of the argument. If $\rho \leq x^\varepsilon$, then the sum over the prime powers ℓ is $\ll x^\varepsilon$. With this bound the argument also proceeds.

Remark. Theorem 25.9 continues to hold if to the final sum in the definition of $(\tilde{\omega}'b)(q,r)$ is adjoined the condition $(\ell, q) = 1$.

26

Exercises: Probability seen from number theory

Let H_j, $j = 1, \ldots, k$, be independent random variables of mean zero. In the previous chapter I employed the inequality

$$E\left(\left|\sum_{j=1}^{k} H_j\right|^{\alpha}\right) \ll \left(\sum_{j=1}^{k} E(H_j^2)\right)^{\alpha/2} + \sum_{j=1}^{k} E(|H_j|^{\alpha}), \quad \alpha \geq 2,$$

of Rosenthal. The similarity in form between this inequality and the cases $\alpha \geq 2$ of Theorem 2.1 suggests that the arithmetic argument of Chapter 2 might adapt to the theory of probability, the more so since the free semi-group property of the integers upon which that argument rests is less immediate than the notion of stochastic independence. As usual, $E(X) = \bar{X}$ denotes the mean of the random variable X, $\operatorname{var}(S) = E((X - \bar{X})^2)$ its variance.

Let

$$D = \left(\sum_{j=1}^{k} E(H_j^2)\right)^{\frac{1}{2}} > 0,$$

and define random variables

$$X_j = \begin{cases} H_j & \text{if } |H_j| \leq D, \\ 0 & \text{if } |H_j| > D, \end{cases} \qquad Y_j = H_j - X_j, \ j = 1, \ldots, k.$$

Since $\bar{X}_j = \bar{Y}_j$,

$$\left|\sum_{j=1}^{k} \bar{X}_j\right| = \left|-\sum_{j=1}^{k} \bar{Y}_j\right| \leq \sum_{j=1}^{k} D^{-1} E(Y_j^2) \leq D.$$

To obtain Rosenthal's inequality it will suffice to establish

$$E\left(\left|\sum_{j=1}^{k} (X_j - \bar{X}_j)\right|^{\alpha}\right) \ll \left(\sum_{j=1}^{k} \operatorname{var}(X_j)\right)^{\alpha/2} \tag{26.1}$$

and

$$E\left(\left|\sum_{j=1}^{k} Y_j\right|^{\alpha}\right) \ll \sum_{j=1}^{k} E(|Y_j|^{\alpha}). \tag{26.2}$$

1. Let S denote the sum of independent random variables Z_j, $j = 1, \ldots, k$, each of mean zero and bounded by M. Prove that

$$|E(e^{wS})| \le \exp\left(e^{|w|M} \operatorname{var}(S)\right)$$

uniformly in complex w.

Hint. $E(e^{wS})$ is the product of the $E(e^{wZ_j})$, $1 \le j \le k$, and $|e^s - 1 - s| \le \frac{1}{2}|s|^2 e^{|s|}$ for complex s.

2. Establish inequality (26.1).

Hint. For positive integers m,

$$E(S^{2m}) = \frac{(2m)!}{2\pi i} \int_{|w|=1} w^{-2m-1} E(e^{wS}) dw.$$

3. Prove that if the independent random variables λ_j, $j = 1, \ldots, k$, satisfy $0 \le \lambda_j \le M$, then

$$E\left(\left(\sum_{j=1}^{n} \lambda_j\right)^m\right) \le \left(mM + \sum_{j=1}^{n} E(\lambda_j)\right)^m$$

holds for all positive integers m.

Hint. Apply the binomial theorem and argue by induction on the number of random variables.

4. Let $\lambda_j = 1$ if $Y_j \ne 0$, $\lambda_j = 0$ otherwise, $j = 1, \ldots, k$. Prove that

$$E\left(\sum_{i=1}^{k} \lambda_i\right) \le 1,$$

and

$$E\left(\left(1 + \sum_{\substack{i=1\\ i\ne j}}^{k} \lambda_i\right)^{\alpha-1}\right) \le 2(\alpha+1)^{\alpha-1}$$

uniformly for $\alpha \ge 1$, $1 \le j \le k$.

5. Prove that

$$\left|\sum_{j=1}^{k} Y_j\right|^{\alpha} \leq \sum_{j=1}^{k} |Y_j|^{\alpha}\left(1+\sum_{\substack{i=1\\ i\neq j}}^{k} \lambda_i\right)^{\alpha-1}, \quad \alpha \geq 1,$$

and establish (26.2).

6. Prove that $(64(\alpha+2))^{\alpha}$ will serve as the implied constant in the inequality of Rosenthal.

If we abandon the definition of D, then an inequality of Manstavičius, [122], asserts that for each α, $1 \leq \alpha < 2$,

$$E\left(\left|\sum_{j=1}^{k} H_j\right|^{\alpha}\right) \ll \left(\sum_{j=1}^{k} E(X_j^2)\right)^{\alpha/2} + \sum_{j=1}^{k} E(|Y_j|^{\alpha})$$

uniformly in D. Our modifications for the cases $0 < \alpha \leq 2$ of Theorem 2.1 adapt here, too.

For distribution functions $F_j(z)$, $j = 1, \ldots, k$, and real t define

$$\delta(t) = \left(\sum_{j=1}^{k} \int_{|z|\leq t} z^2 dF_j(z)\right)^{\frac{1}{2}} + \left(\sum_{j=1}^{k} \int_{|z|>t} |z|^{\alpha} dF_j(z)\right)^{\frac{1}{\alpha}}.$$

7. Let $0 < \alpha \leq 2$. Prove that if $\delta(t)$ exists for all t, then $\delta(\delta(t)) \leq 2\delta(t)$.

8. Establish the inequality of Manstavičius.

9. Let $0 < \alpha < 1$. Prove that

$$E\left(\left|\sum_{j=1}^{k} H_j\right|^{\alpha}\right) = \left|\sum_{j=1}^{k} \bar{X}_j\right|^{\alpha} + O(\delta(t)^{\alpha}), \quad t > 0,$$

without any assumption concerning the mean of H_j.

All implied constants may be chosen independent of the variables H_j, k and t. A version of these inequalities for complex valued random variables may be obtained from the representation

$$\frac{1}{(m!)^2} E(|S|^{2m}) = -\frac{1}{4\pi^2} \int_{|w|=1} \frac{1}{w^{m+1}} \int_{|z|=1} \frac{1}{z} E\left(\exp\left(zS + \frac{w}{z}\bar{S}\right)\right) dz dw,$$

replacing the use of X_j by that of $zX_j + \bar{z}w\bar{X}_j$.

I sketched this proof of Rosenthal's inequality at the probability meeting workshop held in Northwestern University, Illinois, October 14, 1993.

Characteristically, I learned of the inequality of Manstavičius long after I had established the inequalities of Chapters 2 and 3.

27

Additive functions on arithmetic progressions: Small moduli

Theorem 27.1. *Let $1 < \alpha \le 2$, $A > 0$, $w = (\log x)^A$. Then*

$$\sum_{q \le w} q^{\alpha-1} \max_{(r,q)=1} \left| \sum_{\substack{n \le x \\ n \equiv r (\operatorname{mod} q)}} f(n) - \frac{1}{\phi(q)} \sum_{\substack{n \le x \\ (n,q)=1}} f(n) \right|^{\alpha}$$

$$\ll \frac{x^{\alpha}(\log\log x)^3}{(\log x)^{\alpha-1}} \sum_{\ell \le x} \frac{|f(\ell)|^{\alpha}}{\ell},$$

where q, ℓ denote prime powers, is valid uniformly in $x \ge 2$ and complex f.

Theorem 27.1, chosen for its simplicity of form, will play a rôle that in the theory of primes is usually assigned to the Siegel–Walfisz theorem. As the proof shows, a more natural version of Theorem 27.1 would employ the norm on $(L^2 \cap L^{\alpha'}(\mathbb{C}^s))'$. In particular, the power of $\log\log x$ may be reduced.

I shall obtain a version of Theorem 19.1 with the moduli q taken to be a fixed power of their prime q_0, and confined to an interval $Q^{1/2} < q \le Q$. Since there are $\ll \log\log x$ choices for the fixed power, and $\ll \log\log x$ intervals, it will then suffice to obtain $\log\log x$ rather than $(\log\log x)^3$ in the upper bound.

Consider the space F, defined immediately preceding the statement of Theorem 25.6. Let d denote the number of prime powers $q = q_0^k$ in the interval $(Q^{1/2}, Q]$, and let $V^{\beta}(\mathbb{C}^d)$ be the space of d-tuples g topologised by

$$\|g\|_{\beta} = \left(\sum_{Q^{1/2} < q \le Q} \frac{|g(q)|^{\beta}}{q} \left(1 - \frac{1}{q_0}\right) \right)^{1/\beta}.$$

Note that if $Q \le x$, then $V^{\beta}(\mathbb{C}^d)$ is a projection of $L^{\beta}(\mathbb{C}^s)$, with the induced norm. I consider certain functions which collapse F into the smaller space $\mathbb{C}^d$.

Let $S : F \to \mathbb{C}^d$ be given by

$$(h(q,r)) \to \left(\sum_r \theta_{q,r} h(q,r) \right).$$

Lemma 27.2. *Let $\alpha \geq \beta \geq 1$. Considering S between*

$$(\lambda^\alpha \cap \lambda^\beta(F))' \quad \text{and} \quad (V^\alpha \cap V^\beta(\mathbb{C}^d))'$$

we have

$$\|S\| \leq \max_{\tau=\alpha,\beta} \max_q \left(2 \sum_r |\theta_{q,r}|^\tau \right)^{1/\tau}.$$

Proof. It suffices to obtain a bound for the dual (and inflation) map

$$S' : V^\alpha \cap V^\beta(\mathbb{C}^t) \to \lambda^\alpha \cap \lambda^\beta(F),$$

given by

$$(g(q)) \to (\theta_{q,r} g(q)).$$

For any $\tau \geq 1$,

$$\left(\sum_{Q^{1/2} < q \leq Q} \sum_r \frac{1}{q} |\theta_{q,r} g(q)|^\tau \right)^{1/\tau}$$

$$\leq \left(\sum_{Q^{1/2} < q \leq Q} \frac{1}{q} \left(1 - \frac{1}{q_0} \right) |g(q)|^\tau 2 \sum_r |\theta_{q,r}|^\tau \right)^{1/\tau}.$$

We apply this with $\tau = \alpha, \beta$ to obtain the asserted inequality.

Proof of Theorem 27.1. For each prime power q not exceeding x, choose an integer r_q, $1 \leq r_q \leq q$, $(r_q, q) = 1$. Define the operator $D(x) : \mathbb{C}^s \to \mathbb{C}^s$ by

$$(D(x)f)(q) = \frac{q}{[x]} \left(\sum_{\substack{n \leq x \\ n \equiv r_q (\operatorname{mod} q)}} f(n) - \frac{1}{\phi(q)} \sum_{\substack{n \leq x \\ (n,q)=1}} f(n) \right).$$

Let P be the natural projection of $\mathbb{C}^s$ onto $\mathbb{C}^d$.

Let y be a positive real number, not exceeding x. We decompose f in $\mathbb{C}^s$ into the sum of two functions f_j, where $f_1(q) = f(q)$ if $q \leq y$, and $f_1(q) = 0$ otherwise.

On the one hand, a typical component of $D(x)f_1$ has the representation

$$\frac{q}{[x]} \sum_{\substack{\ell \le y \\ (\ell,q)=1}} f(\ell) \left(\sum_{\substack{m \le x/\ell \\ m\ell \equiv r_q (\mathrm{mod}\, q) \\ (m,\ell)=1}} 1 - \frac{1}{\phi(q)} \sum_{\substack{m \le x/\ell \\ (m,q)=1 \\ (m,\ell)=1}} 1 \right)$$

where in the two inner(most) sums m runs over positive integers. This innermost difference is bounded uniformly in ℓ, and by Hölder's inequality

$$\|PD(x)f_1\|_\alpha \ll \|f_1\|_\alpha \frac{wy}{x(\log w)^{1/\alpha}(\log y)^{1/\alpha'}},$$

with norms considered in $V^\alpha(\mathbb{C}^d)$ or $L^\alpha(\mathbb{C}^s)$.

On the other hand, a typical component of $D(x)f_2$ has the representation

$$\frac{q}{[x]} \sum_{\substack{m \le x/y \\ (m,q)=1}} \left(\sum_{\substack{\ell \le x/m \\ \ell m \equiv r_q (\mathrm{mod}\, q) \\ (\ell,m)=1}} f_2(\ell) - \frac{1}{\phi(q)} \sum_{\substack{\ell \le x/m \\ (\ell,q)=1 \\ (\ell,m)=1}} f_2(\ell) \right) = \sum_{m \le x/y} (\eta_m f_2)(q),$$

say. In terms of the norm on $(V^2 \cap V^{\alpha'}(\mathbb{C}^t))'$,

$$\|PD(x)f_2\| \le \sum_{m \le x/y} \|P\eta_m f_2\|.$$

For each integer m and prime power q, $(q,m)=1$, define $\bar{m}(=\bar{m}(q))$ by $m\bar{m} \equiv 1 (\mathrm{mod}\, q)$, $1 \le \bar{m} \le q$.

I approach the operator $P\eta_m$ from Theorem 25.9. The composition

$$(\mu^{\alpha'}(G))' \xrightarrow{\tilde{\omega}'} (\lambda^2 \cap \lambda^{\alpha'}(F))' \xrightarrow{\theta} (V^2 \cap V^{\alpha'}(\mathbb{C}^t))'$$

where θ is defined by $\theta(q,r)=1$ if $(q,m)=1$, $r \equiv \bar{m}(q)r_q (\mathrm{mod}\, q)$, $1 \le r \le q$, and $\theta(q,r)=0$ otherwise, takes a vector (b_ℓ), $x-\rho < \ell \le x$ to a vector $\mathbf{d}$ with q-coordinate

$$\frac{q \log x}{\rho} \left(\sum_{\substack{x-\rho<\ell \le x \\ \ell \equiv \bar{m} r_q (\mathrm{mod}\, q)}} b_\ell - \frac{1}{\phi(q)} \sum_{x-\rho<\ell\le x} b_\ell \right).$$

I set $\rho = x$. The circumstances of Lemma 27.2 hold with $\|S\| \le 2$ and, in $(V^2 \cap V^{\alpha'}(\mathbb{C}^d))'$,

$$\|\mathbf{d}\| \ll (1 + x^{-1+\varepsilon} Q^{(2\alpha'+3)\alpha'})^{1/\alpha'} \left(\frac{\log x}{x} \sum_{\ell \le x} |b_\ell|^\alpha \right)^{1/\alpha}.$$

In our present situation Q is at most a power of $\log x$, and

$$\|\mathbf{d}\| \ll (\log x)^{1/\alpha} \left(\sum_{\ell \le x} \ell^{-1} |b_\ell|^\alpha \right)^{1/\alpha}.$$

Replacing x in this argument by x/m, and bearing in mind the remark made immediately following the proof of Theorem 25.9, we see that

$$\|P\eta_m f_2\| \ll m^{-1} (\log x)^{-1/\alpha'} \|f_2\|_\alpha,$$

uniformly for $m \le x^{2/3}$ (say). If $y \ge x^{1/3}$, then

$$\|PD(x) f_2\| \ll (\log x)^{-1/\alpha'} \log(x/y) \|f_2\|_\alpha.$$

For a suitably large value of c, the choice $y = x(\log x)^{-c}$ gives

$$\|PD(x) f\| \ll (\log x)^{-1/\alpha'} \log\log x \|f\|_\alpha.$$

Since every vector g in $(V^2 \cap V^{\alpha'}(\mathbb{C}^d))'$ satisfies

$$\|g\|_\alpha \ll \|g\| \left(\sum_{Q^{1/2} < q \le Q} \frac{1}{q} \right)^{1/\alpha - 1/2},$$

Theorem 27.1 is established.

A closer examination shows that the proof of Theorem 27.1 yields a non-trivial result only so long as $w < \exp(2(\log x)^{1/\alpha'})$.

As I indicate in Chapter 29, it is possible to introduce into Theorem 27.1 an extra uniformity.

28

Additive functions on arithmetic progressions: Large moduli

Theorem 28.1. *Let $1 < \alpha \le 2$. Then there is a positive δ so that*

$$\sum_{q \le x^\delta} q^{\alpha-1} \max_{(r,q)=1} \left| \sum_{\substack{n \le x \\ n \equiv r (\mathrm{mod}\, q)}} f(n) - \frac{1}{\phi(q)} \sum_{\substack{n \le x \\ (n,q)=1}} f(n) \right|^\alpha$$

$$\ll \frac{x^\alpha (\log\log x)^{3\alpha/2+2}}{(\log x)^{\alpha-1}} \sum_{\ell \le x} \frac{|f(\ell)|^\alpha}{\ell}$$

uniformly in $x \ge 2$ and complex valued additive functions f. Any δ which satisfies $\delta\alpha'^2(2\alpha'+3) < 1$ is valid. For $\alpha > 2$ there is a similar result, with upper bound factor $x^\alpha (\log\log x)^{3+\alpha}(\log x)^{-1}$, valid whenever $28\delta < 1$.

No attempt will be made to obtain larger values of δ. The case $\alpha = 2$ has been studied in some detail; $2\delta < 1$ already guarantees the validity of a version of Theorem 28.1, [60]. In particular, the power of $\log\log x$ may be reduced. However, the investigation of an underlying operator (for example, see [56] p. 139) shows that if $\alpha = 2$, then no matter the value of δ, there exists an additive function f, not identically zero, so that the upper bound factor in Theorem 28.1 exceeds a constant multiple of $x^2(\log x)^{-1}$. Moreover, such a restriction holds for every sufficiently large x.

I reappraise Theorem 28.1 in terms of shift operators, in Chapter 30.

In part the proof of Theorem 28.1 employs Fourier analysis in the complex plane, in particular the Mellin transform

$$\frac{1}{2\pi i} \int_{\theta - i\infty}^{\theta + i\infty} \frac{y^z}{z}\, dz = \begin{cases} 1 & \text{if} \quad y > 1, \\ 0 & \text{if} \quad 0 < y < 1, \end{cases}$$

valid for real $y > 0$, and complex z integrated along the line $\mathrm{Re}\,(z) = \theta > 0$.

It is convenient to introduce the kernel

$$K(v, \rho, z) = \frac{v^{z+1} - (v-\rho)^{z+1}}{\rho z(z+1)},$$

defined for $v \ge \rho > 0$, and complex $z = \sigma + i\tau$, $\sigma = \mathrm{Re}\,(z)$, $z \ne 0, -1$.

Lemma 28.2. *The inequality*

$$\int_{\sigma-i\infty}^{\sigma+i\infty} \sup_{\rho\le v\le x} |K(v,\rho,z)| d\tau \ll x^\sigma \log\left(\frac{xe}{\rho\sigma}\right)$$

holds uniformly for $0<\rho\le x$, $x\ge 1$, $\sigma>0$.

Proof. Let $T>0$. Over the range $|\tau|>T$ we employ the direct bound

$$|K(v,\rho,z)| \ll \frac{x^{\sigma+1}}{\rho|z(z+1)|} \ll \frac{x^{\sigma+1}}{\rho\tau^2}, \qquad \rho\le v\le x.$$

Otherwise

$$|K(v,\rho,z)| = \left|\frac{1}{\rho z}\int_{v-\rho}^{v} y^z dy\right| \ll \frac{x^\sigma}{|z|},$$

so that the range $|\tau|\le\sigma$ contributes

$$\ll \int_{|\tau|\le\sigma} \frac{x^\sigma}{\sigma}\, d\tau \ll x^\sigma,$$

whilst the range $\sigma<|\tau|\le T$ (if nonempty) contributes

$$\ll \int_{\sigma<|\tau|\le T} \frac{x^\sigma}{|\tau|}\, d\tau \ll x^\sigma \log\left(\frac{T}{\sigma}\right).$$

The choice $T = x\rho^{-1}$ completes the proof.

Let k and t be positive integers, $1\le t\le k$. Define

$$D(x,t,z) = D(x,k,t,z) = \sum_{\substack{m\le x\\ m\equiv t(\mathrm{mod}\, k)}} m^{-z} - k^{-z}\sum_{m\le x/k} m^{-z} - t^{-z},$$

for complex numbers z, $\beta = \mathrm{Re}\,(z)$, $0<\beta<1$.

Lemma 28.3. *Let* $0<\varepsilon<1$, $\tau>0$. *Then*

$$\sum_{t=1}^{k} |D(x,t,z)|^\tau \ll k(k^{-\beta}|z|^{1-\beta} + x^{-\beta})^\tau$$

uniformly for $\varepsilon\le\beta\le 1-\varepsilon$, $k\ge 1$ *and* $x\ge 1$.

Proof. I shall prove that with the same uniformities each $D(x,t,z)$ is $\ll k^{-\beta}|z|^{1-\beta} + x^{-\beta}$.

Setting $m = t + vk$ in the first of the sums defining $D(x, t, z)$ we obtain for it the representation

$$\sum_{1\le v\le xk^{-1}} (t+vk)^{-z} + t^{-z} + O(x^{-\beta}).$$

For $v \ge 2$

$$(t+vk)^{-z} = (vk)^{-z}\{1+tv^{-1}k^{-1}\}^{-z} = (vk)^{-z} + O(|z|k^{-\beta}v^{1-\beta}).$$

Hence

$$\sum_{|z|<v\le xk^{-1}} \{(t+vk)^{-z} - (vk)^{-z}\} \ll |z|k^{-\beta} \sum_{v>|z|} v^{-1-\beta} \ll k^{-\beta}|z|^{1-\beta}.$$

Moreover, ignoring any cancellation

$$\sum_{v\le|z|} \{(t+vk)^{-z} - (vk)^{-z}\} \ll k^{-\beta} \sum_{v\le|z|} v^{-\beta} \ll k^{-\beta}|z|^{1-\beta}.$$

Lemma 28.3 is now clear.

For each prime power $q \le x$, let r_q be an integer prime to q, $0 \le r_q < q$. For complex z, $\sigma = \mathrm{Re}(z)$, and real y, $0 < y \le x$, define an operator $Y(y) : \mathbb{C}^s \to \mathbb{C}^s$ by taking $(Y(y)f)(q)$ to be

$$\frac{q}{[x]}\left(\sum_{\substack{n\le y\\ n\equiv r_q \,(\mathrm{mod}\, q)}} \frac{f(n)}{n^z} - \frac{1}{\phi(q)} \sum_{\substack{n\le y\\ (n,q)=1}} \frac{f(n)}{n^z}\right).$$

Let $Q \le x^\delta \le x$, and define a projection $P : \mathbb{C}^s \to \mathbb{C}^s$ by $f(q) \to f(q)$ if $Q^{1/2} < q \le Q$, and otherwise $f(q) \to 0$.

The lion's share of Theorem 28.1 is contained in

Lemma 28.4. *Let $0 < \varepsilon < 1$, $1 < \alpha \le 2$. If Q exceeds a suitable power of $\log x$, then*

$$\|PY(x)\| \ll \frac{(\log\log x)^{1/\alpha+3/2}}{x^\sigma(\log x)^{1/\alpha'}}\,|z|^{1-\sigma}$$

uniformly for $(1/\alpha)+\varepsilon \le \sigma \le 1-(2\alpha'+3)\alpha'\delta-\varepsilon$, and $x \ge 2$, the underlying norm(s) being that of $L^\alpha(\mathbb{C}^s)$.

This result is, of course, vacuous unless $\delta(\alpha')^2(2\alpha'+3) < 1$.

Proof. We view f as a convolution of two arithmetic functions, one of which is supported on the prime powers. We separate these functions using Fourier analysis.

Define

$$\Delta_1(q,r,z) = \sum_{\substack{\ell\le x\\ \ell\equiv r(\operatorname{mod} q)}} \frac{f(\ell)}{\ell^z} - \frac{1}{\phi(q)} \sum_{\substack{\ell\le x\\ (\ell,q)=1}} \frac{f(\ell)}{\ell^z},$$

$$\Delta_2(q,r,z) = \sum_{\substack{\ell\ell_0\le x\\ \ell\ell_0\equiv r(\operatorname{mod} q)}} \frac{f(\ell)}{(\ell\ell_0)^z} - \frac{1}{\phi(q)} \sum_{\substack{\ell\ell_0\le x\\ (\ell,q)=1}} \frac{f(\ell)}{(\ell\ell_0)^z},$$

and

$$\Delta(q,r,z) = \frac{q}{[x]}\left(\Delta_1(q,r,z) - \Delta_2(q,r,z)\right).$$

Here, ℓ denotes a prime power. We may regard $\Delta(q,r,z)$ as defining an operator Δ on f in $\mathbb{C}^s$, taking values in $\mathbb{C}^s$.

For $0 < y \le x$ consider the operator $Z(y) : \mathbb{C}^s \to \mathbb{C}^s$, for which $(Z(y)f)(q)$ is

$$\frac{1}{2\pi i}\int_{\theta-i\infty}^{\theta+i\infty} \sum_{\substack{r=0\\ (r,q)=1}}^{q-1} D(x,q,t_r,z+\psi)\Delta(q,r,z+\psi)\frac{y^\psi}{\psi}\, d\psi,$$

where ψ denotes a complex variable, and the integration is along a line $\operatorname{Re}(\psi) = \theta > 0$. Here t_r $(= t_r(q))$ is determined by the conditions

$$rt_r \equiv r_q(\operatorname{mod} q), \qquad 1 \le t_r \le q.$$

Since

$$\sum_{\substack{r=0\\ (r,q)=1}}^{q-1} \Delta(q,r,z+\psi) = 0,$$

the integrand may be replaced by

$$\sum_{\substack{r=0\\ (r,q)=1}}^{q-1} \Delta(q,r,z+\psi)\left(\sum_{\substack{m\le x\\ m\equiv t_r(\operatorname{mod} q)}} \frac{1}{m^{z+\psi}} - \frac{1}{t_r^{z+\psi}}\right)\frac{y^\psi}{\psi}.$$

Interchanging the order of summation(s) and integration shows that

$$Z(y) = Y(y) - \sum_{j=1}^{2} V_j(y),$$

where the q-coordinate of $V_1(y)f$ is

$$\frac{q}{[x]} \sum_{\substack{r=0 \\ (r,q)=1}}^{q-1} \frac{1}{t_r^z} \left(\sum_{\substack{\ell t_r \leq y \\ \ell \equiv r (\mathrm{mod}\, q)}} \frac{f(\ell)}{\ell^z} - \frac{1}{\phi(q)} \sum_{\substack{\ell t_r \leq y \\ (\ell,q)=1}} \frac{f(\ell)}{\ell^z} \right),$$

and that of $V_2(y)f$ is similar, with every ℓ, save those in the $f(\ell)$, replaced by $\ell\ell_0$.

We smooth by integrating over the interval $x - \rho < y \leq x$:

$$\begin{aligned} &\frac{1}{\rho}\int_{x-\rho}^{x} (Z(y)f)(q)dy \\ &= \frac{1}{2\pi i}\int_{\theta-i\infty}^{\theta+i\infty} K(x,\rho,\psi) \sum_{\substack{r=0 \\ (r,q)=1}}^{q-1} D(x,q,t_r,z+\psi)\Delta(q,r,z+\psi)d\psi. \end{aligned}$$

In terms of the norm on $(L^2 \cap L^{\alpha'}(\mathbb{C}^s))'$, and in an obvious notation,

$$\left\| \frac{1}{\rho}\int_{x-\rho}^{x} PZ(y)fdy \right\| \leq \frac{1}{2\pi}\int_{-\infty}^{\infty} |K| \left\| P\sum_r D_r\Delta_r f \right\| d(-i\mathrm{Im}\,(\psi)).$$

In preparation for an application of Lemma 27.2, we note that if $\varepsilon_0 > 0$, then by Lemma 28.3

$$\left(\sum_{\substack{r=0 \\ (r,q)=1}}^{q-1} |D(x,q,t_r,z+\psi)|^\tau \right)^{1/\tau} \ll q^{\frac{1}{\tau}-\beta}|z+\psi|^{1-\beta}$$

with $\beta = \mathrm{Re}\,(z+\psi)$ holds uniformly for $\varepsilon_0 \leq \beta \leq 1-\varepsilon_0$, $1 \leq q \leq x$. If θ is fixed at a sufficiently small positive value, then the condition $(1/\alpha) + \varepsilon \leq \sigma \leq 1-\varepsilon$ on σ is enough to ensure that this upper bound is $\ll Q^{-c}|z+\psi|^{1-\sigma}$ for some positive c, uniformly for $\tau = 2$, α, $Q^{1/2} < q \leq Q$, $x \geq 2$.

For each positive integer k let P_k denote the projection $\mathbb{C}^s \to \mathbb{C}^s$ which leaves $f(q)$ invariant if $Q^{1/2} < q \leq Q$ and $q = q_0^k$, and otherwise takes $f(q)$ to zero. Thus P is the sum of the P_k, $k \leq \log Q/\log 2$.

By Lemma 27.2, with the rôle of $\theta_{q,r}$ played by $D(x,q,t_r,z+\psi)$,

$$\left\| P_k\sum_r D_r\Delta_r f \right\| \ll \frac{|z+\psi|^{1-\sigma}}{Q^c} \,\|\Delta(q,r,z+\psi)\|,$$

where this last norm is viewed as in $(\lambda^2 \cap \lambda^{\alpha'}(F))'$. We estimate this norm by applications of Theorem 25.9.

Define j_0 by $2^{j_0} \le x^\beta < 2^{j_0+1}$. For $0 \le j \le j_0$ let Γ_j be the operator defined like Δ but with the range $\ell \le x$ (or $\ell\ell_0 \le x$ as the case may be) replaced by $x2^{-j-1} < \ell \le x2^{-j}$. Γ_{j_0+1} is defined like Δ but with the range $\ell \le x2^{-j_0-1}$. Then Δ is the sum of the Γ_j, $0 \le j \le j_0+1$.

Since $(2\alpha'+3)\alpha'\delta < 1-\beta$, again θ assumed sufficiently small in terms of ε, an application of Theorem 25.9 shows that

$$\begin{aligned}\|\Gamma_j f\| &\ll \frac{1}{2^j \log x}\left(\frac{\log(2^{-j-1}x)}{2^{-j-1}x} \sum_{x2^{-j-1}<\ell\le x2^{-j}} |f(\ell)|^\alpha \ell^{-\alpha\beta}\right)^{1/\alpha} \\ &\ll (2^{j(1-\beta)}x^\beta(\log x)^{1/\alpha'})^{-1}\|f\|_\alpha,\end{aligned}$$

uniformly for $0 \le j \le j_0$. The combined contribution of these $\Gamma_j f$ to Δf is

$$\ll (x^\beta(\log x)^{1/\alpha'})^{-1}\|f\|_\alpha \sum_{j\ge 0} 2^{-j(1-\beta)}.$$

Moreover,

$$\begin{aligned}\|\Gamma_{j_0+1} f\| &\ll \frac{1}{2^{j_0+1}\log x}\left(\frac{\log(2^{-j_0}x)}{2^{-j_0}x} \sum_{\ell\le x2^{-j_0-1}} |f(\ell)|^\alpha \ell^{-\alpha\beta}\right)^{1/\alpha} \\ &\ll (x^\beta(\log x)^{1/\alpha'})^{-1}\|f\|_\alpha,\end{aligned}$$

the factors $\ell^{-\alpha\beta}$ being abandoned.

Altogether

$$\|\Delta f\| = \|\Delta(q,r,z+\psi)\| \ll (x^\beta(\log x)^{1/\alpha'})^{-1}\|f\|_\alpha,$$

uniformly for $\mathrm{Re}\,(z)=\sigma$, $\mathrm{Re}\,(\psi)=\theta$, $x\ge 2$.

Employing the direct bound for K,

$$\int_{-\infty}^{\infty} |K||z+\psi|^{1-\sigma} d(-i\mathrm{Im}\,\psi) \ll \rho^{-1}x^{1+\theta}|z|^{1-\sigma},$$

and we see that

$$\left\|\frac{1}{\rho}\int_{x-\rho}^{x} P\left(Y(y)-\sum_{j=1}^{2} V_j(y)\right) f\,dy\right\| \ll x^{-\sigma}\left(\frac{x|z|^{1-\sigma}}{\rho Q^c}\right)(\log x)^{1/\alpha}\|f\|_\alpha,$$

the left norm being that of $(L^2 \cap L^{\alpha'}(\mathbb{C}^s))'$. I shall ultimately choose ρ to have the form $x(\log x)^{-d}$ for a fixed positive d. If Q exceeds a suitable power of $\log x$, the power depending upon α, c, d only, then this upper bound will be $\ll (x^\sigma(\log x)^2)^{-1}|z|^{1-\sigma}$, well within the asserted bound of Lemma 28.4.

The operators $V_j(y)$ warrant study in their own right. The treatment which I give here is designed to reach a nontrivial result rapidly.

The average of the operator $V_1(y)$ has a representation

$$\frac{1}{\rho}\int_{x-\rho}^{x} V_1(y)f(q)dy = \frac{1}{2\pi i}\int_{\theta-i\infty}^{\theta+i\infty} K(x,\rho,\psi) \sum_{\substack{r=0\\(r,q)=1}}^{q-1} \frac{1}{t_r^{z+\psi}}\Delta(q,r,z+\psi)d\psi.$$

We may treat the integral of $PV_1(y)$ much as we did that of $PZ(y)$. The rôle of D_r is here played by $t_r^{-z-\psi}$. Then

$$\left(\sum_{\substack{r=0\\(r,q)=1}}^{q-1} |t_r^{-z-\psi}|^\tau\right)^{1/\tau} \ll 1$$

uniformly for $\mathrm{Re}(\psi) \ge 0$, $\mathrm{Re}(z) = \sigma \ge (1/\alpha)+\varepsilon$, $\tau = 2, \alpha$, since as r traverses the reduced residue classes $(\mathrm{mod}\, q)$, so (in possibly another order) does t_r. For each positive integer k we obtain the bounds

$$\begin{aligned}\left\|\frac{1}{\rho}\int_{x-\rho}^{x} P_k V_1(y) f dy\right\| &\ll (x^\beta(\log x)^{1/\alpha'})^{-1}\|f\|_\alpha \int_{-\infty}^{\infty} |K| d(-i\mathrm{Im}\,\psi)\\ &\ll (x^\sigma(\log x)^{1/\alpha'})^{-1}\|f\|_\alpha \log(x/\rho),\end{aligned}$$

the second step by Lemma 28.2.

A similar estimate holds with $V_2(y)$ in place of $V_1(y)$.

Whilst this is a sufficiently good result for an individual P_k, there are about $\log x$ possible values of k, and a multiplicity factor $\log x$ would swamp the improvement factor $(\log x)^{-1/\alpha'}$. We experience here in a different form the difficulty, noted in Chapter 23, of the multiplicities arising when sums over Dirichlet characters are reduced to sums over primitive Dirichlet characters. Let $h = [B \log\log Q] + 1$, where B is a positive number, shortly to be chosen independent of x, Q. I shall show that

$$\left\|\sum_{k=h}^{\infty} P_k V_j(y) f\right\| \ll (x^\sigma(\log x)^{1/\alpha'})^{-1}(\log\log x)^{1/\alpha}\|f\|_\alpha \tag{28.1}$$

uniformly for $j = 1, 2$, $y \le x$. Granted this result we shall have

$$\left\| \frac{1}{\rho} \int_{x-\rho}^{x} PY(y) f \, dy \right\| \ll (x^\sigma (\log x)^{1/\alpha'})^{-1} \|f\|_\alpha (\log\log x)^2 |z|^{1-\sigma}.$$

We can then strip the averaging from the operator $PY(y)$.

$$\left\| PY(x)f - \frac{1}{\rho} \int_{x-\rho}^{x} PY(y) f \, dy \right\| \le \frac{1}{\rho} \int_{x-\rho}^{x} \|P(Y(x) - Y(y))f\| dy.$$

For each y we treat $P(Y(x) - Y(y))$ via Lemma 27.2 and Theorem 25.5. In the present circumstances $x - y \le x(\log x)^{-d}$, and

$$\begin{aligned} \|P(Y(x) - Y(y))f\| &\ll \frac{(x-y)}{x} \left(\frac{1}{(x-y)} \sum_{x-y<n\le x} \frac{|f(n)|^\alpha}{n^{\sigma\alpha}} \right)^{1/\alpha} \\ &\ll x^{-\sigma} (\log x)^{-d/\alpha'} \left(x^{-1} \sum_{n \le x} |f(n)|^\alpha \right)^{1/\alpha}. \end{aligned}$$

With

$$\mu = \sum_{q \le x} \frac{1}{q} \left(1 - \frac{1}{q_0} \right) f(q)$$

the mean α-power of $|f(n)|$ is

$$\ll |\mu| + \left(x^{-1} \sum_{n \le x} |f(n) - \mu|^\alpha \right)^{1/\alpha}.$$

This last norm is that of $A_\alpha f$, where A_α is the uniformly bounded operator $(L^2 \cap L^{\alpha'}(\mathbb{C}^s))' \to M^\alpha(\mathbb{C}^t)$ which underlies the general Turán–Kubilius inequality obtained in Chapter 2. By Hölder's inequality and an estimate from elementary number theory, $\mu \ll \|f\|_\alpha (\log\log(4Q))^{1/\alpha'}$. If d is fixed at a sufficiently large value

$$\|P(Y(x) - Y(y))f\| \ll (x^\sigma (\log x)^2)^{-1} \|f\|_\alpha$$

uniformly for y in $(x - x(\log x)^{-d}, x]$, and

$$\|PY(x)f\| \ll (x^\sigma (\log x)^{1/\alpha'})^{-1} (\log\log x)^2 |z|^{1-\sigma}.$$

Here the norm is on $(L^2 \cap L^{\alpha'}(\mathbb{C}^s))'$ and may be changed to the norm on $L^\alpha(\mathbb{C}^s)$ at the expense of a factor $\ll (\log\log x)^{1/\alpha - 1/2}$.

To complete the proof of Lemma 28.4 it will suffice to establish the bound (28.1).

The sum

$$\frac{q}{[x]} \sum_{\substack{r=0 \\ (r,q)=1}}^{q-1} \frac{1}{t_r^z} \frac{1}{\phi(q)} \sum_{\substack{\ell t_r \le y \\ (\ell,q)=1}} \frac{f(\ell)}{\ell^z}$$

is readily estimated by Hölder's inequality to be

$$\begin{aligned} &\ll \frac{1}{x} \sum_r \frac{1}{t_r^\sigma} \left(\sum_{\ell t_r \le y} (\ell^{1/\alpha-\sigma})^{\alpha'} \right)^{1/\alpha'} \|f\|_\alpha \\ &\ll (x^\sigma (\log x)^{1/\alpha'})^{-1} \|f\|_\alpha. \end{aligned}$$

The corresponding contribution to the norm at (28.1) is

$$\ll (x^\sigma (\log x)^{1/\alpha'})^{-1} \|f\|_\alpha \Big(\sum q^{-1} (\log q)^\alpha \Big)^{1/\alpha}$$

where the prime-power sum is over those $q = q_0^k$ with $k \ge [B \log\log Q] + 1 \ge 2$, and so bounded.

We are reduced to essentially the following. Define the operator $U : \mathbb{C}^s \to \mathbb{C}^s$ by

$$(Uf)(q) = \frac{q}{x} \sum_{\ell t_\ell \le x} \frac{f(\ell)}{(\ell t_\ell)^\sigma},$$

with t_ℓ defined $(\bmod\, q)$ as earlier in this proof. Again $h = [B \log\log Q] + 1$.

Lemma 28.5. *Let $1 < \alpha \le 2$, $\varepsilon > 0$. Then for a suitable $B > 0$*

$$\left\| \sum_{k=h}^\infty P_k U f \right\|_\alpha \ll x^{-\sigma} ((\log x)^{-1} + Q x^{-1+\varepsilon})^{1/\alpha'} (\log\log Q)^{1/\alpha} \|f\|_\alpha$$

uniformly for $x \ge 2$, $Q \ge 4$.

Proof of Lemma 28.5. The following argument is slightly wasteful. I rapidly overfly the functional analytic foundation of the cases $k < h$.

It will suffice to establish the inequality of Lemma 28.5 for the values $2m(2m-1)^{-1}$ of α, $m = 1, 2, \ldots$. For the remaining values it may then be obtained by applying the Riesz–Thorin interpolation theorem.

The case $\alpha = 2m(2m-1)^{-1}$ I deal with by considering the dual inequality

$$\left(\sum_{\ell \le x} \frac{\ell^{2m(1-\sigma)-1}}{x^{2m}} \left| \sum_{q \le Q}{}' \frac{d_q}{t_\ell(q)^\sigma} \right|^{2m} \right)^{1/(2m)} \le \lambda \left(\sum_{q \le Q}{}' \frac{|d_q|^{2m}}{q} \right)^{1/(2m)},$$

where $'$ indicates that the prime powers $q = q_0^k$ satisfy $k \ge h$. The restriction $Q^{1/2} < q$ may be dropped without hazard. This dual inequality I shall establish with a $\lambda \ll x^{-\sigma}((\log x)^{-1} + Qx^{-1+\varepsilon})^{1/(2m)}(\log\log Q)^{1-1/(2m)}$.

I divide the moduli q into h classes, one for each b, $0 \le b < h$. A typical class contains the moduli $q = q_0^{ah+b}$, $a = 1, 2, \ldots$. Within a particular class moduli $q' > q$ are well spaced in the sense that $q'/q \ge q_0^h$. It will suffice to establish the dual inequality with the moduli confined to a particular class, and the sharper value $\lambda \ll x^{-\sigma}((\log x)^{-1} + Qx^{-1+\varepsilon})^{1/(2m)}$.

Without loss of generality I assume the d_q to be real and non-negative.

Expanding the $2m^{\text{th}}$ power of the sum over the $q \le Q$, we are reduced to estimating

$$R = \sum_{q_1 \le Q} \cdots \sum_{q_{2m} \le Q} d_{q_1} \cdots d_{q_{2m}} \sum_{(r_1, q_1)=1} \cdots$$

$$\cdots \sum_{(r_{2m}, q_{2m})=1} \prod_{j=1}^{2m} \frac{1}{t_{r_j}(q_j)^\sigma} \sum_{\ell} \frac{\ell^{2m(1-\sigma)-1}}{x^{2m}},$$

where the innermost sum is over the prime powers ℓ which satisfy $\ell \equiv r_j (\operatorname{mod} q_j)$, $j = 1, \ldots, 2m$, and

$$\ell \max_{1 \le j \le 2m} t_{r_j}(q_j) \le x.$$

In view of the remarks following Lemma 25.7, the innermost sum is

$$\ll x^{-2m\sigma}\{(\log x)^{-1} + Qx^{-1+\varepsilon}\}([q_1, \ldots, q_{2m}] \max_j t_{r_j}(q_j)^{2m(1-\sigma)})^{-1}.$$

Employing the wasteful inequality

$$\max_j t_{r_j}(q_j)^{2m} \ge \prod_{i=1}^{2m} t_{r_i}(q_i),$$

we see that

$$R \ll x^{-2m\sigma}\{(\log x)^{-1} + Qx^{-1+\varepsilon}\} T$$

with

$$T = \sum_{\substack{q_j \le Q \\ j=1,\dots 2m}} \cdots \sum d_{q_1} \cdots d_{q_{2m}} \sum_{(r_j,q_j)=1} \cdots \sum \prod_{j=1}^{2m} t_{r_j}(q_j)^{-1}[q_1, \dots q_{2m}]^{-1}.$$

I estimate T using the theory of probability.

I construct a model on the lines of the model employed in Lemma 25.2. D will be the product of the prime powers $q_0^{k_0}$ with $k_0 = [\log Q/\log q_0]$. The probability space consists of the residue classes prime to D, with measure $\phi(D)^{-1}$ assigned to each class. This may be interpreted as the natural density

$$\lim_{y\to\infty} \frac{\log y}{y} \sum_{\substack{p\le y \\ p\equiv r (\text{mod } D)}} 1, \qquad (r, D) = 1.$$

The arithmetic function $\psi(n, q, r)$ is defined to be 1 if $n \equiv r (\text{mod}\, q)$, and zero otherwise. The random variable

$$M(q) = d_q \sum_{\substack{r=1 \\ (r,q)=1}}^{q} \psi(n, q, r) t_r(q)^{-1}$$

assumes the value $d_q t_r(q)^{-1}$ with probability $\phi(q)^{-1}$, for each r prime to q, $1 \le r \le q$. For sets of distinct primes q_0 the random variables

$$L_{q_0} = \sum_i M(q_0^{k_i}),$$

taken over a selection of the integers k in the interval $[1, [\log Q/\log q_0]]$, are independent. In the present situation I shall confine the powers $q_0^{k_i}$ to one of the given classes defined earlier.

In terms of the expectation on this probability space,

$$T = E\left(\left(\sum_{q_0\le Q} L_{q_0}\right)^{2m}\right).$$

This expression in turn is

$$\ll E\left(\left(\sum_{q_0\le Q} \{L_{q_0} - E(L_{q_0})\}\right)^{2m}\right) + \left(\sum_{q_0\le Q} E(L_{q_0})\right)^{2m}. \tag{28.2}$$

Again $t_r(q)$ traverses the reduced residue classes $(\bmod q)$ with r. Therefore

$$\sum_{q_0 \le Q} E(L_{q_0}) \ll \sum_{q \le Q}{}' d_q q^{-1} \log q \ll \left(\sum |d_q|^{2m} q^{-1} \right)^{1/(2m)},$$

the last step by Hölder's inequality, noting that the expression

$$\left(\sum_{q \le Q}{}' (\log q)^\gamma q^{-1} \right)^{1/\gamma}$$

with $1/\gamma + 1/\alpha = 1$, is bounded. Here I again employ the restriction $h \ge 2$. To the random variables $L_{q_0} - E(L_{q_0})$ I apply Lemma 25.3. The first sum at (28.2) is thus

$$\ll \sum_{q_0} E(\{L_{q_0} - E(L_{q_0})\}^{2m}) + \left(\sum_{q_0} E(\{L_{q_0} - E(L_{q_0})\}^2) \right)^m.$$

Typically $E(L_{q_0}^{2m})$ has the same form as the sum T, save that the q_j are all powers of the same prime q_0. If the q_i are equal, then we get a contribution of

$$\ll \sum d_q^{2m} q^{-1}.$$

For the congruence conditions $n \equiv r_j (\bmod q_j)$ are then only compatible if all the r_j are equal, so that

$$\sum_{\substack{(r_j, q_j)=1 \\ j}} \cdots \sum \prod_{j=1}^{2m} t_{r_j}(q_j)^{-1} \le \sum_{\substack{r=1 \\ (r,q)=1}}^{q} r^{-2m} \ll 1.$$

Otherwise we may assume that $q_1 \le q_2 \le \cdots \le q_{2m}$, and that not all the q_j are equal. Typically

$$\sum_{r_j} \cdots \sum \prod_{j=1}^{2m} t_{r_j}(q_j)^{-1} \le \left(\sum_{r < q_{2m}} r^{-1} \right)^{2m} \ll (\log Q)^{2m}.$$

Here $[q_1, \dots, q_{2m}] = q_{2m} \ge q_0^{h/(2m)} \prod_{j=1}^{2m} q_j^{1/(2m)}$. The corresponding terms of $E(L_{q_0}^{2m})$ contribute

$$\ll 2^{-h/(2m)} \left(\sum_{q \le Q}{}' d_q q^{-1/(2m)} \right)^{2m} (\log Q)^{2m}$$

$$\ll 2^{-h/(2m)} (\log Q)^{4m-1} \sum_{q \le Q}{}' d_q^{2m} q^{-1},$$

the q all powers of the same q_0. Since $2h \log 2 > h > B \log\log Q$, any choice $B \geq 4m(4m-1)$ ensures that the leading factor in this upper bound is at most 1.

The terms

$$(E(L_{q_0}))^{2m}, \qquad (E(\{L_{q_0} - E(L_{q_0})\}^2))^m$$

may be similarly but more simply dealt with.

This completes the proof of Lemma 28.5, and with it the proof of Lemma 28.4.

Remark. Given $A > 0$, if Q exceeds a certain power of $\log x$, then the proof of Lemma 28.4 yields

$$\|PY(x)\| \ll \frac{(\log\log x)^{1/\alpha+3/2}}{x^\sigma (\log x)^{1/\alpha'}} + \frac{|z|^{1-\sigma}}{x^\sigma (\log x)^A},$$

so that the dependence of the estimate upon z is rather weak.

A more natural version of Lemma 28.4 would involve the norm on $(L^2 \cap L^{\alpha'}(\mathbb{C}^s))'$, and the proof suggests this.

Note that the proof of Lemma 28.5 does not need the good distribution of primes in residue classes to large moduli, only a uniform upper bound of the correct order.

Proof of Theorem 28.1; the cases $1 < \alpha \leq 2$. I begin with an argument based on general principles. Consider the composition of maps

$$(L^2 \cap L^{\alpha'}(\mathbb{C}^s))' \xrightarrow{A_\alpha} M^\alpha(\mathbb{C}^t) \simeq (K^{\alpha'}(\mathbb{C}^v))' \longrightarrow$$

$$\longrightarrow (J^2 \cap J^{\alpha'}(\mathbb{C}^u))' \longrightarrow (V^2 \cap V^{\alpha'}(\mathbb{C}^d))'.$$

Here we identify $\mathbb{C}^t$ and $\mathbb{C}^v$ in the second and third spaces by setting $\rho = x$ in the definition of the $K^\beta(\mathbb{C}^v)$. The map into the dual space on $\mathbb{C}^u$ is the $\widetilde{W}'$ defined immediately preceding Theorem 25.5, modified according to the remark made following the statement of that theorem. The final map is a collapse onto the appropriate residue classes $r_q(\operatorname{mod} q)$, constructed as in Lemma 27.2. We may regard this collapse as involving a projection from $(J^2 \cap J^{\alpha'}(\mathbb{C}^u))'$ to $(\lambda^2 \cap \lambda^{\alpha'}(F))'$ and modify Lemma 27.2, or insert such a projection into the above chain of maps and apply Lemma 27.2 directly.

For $(2\alpha' + 3)\alpha'\delta < 1$, each of these maps represents a linear operator bounded uniformly in $x \geq 2$. Moreover, since the sum $\sum q^{-1}$ taken over the prime powers $q = q_0^k$ in the interval $(Q^{1/2}, Q]$ is bounded uniformly in Q and k, for functions g in $\mathbb{C}^s$ supported on such prime powers only, $\|g\|_\alpha \ll \|g\|$, where the upper bound norm is that of $(L^2 \cap L^{\alpha'}(\mathbb{C}^s))'$.

With the mean μ defined in the proof of Lemma 28.4, we may sum up the outcome of this map diagram with the inequality

(28.3)

$$\sum_{Q^{1/2}<q\le Q} q^{\alpha-1} \max_{(r,q)=1} \left| \sum_{\substack{n\le x \\ n\equiv r (\mathrm{mod}\, q)}} (f(n)-\mu) - \frac{1}{\phi(q)} \sum_{\substack{n\le x \\ (n,q)=1}} (f(n)-\mu) \right|^{\alpha}$$

$$\ll x^{\alpha} \sum_{\ell\le x} \frac{|f(\ell)|^{\alpha}}{\ell}.$$

Again $\mu \ll \|f\|_{\alpha}(\log\log x)^{1/\alpha'}$, and since

$$\sum_{\substack{n\le x \\ n\equiv r (\mathrm{mod}\, q)}} 1 - \frac{1}{\phi(q)} \sum_{\substack{n\le x \\ (n,q)=1}} 1 \ll 1$$

uniformly in $(r,q)=1$ and prime powers q, we may safely omit μ from the inequality.

The q are confined to a fixed power of their prime q_0. Removing this condition and replacing the interval $(Q^{1/2}, Q]$ by $(1, Q]$ introduces a factor of $\ll \log x \log\log x$ into the upper bound at (28.3).

It follows that the contribution towards the expression in Theorem 28.1 which arises from the integers n not exceeding $x(\log x)^{-2}$ in the innermost sums, is $\ll x^{\alpha}(\log x)^{-2}\|f\|_{\alpha}^{\alpha}$, and well within the asserted bound. I shall confine my attention to the integers $x(\log x)^{-2} < n \le x$ which remain.

For $y \ge x(\log x)^{-2}$, $0<\sigma<1$, define

$$\theta(y,\sigma) = \sum_{\substack{x(\log x)^{-2}<n\le y \\ n\equiv r_q (\mathrm{mod}\, q)}} f(n)n^{-\sigma} - \frac{1}{\phi(q)} \sum_{\substack{x(\log x)^{-2}<n\le y \\ (n,q)=1}} f(n)n^{-\sigma}.$$

We require a bound for

$$\sum_{q\le x^{\delta}} q^{\alpha-1}|\theta(x,0)|^{\alpha}.$$

Integration by parts gives

$$\theta(x,0) = x^{\sigma}\theta(x,\sigma) - \sigma \int_{x(\log x)^{-2}}^{x} y^{\sigma-1}\theta(y,\sigma)dy.$$

Applying Hölder's inequality

$$|\theta(x,0)|^{\alpha} \ll x^{\alpha\sigma}|\theta(x,\sigma)|^{\alpha} + x^{\alpha-1}\int_{x(\log x)^{-2}}^{x} y^{\alpha\sigma-\alpha}|\theta(y,\sigma)|^{\alpha}dy.$$

If $\delta\alpha'^2(2\alpha'+3) < 1$, then we may choose a value of σ so that Lemma 28.4 is nontrivially valid. The proof of Theorem 28.1 in the cases $1 < \alpha \leq 2$ is then almost immediate. Note that the restriction on δ guarantees the existence of a number $\zeta > \delta$ which also satisfies $\zeta\alpha'^2(2\alpha' + 3) < 1$, and such that $Q \leq x^\delta$ implies $Q \leq (x(\log x)^{-2})^\zeta$ for all sufficiently large values of x.

Proof of Theorem 28.1; the cases $\alpha > 2$. Let O denote the operator $\mathbb{C}^s \to \mathbb{C}^s$ given by

$$(Of)(q) = \frac{q}{[x]}\left(\sum_{\substack{n\leq x\\ n\equiv r_q (\mathrm{mod}\, q)}} f(n) - \frac{1}{\phi(q)}\sum_{\substack{n\leq x\\ (n,q)=1}} f(n)\right).$$

On $L^\infty(\mathbb{C}^s)$, O is almost uniformly bounded. A typical coefficient of Of satisfies

$$|(Of)(q)| \leq \frac{2q}{x}\left(\sum_{\substack{n\leq x\\ n\equiv r_q (\mathrm{mod}\, q)}} |f(n)| + \frac{1}{q}\sum_{\substack{n\leq x\\ (n,q)=1}} |f(n)|\right)$$

$$\leq 2\|f\|_\infty\left(\frac{q}{x}\sum_{\substack{n\leq x\\ n\equiv r_q (\mathrm{mod}\, q)}} \nu(n) + \frac{1}{x}\sum_{n\leq x}\nu(n)\right).$$

Here $\nu(n)$ does not exceed the number of prime factors of n up to $x^{1/4}$ by more than three, and

$$\sum_{\substack{n\leq x\\ n\equiv r_q (\mathrm{mod}\, q)}} \nu(n) \ll \frac{x}{q} + \sum_{p\leq x^{1/4}}\sum_{\substack{m\leq x/p\\ mp\equiv r_q (\mathrm{mod}\, q)}} 1$$

$$\ll \frac{x}{q}\left(1 + \sum_{p\leq x^{1/4}}\frac{1}{p}\right) \ll \frac{x}{q}\log\log x,$$

uniformly for $x^{1/2} \geq q \geq 2$. Altogether $\|O\|_\infty$ is $\ll \log\log x$ when $\delta \leq 1/2$.

A similar bound holds under the wider condition $\delta < 1$.

On $L^2(\mathbb{C}^s)$, $\|O\|_2 \ll (\log\log x)^{5/2}(\log x)^{-1/2}$ provided $18\delta < 1$. The assertion of Theorem 28.1 for $\alpha > 2$ now follows from an application of the Riesz–Thorin interpolation theorem.

As remarked earlier, when $\alpha = 2$, Theorem 28.1 indeed holds if $2\delta < 1$. The above argument shows that this weaker restraint upon δ also suffices in the cases $\alpha > 2$.

29

Exercises: Maximal inequalities

Arguing through a dual, in the manner that Theorem 25.1 is approached by way of Lemma 25.2, emphasises form over precision.

Without characters

1. Prove that

$$\sum_{p\le Q} p \max_{v-u\le H} \sum_{r=0}^{p-1} \left| \sum_{\substack{u<n\le v \\ n\equiv r (\mathrm{mod}\, p)}} a_n - \frac{1}{p} \sum_{u<n\le v} a_n \right|^2 \ll \left(H + \frac{Q^3}{\log Q} \right) \sum_{n=-\infty}^{\infty} |a_n|^2$$

uniformly for $Q \ge 2$ and all square-summable sequences of complex numbers, a_n.

Norm form inequalities so uniform in the interval over which an inner summation is performed, I call *maximal-gap*.

2. Establish a maximal-gap version of Theorem 25.1.

Hint: The expression corresponding to (25.2) does not exceed $E(Y^{2m})$, where Y is the sum of the $|U(q) - E(U(q))|$ taken over the q not exceeding Q.

After the results of Chapter 4, one might hope for a factor more nearly $H + Q^2$ in the bound of exercise 1. Calculating the traces of the underlying operators, as in Chapter 23, shows that hardly more is to be expected. We may return to Chapter 4 in search of methods robust enough to provide such an improvement.

With additive characters

We begin with a maximal gap version of the basic inequality of the Large Sieve.

3. In the notation of Chapter 4, exercise 17, let x_j, $j = 1, \ldots, J$ be real numbers which satisfy $\|x_j - x_k\| \ge \delta > 0$ for all $j \ne k$. Prove that

$$\sum_{j=1}^{J} \max_{v-u\le H} \left| \sum_{u<n\le v} a_n e^{2\pi i x_j} \right|^2 \ll (H + \delta^{-1} \log \delta^{-1}) \sum_{n=-\infty}^{\infty} |a_n|^2$$

uniformly for $H \geq 1$, $\delta > 0$ and all mean-square summable sequences of complex a_n.

Hint. Geometric progression.

4. Prove that

$$\sum_{q \leq Q} \sum_{\substack{a=1 \\ (a,q)=1}}^{q} \max_{v-u \leq H} \left| \sum_{u < n \leq v} b_n e^{2\pi i a/q} \right|^2 \ll (H + Q^2 \log Q) \sum_{n=-\infty}^{\infty} |b_n|^2$$

uniformly for $H \geq 1$, $Q \geq 2$ and square-summable b_n.

5. Prove that for a multiplicative function g with values in the complex unit disc,

$$\sum_{(\log x)^2 < p \leq x^{\frac{1}{2}-\varepsilon}} \phi(p) \max_{(r,p)=1} \max_{y \leq x} \left| \sum_{\substack{n \leq y \\ n \equiv r (\mathrm{mod}\, p)}} g(n) - \frac{1}{\phi(p)} \sum_{\substack{n \leq y \\ (n,p)=1}} g(n) \right|^2$$

is $\ll (x \log\log x / \log x)^2$, the implied constant depending at most upon ε.

With multiplicative characters

A maximal gap version of the Large Sieve inequality for Dirichlet characters is a little more troublesome.

Formally define the Fourier transform $\hat{h}$ of a complex valued function h on $\mathbb{R}$, by

$$\hat{h}(y) = \frac{1}{\sqrt{2\pi}} \int_{\mathbb{R}} h(x) e^{ixy} dx.$$

6. Let $\delta > 0$. Prove that the function $F_\delta(x)$ which is δ^{-1} when $|x| \leq \delta/2$, and zero on all other reals, has Fourier transform $(\frac{1}{2}\delta y)^{-1} \sin \frac{1}{2}\delta y$.

For an increasing sequence of reals ν, and complex $c(\nu)$, $\sum |c(\nu)| < \infty$, define

$$S(y) = \sum_{\nu} c(\nu) e^{i\nu y}, \quad y \in \mathbb{R},$$

$$C_\delta(x) = \delta^{-1} \sum_{|\nu - x| \leq \delta/2} c(\nu) = \sum_{\nu} c(\nu) F_\delta(x - \nu), \quad x \in \mathbb{R}.$$

7. Prove that $\widehat{C_\delta} = S\widehat{F_\delta}$.

8. Prove that

$$\int_{-\pi/\delta}^{\pi/\delta} |S(y)|^2 dy \leq \frac{\pi^2}{4} \int_{\mathbb{R}} \left| \delta^{-1} \sum_{x \leq \nu \leq x+\delta} c(\nu) \right|^2 dx.$$

Hint. Plancherel.

Let a_n be a sequence of complex numbers, $\sum |a_n| < \infty$. An inequality of Gallagher, [90], provides a finite form of Plancherel's relation convenient for the study of Dirichlet series:

9. Prove that with $\tau = e^{1/T}$,

$$\int_{-T}^{T} \left| \sum_{n=1}^{\infty} a_n n^{it} \right|^2 dt \le \left(\frac{\pi T}{2} \right)^2 \int_0^{\infty} \left| \sum_{y<n\le y\tau} a_n \right|^2 \frac{dy}{y},$$

uniformly for $T > 0$.

10. In the notation of Chapter 4, prove that

$$\sum_{\substack{qr\le Q\\(q,r)=1}} \frac{q}{\phi(qr)} \sum_{\chi(\bmod q)}^{*} \int_{-T}^{T} \left| \sum_{n\le x} a_n \chi(n) c_r(n) n^{it} \right|^2 dt \ll \sum_{n\le x} |a_n|^2 (n + Q^2 T)$$

uniformly for $T \ge 1$, $Q \ge 1$, $x \ge 1$, complex a_n.

In the following ten examples, I shall denote

$$\sum_{\substack{qr\le Q\\(q,r)=1}} \sum_{\chi(\bmod q)}^{*} \text{ by } \widehat{\sum},$$

and $\sum_{n\le x} |a_n|^2$ by $|\mathbf{a}|^2$.

11. If $s = \sigma + it$, $\sigma = \operatorname{Re}(s)$ denotes a complex variable, prove that

$$\int_{-1}^{1} \left| \frac{K^s - 1}{s} \right| dt \le 4K^{\sigma}(1 + \log\log K)$$

uniformly for $K \ge 2$, $\sigma \ge 0$. The integrand is defined to be $\log K$ when $\sigma = 0 = t$.

Hint. Consider the ranges $|t| \le (\log K)^{-1}$, $(\log K)^{-1} < |t| \le 1$ separately.

As a waystation we establish

Lemma 29.1. *Let $\varepsilon > 0$. Then*

$$\widehat{\sum} \max_{0<y\le x} \left| \sum_{y/K<n\le y} a_n \chi(n) c_r(n) \right|^2 \ll (x + Q^2 (\log x)^{2+\varepsilon}) |\mathbf{a}|^2 (1 + \log\log K)^2$$

uniformly for $2 \le K \le x$, $Q \ge 1, \mathbf{a}$.

Let w be half an odd integer, $1/2 \le w \le 2x$, and n a positive integer not exceeding x. Then for $\sigma = (\log x)^{-1}$

$$\frac{1}{2\pi i}\int_{\sigma-i\infty}^{\sigma+i\infty}\left(\frac{w}{n}\right)^s \frac{ds}{s} = \begin{cases} 1 & \text{if} \quad n \le w, \\ 0 & \text{if} \quad n > w. \end{cases}$$

12. Show that the range $|t| \ge T \ge 1$ contributes $\ll (w/n)^\sigma |\log w/n|^{-1} T^{-1}$ to the integral, in particular $\ll T^{-1}x$.

13. Prove that a typical character sum in Lemma 29.1 is

$$\ll \left|\int_{\sigma-iT}^{\sigma+iT} A(\chi)\left(\frac{y^s - (y/K)^s}{s}\right) ds\right|^2 + \frac{x}{T^2}\sum_{n \le x} |\chi(n)c_r(n)|^2 |\mathbf{a}|^2,$$

where

$$A(\chi) = \sum_{n \le x} a_n \chi(n) c_r(n) n^{-s}.$$

14. Prove that $\widehat{\sum} |\chi(n)c_r(n)|^2 \ll Q^2$ uniformly in $Q \ge 1$, $n \ge 1$. Deduce that with $T = x^2$, the second term in exercise 13 contributes $\ll x^{-1}Q^2|\mathbf{a}|^2$ to the bound in Lemma 29.1.

15. Let $\varepsilon > 0$. Prove that

$$\widehat{\sum}\int_{1 \le |t| \le T} |A(\chi)|^2 \frac{(\log(2+|s|))^{1+\varepsilon}}{|s|}\, dt \ll \sum_{n \le x} |a_n|^2 (n + Q^2(\log T)^{2+\varepsilon}),$$

the implied constant absolute.

Hint. Cover the range $1 \le |t| \le T$ with pairs of intervals $2^\nu \le |t| \le 2^{\nu+1}$, $1 \le \nu \le \log T/\log 2$.

16. Prove that

$$\widehat{\sum}\int_{-1}^{1} |A(\chi)|^2 \left|\frac{1-K^{-s}}{s}\right| dt \ll (x+Q^2)|\mathbf{a}|^2(1+\log\log K).$$

17. Establish Lemma 29.1.

For real y define

$$B(y) = \sum_{n \le y} a_n \chi(n) c_r(n).$$

Let $\varepsilon > 0$ and define the integer k by $2^k \le (\log x)^\varepsilon < 2^{k+1}$.

18. Prove that for $y \le x$,

$$|B(y)|^2 \ll \max_{w \le x(\log x)^{-\varepsilon}} |B(w)|^2 + \sum_{\nu=0}^{k} 2^{\nu/2} \max_{w \le x2^{-\nu}} \left| \sum_{w/2 < n \le w} a_n \chi(n) c_r(n) \right|^2,$$

the implied constant absolute.

19. Let $\varepsilon > 0$. Prove that

$$\sum_{\substack{qr \le Q \\ (q,r)=1}} \frac{q}{\phi(qr)} \sum_{\chi (\bmod q)}^{*} \max_{y \le x} \left| \sum_{n \le y} a_n \chi(n) c_r(n) \right|^2 \ll (x + Q^2 (\log x)^{2+\varepsilon}) \sum_{n \le x} |a_n|^2,$$

uniformly in $Q \ge 1$, $x \ge 2$ and complex a_n.

This gives a version of the inequality in Chapter 4 exercise 51 that is maximal but not maximal gap. There is a corresponding version of Chapter 4 exercise 54:

20. Let $\varepsilon > 0$. Prove that

$$\sum_{q \le Q} \sum_{\chi (\bmod q)}^{*} \max_{y \le x} \left| \sum_{p \le y} a_p \chi(p) \right|^2 \ll \left(\frac{x}{\log x} + Q^{2+\varepsilon} \right) \sum_{p \le x} |a_p|^2,$$

where p denotes a prime number, and the implied constant is uniform in $Q \ge 2$, $x \ge 2$, complex a_p.

Let $w \ge 2$. I continue with the convention $\widehat{\sum}$, but $|\mathbf{a}|^2$ will denote $\sum |a_n|^2$ taken over the interval $w < n \le 3w$. Our next waystation is

Lemma 29.2. *The inequality*

$$\widehat{\sum} \max_{\substack{0 \le v-u \le y \\ w < u \le 2w}} \left| \sum_{u < n \le v} a_n \chi(n) c_r(n) \right|^2 \ll (y + Q^2 (\log y)^2) \sum_{w < n \le 3w} |a_n|^2$$

holds uniformly for $w \ge y \ge 2$, a_n.

This time we employ the representation

$$\sum_{u < n \le v} a_n \chi(n) c_r(n) = \frac{1}{2\pi i} \int_{\sigma - i\infty}^{\sigma + i\infty} \sum_{w < n \le 3w} a_n \chi(n) c_r(n) n^{-s} K(s) ds$$

with $K(s) = K(u, v, s) = s^{-1}(v^s - u^s)$, $\sigma = (\log w)^{-1}$, v, u half odd integers. Note that since u belongs to the interval $(w, 2w]$, $v \le 3w$. We break the integral into four pieces I_j, $j = 1, 2, 3, 4$, corresponding to the ranges $|t| \le 2wy^{-1}$, $2wy^{-1} < |t| \le 2wy^{-1}\log y$, $2wy^{-1}\log y < |t| \le w$, $|t| > w$, of the variable $t = -i\mathrm{Im}\,(s)$.

21. Prove that

$$\max_{w<u\le 2w} \sum_{m\le 3w} \left|\log \frac{u}{m}\right|^{-2} \ll w^2.$$

22. Prove that

$$\widehat{\sum} \max_{u,v} |I_4|^2 \ll Q^2|\mathbf{a}|^2.$$

23. Prove that

$$\int_{|t|\le 2wy^{-1}} \widehat{\sum} \left| \sum_{w<n\le 3w} a_n\chi(n)c_r(n)\right|^2 dt \ll (w + Q^2wy^{-1})|\mathbf{a}|^2,$$

hence that

$$\widehat{\sum} \max_{u,v} |I_1|^2 \ll (y + Q^2)|\mathbf{a}|^2.$$

24. Prove that

$$\widehat{\sum} \int |s|^{-1/2} \left| \sum_{w<n\le 3w} a_n\chi(n)c_r(n)\right|^2 dt \ll \left(\frac{w}{y}\right)^{1/2} (y + Q^2(\log y)^{1/2})|\mathbf{a}|^2,$$

the integral taken over $2wy^{-1} < |t| \le 2wy^{-1}\log y$.

25. Prove that

$$\widehat{\sum} \int |s|^{-1} \left| \sum_{w<n\le 3w} a_n\chi(n)c_r(n)\right|^2 dt \ll \left(\frac{y}{\log y} + Q^2\log y\right) |\mathbf{a}|^2,$$

the integral taken over $2wy^{-1}\log y < |t| \le w$.

26. Establish Lemma 29.2.

27. Let $\varepsilon > 0$. Prove that

$$\sum_{\substack{qr\le Q\\(q,r)=1}} \frac{q}{\phi(qr)} \sum_{\chi(\mathrm{mod}\, q)}^{*} \max_{0\le v-u\le y} \left| \sum_{u<n\le v} a_n\chi(n)c_r(n)\right|^2$$

$$\ll (y + Q^2(\log y)^{2+\varepsilon}) \sum_{n=-\infty}^{\infty} |a_n|^2$$

uniformly in $Q \ge 1$, $y \ge 2$ and square-summable sequences of complex numbers a_n.

This provides a maximal gap version of the Selberg inequality asserted in exercise 51 of Chapter 4. Note that the lines in the complex plane over which we carry out an inverse Mellin transform, vary their position with u.

28. Prove that

$$\sum_{q\le Q}\ \sideset{}{^*}\sum_{\chi(\bmod q)} \max_{0\le v-u\le y} \left| \sum_{u<p\le v} a_p \chi(p) \right|^2 \ll \left(\frac{y}{\log y} + Q^{2+\varepsilon} \right) \sum_{p\ge 2} |a_p|^2,$$

where p denotes a prime number, uniformly for $Q \ge 1$, $y \ge 2$, square-summable complex a_p.

Again without characters

Choose a real $y_p \ge 1$ for each prime $p \ge 2$. Define $\theta_{n,p}$ to be 1 if the positive integer n does not exceed y_p, to be zero otherwise. For a complex-valued function f on the primes, and a real $Q \ge 2$, let S denote the sum

$$\sum_{p,q\le Q} \frac{f(p)\overline{f(q)}}{pq} \frac{\min(y_p, y_q)}{\sqrt{y_p y_q}}$$

taken over pairs of primes not exceeding Q.

29. Prove that

$$\sum_{n=1}^{\infty} \left| \sum_{\substack{p|n \\ p\le Q}} \frac{f(p)\theta_{n,p}}{\sqrt{y_p}} \right|^2 \le \sum_{p\le Q} \frac{|f(p)|^2}{p} \left(1 - \frac{1}{p}\right) + S + O\left(\sum_{p\le Q} \frac{|f(p)|^2}{p} \sum_{p\le Q} \frac{p}{y_p} \right).$$

30. Prove that

$$-2\mathrm{Re} \sum_{n=1}^{\infty} \sum_{\substack{p|n \\ p\le Q}} \frac{f(p)\theta_{n,p}}{\sqrt{y_p}} \sum_{q\le Q} \frac{\overline{f(q)}\theta_{n,q}}{q\sqrt{y_q}}$$

$$= -2S + O\left(\sum_{p\le Q} \frac{|f(p)|^2}{p} \left\{ \sum_{p,q\le Q} \frac{q}{y_p y_q} \right\}^{1/2} \right).$$

31. Prove that

$$\sum_{n=1}^{\infty}\left|\sum_{\substack{p|n\\ p\le Q}}\frac{f(p)\theta_{n,p}}{\sqrt{y_p}}-\sum_{q\le Q}\frac{f(q)\theta_{n,q}}{q\sqrt{y_q}}\right|^2 \le \sum_{p\le Q}\frac{|f(p)|^2}{p}\left(1-\frac{1}{p}\right)$$

$$+O\left(\sum_{p\le Q}\frac{|f(p)|^2}{p}\left\{\sum_{p\le Q}\frac{p}{y_p}+\left(\sum_{p,q\le Q}\frac{q}{y_p y_q}\right)^{1/2}\right\}\right).$$

32. Prove that

$$\sum_{p\le (T\log T)^{1/2}} p\max_{y\ge T}\frac{1}{y}\left|\sum_{\substack{n\le y\\ n\equiv 0(\text{mod } p)}} a_n-\frac{1}{p}\sum_{n\le y}a_n\right|^2 \ll \sum_{n=1}^{\infty}|a_n|^2$$

holds uniformly for $T\ge 2$ and all square-summable complex sequences a_n.

This may be compared with the inequality

$$\sum_{p\le (T/\log T)^{1/2}} p\max_{v-u\le T}\sum_{j=1}^{p-1}\left|\sum_{\substack{u<n\le v\\ n\equiv j(\text{mod } p)}} a_n-\frac{1}{p}\sum_{u<n\le v}a_n\right|^2 \ll T\sum_{n=-\infty}^{\infty}|a_n|^2$$

implied by the present exercise 4.

Maximal versions of Theorem 28.1, case $\alpha = 2$

In [76] I give the following improved version of the mean square case of Theorem (28.1).

Theorem. *If $\delta < 1/2$, then:*

$$\sum_{q\le x^\delta}\phi(q)\max_{(r,q)=1}\max_{y\le x}\left|\sum_{\substack{n\le y\\ n\equiv r(\text{mod } q)}} f(n)-\frac{1}{\phi(q)}\sum_{\substack{n\le y\\ (n,q)=1}}f(n)\right|^2 \tag{29.1}$$

$$\ll \frac{x^2}{\log x}\sum_{\ell\le x}\frac{|f(\ell)|^2}{\ell},$$

where ℓ denotes a prime power.

It is worthwhile to consider the improvements. Here the moduli q range up almost to $x^{1/2}$, as expected in Chapter 23. In part this is possible because the function $D(Q)$ in the upper bound at (23.6) may be taken $Q^{2+\varepsilon}$ for each fixed $\varepsilon > 0$.

The earlier version of (29.1) that I gave in Chapter 6 of my book, [56], allowed the moduli only up to a smaller power of x. It also involved Cesàro summation; the function f was weighted by the factor $1 - ny^{-1}$. My first draft of the book had no such weight upon the function f. In making precise the results of that Chapter 6 I encountered a difficulty, of an analytic nature, which exacerbated the Wiener phenomenon. I illustrate the difficulty using the present treatment.

The upper bound of Lemma 28.3 depends upon $|z|$. In order to obtain an unweighted version of the sum to be squared in (29.1) it would be natural to consider Mellin integrals of the type

$$\frac{1}{2\pi i}\int_{\theta-i\infty}^{\theta+i\infty} \frac{w^z}{z} \sum_{\substack{n\le x \\ n\equiv r (\bmod q)}} f(n)n^{-z}dz,$$

where $1 \le w \le x$. We might hope to squeeze by the plain Wiener phenomenon that the kernel $z^{-1}w^z$ is only $O(|z|^{-1})$ as $|\mathrm{Im}\,(z)| \to \infty$. However, when multiplied by a factor $|z|^{1-\sigma}$ with $0 < \sigma < 1$, introduced by application of Lemma 28.3, the kernel gives rise to a function far too large to be integrable over the whole complex line $\mathrm{Re}\,(z) = \theta > 0$.

The kernel K employed in the proof of Lemma 29.1 is related to Cesàro summation, the average being over a short interval $(x - \rho, x]$ rather than the full interval $(0, x]$. This kernel is $O(|z|^{-2})$ as $|\mathrm{Im}\,(z)| \to \infty$, and the existence of the relevant integrals is immediate. Again the additive function is weighted.

To remove the weights from the estimates required a Large Sieve inequality involving gaps, and if the ultimate inequality (say (29.1)) were to have the wider generality of a maximal form, then these gaps should be allowed to vary their position. Not having a maximal gap version of the Large Sieve to hand, and not wanting to stop the writing of the book to try and devise one, I introduced the Cesàro summation, and changed the remaining chapters as necessary. Some complication was in the offing!

I note that the version of a mean square case of Theorem 28.1 given in the paper of Hildebrand, [101], is in error: at the foot of p. 399 the minimum can be excessively large; set $M = D = x^{\frac{1}{2}}/4$. Implicitly, Hildebrand has fallen into the trap that I did.

A sufficient form of maximal gap Large Sieve inequality I obtained in [60]. This allowed me to establish a version of (29.1) weaker by only a factor $(\log\log x)^4$ in the upper bound. A nearly best possible treatment of maximal gap Large Sieve inequalities involving Dirichlet characters took me a little longer, [73], [74], it is expounded in the foregoing exercises 11–20 and 21–28 respectively, and enabled me to obtain the inequality (29.1) stated.

I shall return to discuss methods for obtaining a maximal version of the general inequality in Theorem 28.1.

As mentioned before, the upper bound factor $x(\log x)^{-1}$ in (29.1) is best possible. For moduli $q = q_0^k$ with k fixed, this upper bound is obtained using Fourier analysis and character sum estimates. The implied constant in (29.1) then depends upon k.

A similar situation prevails in the proof of Lemma 28.4, save that the moduli q are confined to an interval $Q^{1/2} < q \le Q$. This restriction is a little artificial. It arises from the mechanics of the Selberg sieve, applied to avoid appeal to estimates for the number of primes in residue classes to large moduli. There is an alternative procedure, more messy but more accurate; I continue in the notation of Lemma 28.4.

Given a positive c, to be thought *large but fixed*, the contribution to the operator $V_1(y)$ from those t_r exceeding $(\log x)^c$ may be deemed negligible. Ultimately they occur only in sums of the form $\sum t_r^{-\beta}$ with $\beta > 1$, where they contribute $\ll (\log x)^{-c(\beta-1)}$. It will therefore suffice to consider the functions $\Delta_j(q, r, z)$ for at most $(\log x)^c$ different residue classes $r(\operatorname{mod} q)$. Of course these classes may have any of $\phi(q)$ possible values. The operators bounded in Chapter 25 may then be modified to take this restriction on r into account, and instead of applying the Selberg sieve method, we may appeal to the Bombieri–Vinogradov theorem mentioned in Chapter 23.

Let $1 < \alpha \le 2$. For moduli $q = q_0^k$ with k fixed, but otherwise restricted only by $q \le x^\delta$, we may reformulate the inequality of Theorem 28.1 in terms of the appropriate norm on $(L^2 \cap L^{\alpha'})'$, and strip the powers of $\log\log x$ from the upper bound factor.

In the present treatment, the contributions to the sum(s) appearing in Theorem 28.1 which arise from the moduli $q = q_0^k$ with k large ($\ge h$) are estimated together in (28.1). This also introduces powers of $\log\log x$ into the upper bound. Consider the case $\alpha = 2$.

In order to obtain (29.1), in our present notation we would need

$$\left\| \sum_{k \ge k_0} P_k V_j(y) \right\|_2 \ll (x^\sigma (\log x)^{1/2})^{-1} \|f\|_2,$$

valid for some $k_0 \ge 2$, independent of x, f. In turn, it would suffice to obtain a sharpening of the corresponding assertion of Lemma 28.5:

$$\left\| \sum_{k \ge k_0} P_k U f \right\|_2 \ll x^{-\sigma} ((\log x)^{-1} + Q x^{-1+\varepsilon})^{1/2} \|f\|_2.$$

In these bounds it is to be (now) understood that the projection P_k has the restriction $Q^{1/2} < q \le Q$ in its definition replaced by $q \le Q$.

Largely in the notation of the proof of Lemma 28.5 we are reduced to establishing two bounds. The first asserts that

$$\sum_{\substack{t_j (\bmod q_j),\\ j=1,2}}{}' t_1^{-\sigma} t_2^{-\sigma} \max(t_1, t_2)^{-2(1-\sigma)} \ll 1, \quad 0 < \varepsilon \le \sigma \le 1-\varepsilon, \tag{29.2}$$

where $'$ indicates that the $q_j = q_0^{e_j}$ are powers of the same prime, $1 \le e_i < e_j$, and $t_2 \equiv t_1 w (\bmod q_1)$ for some (fixed) w.

The second asserts that

$$\left| \sum_{1 \le i,j \le m} d_i d_j q_0^{-\max(i,j)} \right| \le 8 \sum_{i=1}^{m} |d_i|^2 q_0^{-i} \tag{29.3}$$

uniformly in real d_i, integers $m \ge 1$, primes $q_0 \ge 2$.

I consider these sums in turn, beginning with that of (29.2). Given t, there can be at most one $t_2 < t_1$. For a second, say t_2', would satisfy the same congruence restriction $(\bmod q_1)$, and so $|t_2' - t_2| \ge q_1 \ge t_1$. The contribution to the sum from pairs (t_1, t_2) with $t_2 < t_1$ is not more than $\sum t_1^{-2+\sigma} \ll 1$.

If now $t_2 \ge t_1$, then t_2 has the form $t_0 + s q_0^{e_i}$, $s = 0, 1, \ldots$, where $t_0 \ge t_1$. There is a corresponding contribution to the sum of

$$\begin{aligned}
&\ll \sum t_1^{-\sigma} \sum_s (t_1 + s q_1)^{-2+\sigma} \\
&\ll \sum t_1^{-\sigma} \left(t_1^{-2+\sigma} + q_1^{-2+\sigma} \sum_{s=1}^{\infty} s^{-2+\sigma} \right) \\
&\ll \sum t_1^{-\sigma} (t_1^{-2+\sigma}) = \sum t_1^{-2} \ll 1.
\end{aligned}$$

The inequality (29.2) is established.

Let Λ denote the quadratic form in (29.3). We may assume the d_i to be non-negative. Then

$$\Lambda \le 2 \sum_i d_i \sum_{j \ge i} d_j q_0^{-j}.$$

Let $D = \sum d_j^2 q_0^{-j}$. An application of the Cauchy–Schwarz inequality gives

$$\Lambda^2 \le 4D \sum_i q_0^i \left| \sum_{j \ge i} d_j q_0^{-j} \right|^2.$$

Expanding the square and interchanging summations gives the upper bound the form

$$4D \sum_{j_1, j_2} d_{j_1} d_{j_2} q_0^{-(j_1+j_2)} \sum_{i \le \min(j_1, j_2)} q_0^i.$$

The coefficient of $d_{j_1} d_{j_2}$ does not exceed

$$q_0^{-(j_1+j_2)} q_0^{\min(j_1, j_2)} \sum_{r=0}^{\infty} q_0^{-r} = q_0(q_0 - 1)^{-1} q_0^{-\max(j_1, j_2)}.$$

Altogether

$$\Lambda^2 \le 4Dq_0(q_0 - 1)^{-1}\Lambda.$$

Inequality (29.3) is established.

I note that the version of Lemma 28.5 given in my paper, [60], is not completely satisfactory. In the formulation of Lemma 28 there, the summation condition $\ell t_\ell \le v$ should be restored in place of $\ell \le v$. The application of that lemma in [60] and in [76] should be modified using the argument given above. The theorems of those papers remain valid.

Maximal variants of Theorem 28.1 can be derived for each of the norms. For the operator O with $\alpha = \infty$ the necessary modifications are straightforward. Interpolation using the bound of (29.1) then covers the cases $\alpha > 2$. When $1 < \alpha < 2$, the proof is more elaborate. Rather than parse the method as a series of exercises, I give an outline. Many of the ideas considered in the present volume appear.

Maximal version of Theorem 28.1, case $1 < \alpha < 2$

I first obtain a coarse maximal version of Theorem 25.5. Consider the composition of maps

$$(K^{\alpha'}(\mathbb{C}^v))' \xrightarrow{N(z)} (K^{\alpha'}(\mathbb{C}^v))' \xrightarrow{\widetilde{W}'} (J^2 \cap J^{\alpha'}(\mathbb{C}^u))',$$

where the first map (typically) takes a_n to $a_n n^{-z}$. For each fixed complex number z with $\operatorname{Re}(z) \ge 0$ the first map has norm at most 1. The norm of the composition therefore does not exceed the estimate for the norm of $\widetilde{W}'$ alone given by Theorem 25.5.

In the notation of Theorem 25.5, using the Kernel $K(v, \frac{1}{4}, z)$ from Lemma 28.2, there is a representation

$$\sum_{\substack{n \le y \\ n \equiv r (\bmod q)}} a_n = \frac{1}{2\pi i} \int_{\theta - i\infty}^{\theta + i\infty} K\left(y, \frac{1}{4}, z\right) \sum_{\substack{n \le x \\ n \equiv r (\bmod q)}} a_n n^{-z} dz,$$

valid for all half-odd-integer y which do not exceed x, and real $\theta > 0$. Set $\theta = (\log x)^{-1}$. If we choose such a $y = y_{q,r}$ for each of the pairs (q, r) which enumerate the coordinates of $\mathbb{C}^u$, setting $\rho = x$ in the plain definition of $\widetilde{W}$, then the map M which takes $\mathbf{a}$ in $\mathbb{C}^v$ to a vector in $\mathbb{C}^u$ with typical coordinate

$$\frac{q(1-q_0^{-1})^{-1}}{[x]}\left(\sum_{\substack{n\le y_{q,r}\\ n\equiv r(\bmod q)}} a_n - \frac{1}{q}\sum_{n\le y_{q,r}} a_n\right)$$

has an integral representation

$$\frac{1}{2\pi i}\int_{\theta-i\infty}^{\theta+i\infty} K\left(y_{q,r}, \frac{1}{4}, z\right)\widetilde{W}'\circ N(z)dz$$

for that coordinate. Bearing in mind Remark (ii) following the proof of Lemma 3.2, we see that

$$\begin{aligned}\|M\| &\le \frac{1}{\pi}\int_{\theta-i\infty}^{\theta+i\infty} \sup_{y\le x}\left|K\left(y, \frac{1}{4}, z\right)\right| \|\widetilde{W}'\circ N(z)\| d\mathrm{Im}\,(z)\\ &\ll \|\widetilde{W}'\|x^\theta \log(4x/\theta) \ll (1 + x^{-1}Q^{(2\alpha'+3)\alpha'})^{1/\alpha'}\log x,\end{aligned}$$

the penultimate step by an appeal to Lemma 28.2. A similar bound holds if we use the variant definition of $\widetilde{W}'$ from the remark following the statement of Theorem 25.5.

Consider next the chain of maps:

$$(K^{\alpha'}(\mathbb{C}^v))' \xrightarrow{M} (J^2\cap J^{\alpha'}(\mathbb{C}^u))' \xrightarrow{\text{collapse}} (L^2\cap L^{\alpha'}(P\mathbb{C}^s))',$$

where the collapse selects the residue class $r_q(\bmod q)$ and abandons the others, and $P\mathbb{C}^s$ denotes the space of vectors one corresponding to each prime power $q = q_0^k$. Here k is fixed. $P\mathbb{C}^s$ may be viewed as a projection of $\mathbb{C}^s$. The norm of the collapse into $(L^2\cap L^{\alpha'}(P\mathbb{C}^s))'$ is by a variant of Lemma 27.2 at most 2. We may lift an estimate for the norm of any element in $(L^2\cap L^{\alpha'}(P\mathbb{C}^s))'$ to an estimate for the norm of the same element in $L^\alpha(P\mathbb{C}^s)$ at the expense of an increasing factor which is $\ll (\log\log x)^{1/\alpha-1/2}$. As mentioned in the discussion following the statement of Theorem 25.1, this device on the larger space $(J^2\cap J^{\alpha'}(\mathbb{C}^u))'$ is too costly. In particular

$$\left(\sum_{\substack{q\le Q\\ q=q_0^k}} q^{\alpha-1}\max_{(r,q)=1}\max_{y\le x}\left|\sum_{\substack{n\le y\\ n\equiv r(\bmod q)}} a_n - \frac{1}{\phi(q)}\sum_{\substack{n\le y\\ (n,q)=1}} a_n\right|^\alpha\right)^{1/\alpha}$$

$$\ll x(1+x^{-1}Q^{(2\alpha'+3)\alpha'})^{1/\alpha'}\log x(\log\log x)^{1/\alpha-1/2}\left(\frac{1}{x}\sum_{n\le x}|a_n|^\alpha\right)^{1/\alpha}.$$

Although not of best form, this estimate will serve us.

We allow k to vary. This introduces a further factor of $\ll (\log x)^{1/\alpha}$ into the upper bound. Setting $a_n = f(n)$, and noting that

$$\left(\frac{1}{[x]}\sum_{n\le x}|f(n)|^\alpha\right)^{1/\alpha} \ll \|A_\alpha f\| + |\mu| \ll \|f\| + |\mu|$$

with the upper bound norm in $(L^2 \cap L^{\alpha'}(\mathbb{C}^s))'$, we reach

$$\sum_{q\le Q} q^{\alpha-1} \max_{(r,q)=1}\max_{y\le x}\left|\sum_{\substack{n\le y\\ n\equiv r(\bmod q)}} f(n) - \frac{1}{\phi(q)}\sum_{\substack{n\le y\\ (n,q)=1}} f(n)\right|^\alpha \tag{29.4}$$

$$\ll x^\alpha(1+x^{-1}Q^{(2\alpha'+3)\alpha'})^{\alpha-1}(\log x)^{\alpha+1}\log\log x \sum_{\ell\le x}\ell^{-1}|f(\ell)|^\alpha.$$

Note that if the value of r_q is independent of q, x, then a shorter argument is possible. For example:

$$\sum_{q\le x} q^{\alpha-1}\max_{y\le x}\left|\sum_{\substack{|r|<n\le y\\ n\equiv r(\bmod q)}} f(n)\right|^\alpha$$

$$\ll x^{\alpha-1}\sum_{q\le x}\sum_{\substack{|r|<n\le x\\ n\equiv r(\bmod q)}}|f(n)|^\alpha \ll x^{\alpha-1}\sum_{|r|<n\le x}|f(n)|^\alpha\sum_{q|(n-r)}1.$$

Each inner sum is $\ll (\log x)^2$; and so on.

From the inequality (29.4) I obtain a coarse maximal-gap version of Theorem 28.1. For each prime power q not exceeding x, let x_q, y_q, $0 \le y_q \le x_q \le x$, satisfy $x_q - y_q \le h$. Consider the map $Z : \mathbb{C}^s \to \mathbb{C}^s$ which takes f to the vector whose q-coordinate is

$$\frac{q}{x}\left(\sum_{\substack{y_q<n\le x_q\\ n\equiv r_q(\bmod q)}} f(n) - \frac{1}{\phi(q)}\sum_{\substack{y_q<n\le x_q\\ (n,q)=1}} f(n)\right).$$

Let $\|Z\|_\alpha$ denote the norm of $\mathbb{Z}$ considered between $L^\alpha(\mathbb{C}^s)$ and itself.

Let $1 < \beta < \alpha \le 2$. From (29.4)

$$\|Z\|_\beta \ll (1+x^{-1}Q^{(2\beta'+3)\beta'})^{1/\beta'}(\log x)^{2-1/\beta'}(\log\log x)^{1-1/\beta'}.$$

However, the sharper bound

$$\|Z\|_2 \ll (x^{-1}(h + Q^2 \log Q))^{1/2} \log x$$

is obtained in [60] Lemma 4, and may be deduced from exercise 4 of the present chapter. Applying the Riesz–Thorin theorem to interpolate between the cases β and 2, we see that

$$\|Z\|_\alpha \ll (x^{-1}(h + Q^2 \log Q))^{\theta_1} (\log x)^{\theta_2} \tag{29.5}$$

for certain positive numbers θ_1, θ_2, depending upon α, β, provided $Q^{(2\beta'+3)\beta'} \leq x$. For β near to α this last represents only a slight weakening of the condition $Q^{(2\alpha'+3)\alpha'} \leq x$ appropriate to the direct application of Theorem 25.5.

To obtain a maximal version of Theorem 28.1 we introduce the operator $L : \mathbb{C}^s \to \mathbb{C}^s$ given by

$$(Lf)(q) = \frac{q}{[x]} \left(\sum_{\substack{n \leq x_q \\ n \equiv r_q (\mathrm{mod}\, q)}} f(n) - \frac{1}{\phi(q)} \sum_{\substack{n \leq x_q \\ (r,q)=1}} f(n) \right),$$

where $0 < x_q \leq x$. We seek an integral representation of the form

$$(Lf)(q) = \frac{1}{2\pi i} \int_{\sigma - iT}^{\sigma + iT} \frac{x_q^z}{z} (Y(x)f)(q) dz + \text{small}, \tag{29.6}$$

where $\sigma = \mathrm{Re}\,(z)$ satisfies the condition $1/\alpha + \varepsilon \leq \sigma \leq 1 - (2\alpha'+3)\alpha'\delta - \varepsilon$ of Lemma 28.4, and the integration is taken over the line segment $\mathrm{Re}\,(z) = \sigma$, $|\mathrm{Im}\,(z)| \leq T$ with T a (fixed) power of $\log x$. Only the last requirement causes difficulties. We again encounter Wiener's phenomenon; but this time we are prepared. In the notation of Chapters 3 and 23, we have effectively shown that as x increases, certain projections of the operators $A'_{\alpha'} E^r A_\alpha$ and their maximal variants satisfy approximate Lipschitz conditions; they are in some sense 'smooth'.

If v is confined to half-odd integers, $2 \leq h \leq y$, $x \geq 0$, $T \geq 1$, then

$$\frac{1}{h} \sum_{y-h \leq v \leq y} \left| \int_{\substack{\mathrm{Re}\,(z)=\sigma \\ |\mathrm{Im}\,(z)|>T}} \sum_{n \leq x} a_n \left(\frac{v}{n}\right)^z \frac{dz}{z} \right|$$
$$\ll \frac{1}{T} \sum_{n \leq x} \frac{|a_n|}{n^\sigma} \frac{1}{h} \sum_{y-h \leq v \leq y} \frac{v^\sigma}{|\log v/n|} \ll \frac{y^\sigma}{T} \sum_{n \leq x} \frac{|a_n|}{n^\sigma} \left(1 + \frac{n \log(2n)}{h}\right).$$

For at least one value of v in $[y-h, y]$, the integral does not exceed the second upper bound; cf. [41] Vol. 1, Lemma 9.6; Halász, [92]. Together with an application of the maximal gap inequality (29.5), this yields a representation (29.6) within a suitably small error but with x_q, *in the integrand only*, replaced by a y_q which satisfies $|x_q - y_q| \le h$, $y_q \le x$. This may be achieved with h of the form $x(\log x)^{-c}$ for a suitable positive c, and the value of c in turn controls the necessary size of T.

In terms of the projection P of Lemma 28.4, we so reach

$$\|PL\| \ll \int_{\sigma-iT}^{\sigma+iT} \frac{x^\sigma}{|z|} \|PY(x)\| d\mathrm{Im}\,(z) + (\log x)^{-2},$$

uniformly for $Q^{(2\alpha'+3)\alpha'+\varepsilon} \le x$. The remark made following the proof of Lemma 28.5 shows that this integral is

$$\ll \frac{(\log\log x)^{1/\alpha+3/2}}{(\log x)^{1/\alpha'}} + \frac{1}{(\log x)^A} \int_{\sigma-iT}^{\sigma+iT} |z|^{-\sigma} d\mathrm{Im}\,(z),$$

where any fixed positive value of A may be chosen provided (the underlying) Q exceeds a suitably power of $\log x$. In view of the size of T, a maximal version of Theorem 28.1 certainly holds if the moduli q are confined to an interval $(\log x)^B < q \le x^\delta$ for a suitable positive B, independent of x, f.

To derive a maximal analogue of Theorem 27.1 it will suffice to estimate expressions of the type

$$\frac{1}{\rho} \int_{x_q-\rho}^{x_q} \left(\sum_{\substack{n\le t \\ n\equiv r_q (\mathrm{mod}\, q)}} f(n) - \frac{1}{\phi(q)} \sum_{\substack{n\le t \\ (n,q)=1}} f(n) \right) dt$$

with $x(\log x)^{-E} = \rho \le x_q \le x$, and use the gap bound (29.5) to remove the Cesàro weighting. The treatment of $PD(x)f_1$ in the proof of Theorem 27.1 carries over to apply (in an integrated form) to PLf_1. For $(Lf_2)(q)$, the analogue of $(D(x)f_2)(q)$, there is a representation

$$\frac{q}{[x]} \sum_{\substack{m\le x/y \\ (m,q)=1}} \int \frac{K(x_q m^{-1}, \rho, z)}{2\pi i} \left(\sum_{\substack{\ell\le x/m \\ \ell m\equiv r_q (\mathrm{mod}\, q) \\ (\ell,m)=1}} \frac{f_2(\ell)}{\ell^z} - \frac{1}{\phi(q)} \sum_{\substack{\ell\le x/m \\ (\ell,q)=1 \\ (\ell,m)=1}} \frac{f_2(\ell)}{\ell^z} \right) dz,$$

with the integral taken over the line from $\sigma - i\infty$ to $\sigma + i\infty$. In this case we may set $\sigma = (\log x)^{-1}$. Following the proof of Theorem 27.1 the operators

η_m now act upon vectors with $f(q)q^{-z}$ rather than $f(q)$ as q-coordinate. The resulting upper bound for $P\eta_m f_2$ may be taken the same as before, independent of z. Note that since y will be chosen of the form $x(\log x)^{-d}$, the various m will be small compared to x, and may be deemed small compared to the x_q.

In this way we obtain a maximal version of Theorem 28.1.

33. Realise a maximal version of Theorem 28.1.

30

Shift operators and orthogonal duals

I recast part of Theorem 28.1. Again P_δ denotes the projection of $\mathbb{C}^s$ to $\mathbb{C}^s$ which takes $f(q)$ to zero if $q > x^\delta$, and otherwise leaves $f(q)$ invariant.

Theorem 30.1. *For each $\alpha > 1$ there are positive reals δ, γ so that for every nonzero integer r*

$$\|P_\delta A'_{\alpha'} E^r A_\alpha\| \ll (\log x)^{-\gamma}.$$

The norm is that on $(L^2 \cap L^{\alpha'}(\mathbb{C}^s))'$ if $1 < \alpha \leq 2$, that on $L^2 \cap L^\alpha(\mathbb{C}^s)$ if $\alpha \geq 2$.

For $1 < \alpha \leq 2$ the operator is interpreted as

$$(L^2 \cap L^{\alpha'}(\mathbb{C}^s))' \xrightarrow{A_\alpha} M^\alpha(\mathbb{C}^t) \xrightarrow{E^r} M^\alpha(\mathbb{C}^t) \cong (M^{\alpha'}(\mathbb{C}^t))' \xrightarrow{A'_{\alpha'}} (L^2 \cap L^{\alpha'}(\mathbb{C}^s))',$$

followed by the projection P_δ.

Proof. I consider the case when r is positive. The case of r negative may be similarly dealt with.

It follows from Theorem 28.1 that for suitable $\delta > 0$, $\theta > 0$,

$$(30.1)\quad \sum_{q \leq x^\delta} q^{\alpha-1} \left| \sum_{\substack{n \leq x-r \\ n \cong 0 (\mathrm{mod}\, q)}} f(n+r) - \frac{1}{q} \sum_{\substack{n \leq x-r \\ (n,q)=1}} f(n+r) \right|^\alpha \ll \frac{x^\alpha}{(\log x)^\theta} \|f\|_\alpha^\alpha.$$

Note that $q^{-1} = \phi(q)^{-1} - \phi(qq_0)^{-1}$.

Since

$$\frac{1}{q}\left(1-\frac{1}{q_0}\right)\sum_{n\le x-r} f(n+r) - \frac{1}{q}\sum_{\substack{n\le x-r\\(n,q)=1}} f(n+r)$$

$$=\frac{1}{q}\sum_{\ell\le x} f(\ell)\left\{\left(1-\frac{1}{q_0}\right)\sum_{\substack{r<m\le x\\ m-r\cong 0(\operatorname{mod}\ell)}} 1-\frac{1}{q}\sum_{\substack{r<m\le x\\ m-r\cong 0(\operatorname{mod}\ell)\\(m,q)=1}} 1\right\}$$

$$\ll q^{-1}\sum_{\ell\le x}|f(\ell)|,$$

we may replace

$$\frac{1}{q}\sum_{\substack{n\le x-r\\(n,q)=1}} f(n+r) \quad\text{by}\quad \frac{1}{q}\left(1-\frac{1}{q_0}\right)\sum_{n\le x-r} f(n+r)$$

in the sum at (30.1) at an expense

$$\ll \sum_{q\le x^\delta} q^{\alpha-1}q^{-\alpha}\left(\sum_{\ell\le x}|f(\ell)|\right)^\alpha \ll x^\alpha(\log x)^{1-\alpha}\log\log x\|f\|_\alpha^\alpha.$$

This is acceptible.

Finally, we replace every $f(n+r)$ by $f(n+r)-\mu$, where

$$\mu=\sum_{\ell\le x}\frac{1}{\ell}\left(1-\frac{1}{\ell_0}\right)f(\ell).$$

This introduces an error of

$$\ll |\mu|^\alpha\sum_{q\le x^\delta} q^{\alpha-1} \ll x^\alpha(\log x)^{-1}(\log\log x)^{\alpha-1}\|f\|_\alpha^\alpha,$$

which is also acceptible.

Theorem 30.1 is established.

I conjecture that Theorem 30.1 holds without the projection P_δ. Since $(E^r)'=E^{-r}$, it would suffice to establish this conjecture for only one of the ranges $1\le\alpha\le 2$, $2\le\alpha<\infty$; half the possibilities, so-to-speak.

A slightly weaker conjecture would insert the projection and assert an upper bound that slowly degraded as δ (depending upon x) approached 1 from below.

The present argument shows that

$$\|A'_{\alpha'} E^{\tau} A_{\alpha} P_{\delta}\| \ll (\log x)^{-\tau}, \qquad \tau \neq 0,$$

on each of the spaces, the (positive) value of τ varying with α. As an example

$$\sum_{q \le x} q^2 \left| \sum_{\substack{n \le x \\ n - r \cong o (\mathrm{mod}\, q)}} f_{\delta}(n) - \frac{1}{q}\left(1 - \frac{1}{q_0}\right) \sum_{n \le x} f_{\delta}(n) \right|^3$$
$$\ll \frac{x^3 (\log\log x)^4}{(\log x)^{1/2}} \sum_{\ell \le x^{\delta}} \frac{|f(\ell)|^3}{\ell}$$

holds whenever $r \neq 0$, $162\delta < 1$. Here q, ℓ run through prime powers, and $f_{\delta}(n)$ denotes the truncated additive function

$$\sum_{\substack{\ell \| n \\ \ell \le x^{\delta}}} f(\ell).$$

The restriction on δ can certainly be weakened. The conjecture asserts that it can be removed altogether, possibly after decreasing the power of $\log x$ in the upper bound. Both the conjecture and (loosely speaking) the interchangeability of the support of P and f appear explicitly in my 1987 paper, [63].

For real $y \ge 2$ and positive D, r, let

$$E(y, D, r) = \pi(y, D, r) - \frac{1}{\phi(D)} \int_2^y \frac{du}{\log u}.$$

In a form appropriate to the present volume, a conjecture of Elliott and Halberstam, [80], asserts that for each fixed $A > 0$, there is a z near to x such that

$$\sum_{D \le z} \phi(D) \max_{y \le x} \max_{(r,D)=1} |E(y, D, r)|^2 \ll x^2 (\log x)^{-A}.$$

The choice $z = x^{1/2}(\log x)^{-B}$, for a certain positive B, is guaranteed by the theorem of Bombieri, [6]. Moreover, according to Bombieri, Friedlander and Iwaniec, [8], if we fix r and set $y = x$, so removing two uniformities, then a value $A > 2$ can still be achieved with z raised to $x^{1/2}\exp((\log x)^c)$ for a certain $c < 1$. Recent work of Friedlander, Granville, Hildebrand and Maier, [88], shows that for $A > 1$,

$$z = x\exp\left(-\frac{1}{4}(A-1)(\log\log x)^2(\log\log\log x)^{-1}\right)$$

in the basic inequality, and indeed smaller values of z, cannot be taken. It is widely believed that the Elliott–Halberstam conjecture is valid with z any fixed power x^β, $0 < \beta < 1$.

It might be thought that the work of Friedlander, Granville, Hildebrand and Maier casts doubt upon the possible validity of the conjecture made in this chapter. However, circumstances here are of a particular type. A version of the Elliott–Halberstam conjecture analogous to the above operator conjecture would be that for some positive A and each fixed non-zero r,

$$\sum_{q \le x} \phi(q)E(x,q,r)^2 \ll x^2(\log x)^{-2-A}.$$

Note that the moduli q are confined to prime powers. The terms of this sum corresponding to moduli in the range $x^{1-\beta} < q \le x$ may already be estimated together, using Selberg's sieve (cf. [41], Lemma 4.19). They contribute $\ll \beta x^2(\log x)^{-2}$, uniformly in $0 < \beta \le 1/2$ and $x \ge 2$. The validity of the general Elliott–Halberstam conjecture for every $z = x^\beta$ would give at once an upper bound $o(x^2(\log x)^{-2})$ for this modified Elliott–Halberstam conjecture. Even if the conjecture involving the shifting of the operators A_α fails with the asserted upper bound, it might well hold with an upper bound that is $o(1)$ as $x \to \infty$. Such a version of the conjecture would still suffice for the applications to number theory considered in the present volume. The validity of the Elliott–Halberstam conjecture modified with prime power moduli all the way up to x would itself have interesting number theoretical consequences.

31

Differences of additive functions. Local inequalities

In this chapter f continues to denote a complex valued additive function. Again s is the number of prime powers not exceeding $x(\geq 2)$, and the function analytic notation of Chapters 3, 20 and 21 continues in force.

Theorem 31.1. *For each $\alpha > 1$ there is a constant c so that*

$$\|f - F\log\| \ll \max_{x^2 \leq y \leq x^c} \left(y^{-1} \sum_{x < n \leq y} |f(n) - f(n-1)|^\alpha \right)^{1/\alpha}, \quad x \geq 2,$$

where the norm is that on $(L^2 \cap L^{\alpha'}(\mathbb{C}^s))'$ if $1 < \alpha \leq 2$, that on $L^2 \cap L^\alpha(\mathbb{C}^s)$ if $\alpha \geq 2$. The function F of x is $(f, \log)\|\log\|_2^{-2}$ in terms of the inner product and norm on $L^2(\mathbb{C}^s)$.

If

$$\left(x^{-1} \sum_{2 \leq n \leq x} |f(n) - f(n-1)|^\alpha \right)^{1/\alpha} \tag{31.1}$$

is added to the upper bound, then $x^{-1/\alpha}|F|$ may be added to the norm that is bounded.

The implied constants depend only upon α.

By analogy with Theorem 21.1, an ideal upper bound in Theorem 31.1 would be a constant multiple of the expression (31.1).

Let λ be a complex number,

$$\rho = \sum_{q \leq x} \frac{1}{q} \left(1 - \frac{1}{q_0}\right) (f(q) - \lambda \log q).$$

We may employ the representation

$$\begin{aligned} f(n) - f(n-1) = (f(n) - \lambda \log n - \rho) - (f(n-1) - \lambda \log(n-1) - \rho) \\ - \lambda \log(1 - n^{-1}) \end{aligned}$$

and apply Minkowski's inequality to dominate the norm at (31.1) by

$$2\left(x^{-1}\sum_{2\le n\le x}|f(n)-\lambda\log n-\rho|^{\alpha}\right)^{1/\alpha}+\left(x^{-1}\sum_{2\le n\le x}|\lambda\log(1-n^{-1})|^{\alpha}\right)^{1/\alpha}.$$

In turn, the first of these norms is by Theorem 2.1, $\ll \|f-\lambda\log\|$. The second norm is $\ll x^{-1/\alpha}|\lambda|$. An upper bound of the form (31.1) would render Theorem 31.1 largely best possible.

The upper bound given by Theorem 31.1 introduces the constant $c(>2)$. As a compensation it is shown that the values of f on the interval $(1,x]$ are largely controlled by the values of the difference $f(n)-f(n-1)$ on the interval $(x,x^c]$.

The differences $f(n)-f(n-1)$, $2\le n\le x$, determine the values $f(n)$, $2\le n\le x$. I begin with a quantitative version of this assertion.

Lemma 31.2. *The inequality*

$$x^{-1}\sum_{n\le x}|f(n)|^{\alpha}\ll(\log x)^{\alpha}\sum_{2\le m\le x}m^{-1}|f(m)-f(m-1)|^{\alpha}$$

holds uniformly in f, $x\ge 2$.

Proof. Consider the following reduction algorithm acting upon odd integers n. Choose $j_0=1$ or 3 so that $n-j_0$ is exactly divisible by 2, and take n to $(n-j_0)/2$. Reapply the operation to $(n-j_0)/2$ and continue inductively until 1 is reached. Equivalently, we may establish directly that every odd integer n has a unique representation of the form

$$n=j_0+2j_1+\cdots+2^{r-1}j_{r-1}+2^r,$$

with $r\ge 0$, $j_i=1$ or 3.

An odd integer m will be descended from n, using the above operation, if we can apply inflation step(s) $m\to 2m+u$, $u=1$ or 3, successively to reach n. In J such steps we reach an integer $>2^J m$. In order for n not to exceed x, $2^J\le x/m$ must be satisfied. Since there are two possible images of m in each inflation step, at most x/m odd integers n lie above m and do not exceed x.

From the representation

$$f(n)=\{f(n)-f(n-j_0)\}+f(2)+\left\{f\left(\frac{n-j_0}{2}\right)-f\left(\left(\frac{n-j_0}{2}\right)-j_1\right)\right\}+\cdots$$

with an application of Hölder's inequality we obtain the bound

$$|f(n)|^{\alpha}\ll(|f(2)|\log n)^{\alpha}+(\log n)^{\alpha-1}\sum_{m}|f(m)-f(m-v)|^{\alpha}$$

where the odd integers m run through $n, (n-j_0)/2, \dots$, as obtained by the algorithmic operation, and each v has a value 1 or 3. Summing over n and applying the remark from the previous paragraph gives:

(31.2)

$$\sum_{\substack{n\le x\\ n \text{ odd}}} |f(n)|^\alpha \ll |f(2)|^\alpha x(\log x)^\alpha + x(\log x)^{\alpha-1} \sum_{2\le m\le x} m^{-1}|f(m)-f(m-1)|^\alpha,$$

since if $v=3$, then $|f(m)-f(m-v)| \le |f(m)-f(m-1)| + |f(m-1)-f(m-2)| + |f(m-2)-f(m-3)|$. Note that $f(2)=f(2)-f(1)$.

If n is an even integer, then $f(n) = f(n-1) + \{f(n)-f(n-1)\}$ with $n-1$ odd. We obtain an analogue of (31.2) involving even integers n.

Lemma 31.2 is established.

The following result will be useful. The norm is that on $(L^2 \cap L^{\alpha'}(\mathbb{C}^s))'$, $1<\alpha\le 2$, or that on $L^2\cap L^\alpha(\mathbb{C}^s)$, $\alpha\ge 2$.

Lemma 31.3. *If for some scalar z $\|f - z\log\| \ll \varepsilon$, then a similar inequality holds with the choice $z=z_0$, $z_0\|\log\|_2^2 = (f,\log)$.*

Proof. The projection

$$Q : g \mapsto (g,\log)\|\log\|_2^{-2}\log$$

is uniformly bounded on each of the spaces $L^\beta(\mathbb{C}^s)$, $\beta\ge 1$. This may be seen by applying Hölder's inequality to the inner product $(g,\log)$, noting that $\|\log\|_\beta \sim \beta^{-1/\beta}\log x$ as $x\to\infty$. It is therefore bounded on the spaces $L^2\cap L^\alpha(\mathbb{C}^s)$ and $(L^2\cap L^{\alpha'}(\mathbb{C}^s))'$. Note that Q is self-dual. In particular, $\|Q(f-z\log)\| \ll \varepsilon$.

Proof of Theorem 31.1. As in Chapter 20, $A : \mathbb{C}^s \to \mathbb{C}^t$ will denote the operator associated with the (standard) Turán–Kubilius inequality, $\check{A} : \mathbb{C}^t \to \mathbb{C}^s$ a modified version of its dual. Again

$$\mu = \sum_{q\le x} \frac{1}{q}\left(1-\frac{1}{q_0}\right) f(q).$$

I employ the shift operator E^{-1} from Chapter 23. For $0<\delta<1$, P_δ will denote the projection $\mathbb{C}^s \to \mathbb{C}^s$ which takes $f(q)$ to zero if $q > x^\delta$, and leaves the remaining $f(q)$ alone.

From what we established in Chapter 30, for each $\alpha>1$ there are positive constants $\delta, \gamma (<1)$ so that

$$\|P_\delta \check{A} E^{-1} A\| \ll (\log x)^{-\gamma}.$$

Since the operator $P_\delta \check{A}$ is bounded uniformly in x
$\|P_\delta \check{A}(I-E^{-1})Af\| \ll \|(I-E^{-1})Af\|_\alpha$ in $M^\alpha(\mathbb{C}^t)$

$$\ll \left(x^{-1} \sum_{2\le n\le x} |f(n)-f(n-1)|^\alpha \right)^{1/\alpha} + x^{-1/\alpha}|\mu|.$$

An application of Hölder's inequality shows that

$$\mu \ll \|f\|_\alpha \left(\sum_{q\le x} q^{-1} \right)^{1/\alpha'}.$$

Altogether

$$\|P_\delta \check{A}Af\| \ll \left(x^{-1} \sum_{2\le n\le x} |f(n)-f(n-1)|^\alpha \right)^{1/\alpha} + (\log x)^{-\gamma}\|f\|,$$

the first and last norms in $L^2 \cap L^\alpha(\mathbb{C}^s)$ or $(L^2 \cap L^{\alpha'}(\mathbb{C}^s))'$, as the case may be. I denote the upper bound by $\Delta(f,x)$.

It follows from Lemma 21.8 that in terms of the operator T of Chapter 20, $\|\check{A}A-(I-T)\| \ll (\log x)^{-t}$, for a positive t. Without loss of generality we may assume $t=\gamma$, so that

$$\|P_\delta(I-T)f\| \ll \Delta(f,x). \tag{31.3}$$

We continue with the method of the stable dual applied somewhat in the manner of Chapter 8. The analogue of (8.3) is

$$\sum_{q\le x^\delta}^{**} \frac{1}{q} \left| f(q) + m\left(\frac{x}{q}\right) - m(x) \right|^\alpha \ll \Delta(f,x)^\alpha$$

where

$$m(y) = \sum_{\ell\le y} \frac{1}{\ell}\left(1-\frac{1}{\ell_0}\right) f(\ell),$$

ℓ runs over prime powers, and ** denotes that a certain summation condition upon the q (and depending upon f) is to be satisfied if $1<\alpha\le 2$. There is a similar analogue to (8.4).

If $K\ge 6$, $R\ge 2$, $R^K\le x^\delta$, then applications of Hölder's inequality show that

$$\sum_{R<q\le R^K} \frac{1}{q} \left| f(q) + m\left(\frac{x}{q}\right) - m(x) \right| \ll \Delta(f,x)(\log K)^{1/2}.$$

Let $0 < \eta < \delta/2$. I set $R = x^\eta$, $R^K = x^{\delta/2}$, $\tau = K^{1/2}$, $\psi(w) = \phi(w) = m(x/w)$ and apply Theorem 6.1. For x sufficiently large in terms of η the condition $R \ge 2$ will be satisfied, and (6.2) will hold with an ε which is a multiple of

$$(\log K)^{-1/2} \max_{x^{1-\delta/4} \le y \le x^{1+\delta/4}} \Delta(f, y),$$

the multiple depending at most upon α. Here $K = \delta/(2\eta)$, so that $\log K$ becomes large as η becomes smaller; but I shall not apply this refinement.

For $u \le v$,

$$\psi(v) - \psi(u) = \sum_{x/v < \ell \le x/u} \frac{1}{\ell}\left(1 - \frac{1}{\ell_0}\right) f(\ell),$$

which an application of Hölder's inequality shows not to exceed

$$\|f\|_\alpha \left(\sum_{x/v < \ell \le x/u} \frac{1}{\ell} \right)^{1/\alpha'}.$$

Under the conditions upon u, v in the statement of Theorem 6.1, the sum over the reciprocals of ℓ has the estimate

$$\log\left(\frac{\log x/u}{\log x/v}\right) + O((\log(x/v))^{-1})$$

which is $\ll (\log x)^{-1}$ for all x sufficiently large in terms of δ. The continuity condition (6.3) on ψ is amply satisfied.

There is a representation

$$m(x/w) = -D \log w + m(x) + O(\varepsilon)$$

valid uniformly for $1 \le w \le R^\tau$. In particular, for any positive θ not exceeding $\beta = \eta(\delta/(2\eta))^{1/2}$,

$$\left(\sum_{x^\theta < q \le x^\beta} \frac{1}{q} |f(q) - D \log q|^\alpha \right)^{1/\alpha} \ll \varepsilon,$$

the implied constant depending upon α, β, θ. In function analytic terms

$$\|(P_\beta - P_\theta)(f - D\log)\| \ll \max \Delta(f, y), \tag{31.4}$$

with the maximum taken over the interval $[x^{1-\delta/4}, x^{1+\delta/4}]$.

In addition, Theorem 6.1 guarantees the representation

$$D\tau\eta \log x = m(x) - m(xR^{-\tau}). \tag{31.5}$$

If in the inequality (31.4) we replace x by $x^{1/\beta}$, then for any $\zeta < 1$,

$$\|(I - P_\zeta)(f - D_1 \log)\| \ll \max \Delta(f, y), \tag{31.6}$$

with the maximum now taken over the interval $[x, x^{5/(4\beta)}]$, say.

To remove the projection P_ζ in this inequality, let us temporarily denote $f - D_1 \log$ by g. We apply the inequality (31.3) with g in place of f. Since

$$\begin{aligned}\Delta(\log, x) &\ll \left(x^{-1} \sum_{2 \le n \le x} \left| \log\left(1 - \frac{1}{n}\right) \right|^\alpha \right)^{1/\alpha} + (\log x)^{-\gamma} \|\log\| \\ &\ll (\log x)^{1-\gamma},\end{aligned}$$

and from (31.5)

$$D_1 \log x \ll \left(\sum_{x^{1/\beta_1} < q \le x^{1/\beta}} q^{-1} |f(q)|^\alpha \right)^{1/\alpha}, \quad \beta_1^{-1} = \beta^{-1}(1 - \eta\tau),$$

we see that

$$\|P_\delta(I - T)g\| \ll \Delta(f, x) + \Delta(f, x^{1/\beta}).$$

Moreover, $P_\delta(I - T)$ is a (uniformly) bounded operator, so from (31.6)

$$\|P_\delta(I - T)(I - P_\zeta)g\| \ll \max \Delta(f, y),$$

with the larger range of y. Combined with the previous inequality this yields

$$\|P_\delta(I - T)P_\zeta g\| \ll \max \Delta(f, y),$$

the maximum now taken over $[x^{1-\delta/4}, x^{5/(4\beta)}]$.

We fix ζ at a value less than $\min(1 - \delta, \delta)$. The value of $TP_\zeta g$ at a prime power $q \le x^\delta$ is

$$\sum_{\substack{x/q < \ell \le x \\ \ell \le x^\zeta}} \frac{g(\ell)}{\ell} \left(1 - \frac{1}{\ell_0}\right) = 0.$$

The operator $P_\delta T P_\zeta$ is trivial, and

$$\|P_\zeta g\| \ll \max \Delta(f, y).$$

At the expense of widening the interval of y to include the range $[x^{1-\delta/4}, x]$ we may omit the projection P_ζ from (31.6).

The following remark allows a simplification of this result. Suppose that $y_1 < y_2$ and that there are σ_j prime powers not exceeding y_j, $j = 1, 2$. There is a natural projection π from $\mathbb{C}^{\sigma_2}$ onto $\mathbb{C}^{\sigma_1}$, and for every vector f in $\mathbb{C}^{\sigma_2}$, πf may be considered in $\mathbb{C}^{\sigma_1}$ and in $\mathbb{C}^{\sigma_2}$. In particular, after Lemma 3.2.

$$\begin{aligned}
&\|\pi f\| \quad \text{considered in} \quad (L^2 \cap L^{\alpha'}(\mathbb{C}^{\sigma_1}))' \\
&\le \inf_{z>0} \left\{ \left(\sum_{\substack{q \le y_1 \\ |(\pi f)(q)| \le z}} \frac{1}{q}\left(1 - \frac{1}{q_0}\right) |(\pi f)(q)|^2 \right)^{1/2} \right. \\
&\qquad \left. + \left(\sum_{\substack{q \le y_1 \\ |(\pi f)(q)| > z}} \frac{1}{q}\left(1 - \frac{1}{q_0}\right) |(\pi f)(q)|^\alpha \right)^{1/\alpha} \right\} \\
&\le \text{a similar expression with } y_2 \text{ in place of } y_1 \\
&\le 2\|\pi f\| \quad \text{considered in} \quad (L^2 \cap L^{\alpha'}(\mathbb{C}^{\sigma_2}))'.
\end{aligned}$$

With this in mind we have established:
(31.7)

$$\|f - D_1 \log\| \ll \max_y \left(y^{-1} \sum_{2 \le n \le y} |f(n) - f(n-1)|^\alpha \right)^{1/\alpha} + (\log x)^{-\gamma} \|f\|$$

where the maximum is taken over $x^{1-\delta/4} \le y \le x^{5/(4\beta)}$, the initial norm is that of $L^2 \cap L^\alpha(\mathbb{C}^s)$, or $(L^2 \cap L^{\alpha'}(\mathbb{C}^s))'$, and the final norm is of a similar nature with s replaced by s_1, the number of prime powers not exceeding $x^{5/(4\beta)}$.

The inequality (31.7) lends itself to iteration. Let $\nu = 5/(4\beta)$. The function $D_1 = D_1(x)$ is given by (31.5) with x replaced by $x^{1/\beta}$. If in (31.7) we replace f (on $\mathbb{C}^{s_1}$) by $f - D_1(x^\nu) \log$, then for some function H of x,

$$\begin{aligned}
\|f - H \log\| &\ll \max_y \left(y^{-1} \sum_{2 \le n \le y} |f(n) - f(n-1)|^\alpha \right)^{1/\alpha} \\
&\quad + \max_y \left(y^{-1} \sum_{2 \le n \le y} \left| \log\left(1 - \frac{1}{n}\right) \right|^\alpha \right)^{1/\alpha} |D_1(x^\nu)| \\
&\quad + (\log x)^{-\gamma} \|f - D_1(x^\nu) \log\|
\end{aligned}$$

with the same convention as to the norms. Here

$$D_1(x^\nu)\log x \ll \left(\sum_{x^{\nu/\beta_1} < q \le x^{\nu/\beta}} q^{-1}|f(q)|^\alpha \right)^{1/\alpha}.$$

We now apply (31.7) to $\|f - D_1(x^\nu)\log\|$ and continue inductively. After k steps we reach a bound

$$\|f - S\log\| \ll \max_y \left(y^{-1} \sum_{2\le n\le y} |f(n) - f(n-1)|^\alpha \right)^{1/\alpha} + (\log x)^{-k\gamma}\|f\|_*$$

where the maximum is taken over an interval $x^{1-\delta/4} \le y \le x^d$ for some $d > 0$, and the * norm is of the appropriate form on $\mathbb{C}^{s_k}$, with s_k the number of prime powers not exceeding $x^{\nu^k/\beta}$.

Let $c \ge 4\max(d, \nu^k/\beta)$. The bound of Lemma 31.2 together with an integration by parts shows that

$$\sum_{q\le x^{c/4}} q^{-1}|f(q)|^\alpha \ll (\log x)^{\alpha+1} \sum_{2\le m\le x^{c/4}} m^{-1}|f(m) - f(m-1)|^\alpha.$$

A further integration by parts shows this upper bound to be

$$\ll (\log x)^{\alpha+2} \max_{z\le x^{c/4}} z^{-1} \sum_{2\le m\le z} |f(m) - f(m-1)|^\alpha.$$

If we fix k to satisfy $k\gamma > 2 + (2/\alpha)$, and replace x by x^2, then:
(31.8)

$$\|f - S\log\| \ll \max_{x\le y\le x^{c/2}} \left(y^{-1} \sum_{2\le n\le y} |f(n) - f(n-1)|^\alpha \right)^{\frac{1}{\alpha}}$$
$$+ (\log x)^{-1} \max_{z\le x} \left(z^{-1} \sum_{2\le m\le z} |f(m) - f(m-1)|^\alpha \right)^{\frac{1}{\alpha}}.$$

The function S (of x and f) is not specified, but according to Lemma 31.3 we may take for it $\|\log\|_2^{-2}(f, \log)$.

The second bounding term in (31.8) is by Theorem 2.1, $\ll (\log x)^{-1}\|f\|$. At a similar expense we may cut the range of summation in the first bounding term from $2 \le n \le y$ to $x^{2/3} < n \le y$. Replacing f by $f - S\log$, with the value of S specified in the previous paragraph, takes S to zero, and gives

$$\|f - S\log\| \ll$$
$$\max_{x\le y\le x^{c/2}} \left(y^{-1} \sum_{x^{2/3} < n\le y} |f(n) - f(n-1)|^\alpha \right)^{1/\alpha} + x^{-2/3-1/(3\alpha)}|S|.$$

To remove from this upper bound the blemish involving $|S|$, define

$$J(f) = \left(x^{-1} \sum_{x^{1/2}<n\le x} |f(n) - f(n-1)|^\alpha \right)^{1/\alpha}.$$

An elementary calculation shows that $J(\log) \gg x^{-1/2-1/(2\alpha)}$. Hence

$$x^{-1/2-1/(2\alpha)}|S| \ll J(S\log) \le J(S\log - f) + J(f).$$

Here

$$\begin{aligned} J(S\log - f) &\le \|(I - E^{-1})A(f - S\log)\| \quad \text{in} \quad M^\alpha(\mathbb{C}^t) \\ &\ll \|f - S\log\|. \end{aligned}$$

Therefore

$$x^{-2/3-1/(3\alpha)}|S| \ll x^{-1/6+1/(6\alpha)}\{\|f - S\log\| + J(f)\}.$$

With what we have already established,

$$\|f - S\log\| \ll \max_{x<y\le x^{c/2}} \left(y^{-1} \sum_{x^{2/3}<n\le y} |f(n) - f(n-1)|^\alpha \right)^{1/\alpha}. \tag{31.9}$$

To obtain the first assertion of Theorem 31.1, replace x in the inequality (31.9) by x^2. This lifts $\|f - S\log\|$ up to be taken on a larger space than $\mathbb{C}^s$. We project back down to $\mathbb{C}^s$, bearing in mind the note concerning norms on the spaces $\mathbb{C}^{\sigma_j}$ made earlier in this proof. This allows us to consider $\|f - S\log\|$ to be taken on $\mathbb{C}^s$, at the expense of an extra factor 2 in the implied constant of the final inequality. Of course, the value of S has been thereby changed, but we fix it anew using Lemma 31.3.

If $K(f)$ is defined like $J(f)$ but with the summation condition $x^{1/2} < n \le x$ replaced by $2 \le n \le x$, then arguing as we did with $J(f)$

$$x^{-1/\alpha}|F| \ll K(F\log) \le K(F\log - f) + K(f).$$

The first of the two ultimate bounding terms is by Theorem 2.1 at most a constant times $\|f - F\log\|$, for which we already have a satisfactory estimate.

The proof of Theorem 31.1 is complete.

As is suggested by the form of Lemma 31.2, other forms of Theorem 31.1 are possible. An example is

$$\|f - F\log\| \ll \left(\frac{1}{\log x}\sum_{2\le n\le x^c}\frac{|f(n)-f(n-1)|^\alpha}{n}\right)^{1/\alpha}.$$

To argue this particular case the norm on $M^\alpha(\mathbb{C}^t)$ would be replaced by

$$\left(L^{-1}\sum_{n\le x} n^{-1}|z_n|^\alpha\right)^{1/\alpha},$$

with $L = \sum n^{-1}$, $1 \le n \le x$.

Each $y > x(> 0)$ lies in an interval $(2^j x, 2^{j+1}x]$, j a non-negative integer. For $b_n \ge 0$,

$$\sum_{x<n\le y} b_n \le \sum_{\nu=0}^{j} 2^\nu x(2^\nu x)^{-1}\sum_{2^\nu x<n\le 2^{\nu+1}x} b_n \le 2y\max_{x\le w\le y}\left(w^{-1}\sum_{w<n\le 2w} b_n\right),$$

since the sum of the powers 2^ν is not more than $2yx^{-1}$. At little loss we may replace the upper bound of Theorem 31.1 by a (constant) multiple of

$$\max_{x^2\le y\le x^c}\left(w^{-1}\sum_{w<n\le 2w}|f(n)-f(n-1)|^\alpha\right)^{1/\alpha}.$$

If $a > 0$, b, $A > 0$, B are integers for which $\Delta = aB - Ab \ne 0$, then one can establish a form of Theorem 31.1 involving two (not necessarily distinct) additive functions f_j. The difference $f(n) - f(n-1)$ is replaced by $f_1(an+b) + f_2(An+B)$, and the norm $\|f - F\log\|$ by the sum of norms $\|P(f_j - F_j\log)\|$ in the same space, with a projection P that takes $f(q)$ to zero when $(q, aA\Delta) > 1$, and leaves it invariant otherwise. Thus P is almost the identity. To this end one can employ the arguments, involving Kloosterman sums and Selberg's sieve, given in the first three chapters of my book, [56].

A form of Theorem 31.1 involving sums of translations of a single additive function: $\lambda_1 f(n+a_1)+\cdots+\lambda_k f(n+a_k)$ can be obtained. Here the algebraic use of shift operators, demonstrated in the next chapter, is helpful.

No attempt has been made to keep the value of c in Theorem 31.1 small. When $\alpha = 2$, the possibility of every fixed $c > 8$ is sketched in [59]. An improvement allowing every $c > 4$ is given in Hildebrand, [102]. There he notes that the support of f on the prime powers $q \le x(\log x)^{-D}$ with D suitably large but fixed, may be neglected. Note, however, that according to a remark made in Chapter 29 of the present volume, Hildebrand's Lemma 2, which plays an essential rôle in his argument, needs a substantial adjustment to its proof.

32

Linear forms in shifted additive functions

Theorem 32.1. *Let $a_1 < a_2 < \cdots < a_k$ be integers and $\lambda_1, \ldots, \lambda_k$ real numbers not all zero. Then*

$$\lambda_1 f(n + a_1) + \cdots + \lambda_k f(n + a_k) \tag{32.1}$$

belongs to $\mathcal{L}^\alpha$, $1 < \alpha < 2$, if and only if there is a constant D for which the series :

$$\sum_{|f(q)-D\log q|\le 1} q^{-1}|f(q) - D\log q|^2, \qquad \sum_{|f(q)-D\log q|>1} q^{-1}|f(q) - D\log q|^\alpha \tag{32.2}$$

converge. When $\lambda_1 + \cdots + \lambda_k \neq 0$, D will be zero. For $\alpha \ge 2$ a similar result is valid without the summation restrictions upon the prime powers q.

As usual, $f(n)$ is defined to be zero when $n \le 0$. Theorem 32.1 may be compared with Theorem 8.1, which it generalises. It follows from Theorem 8.1 that the series at (32.2) converge if and only if $f - D\log$ belongs to $\mathcal{L}^\alpha$.

I first treat the difference $f(n) - f(n-1)$. I shall reduce all other cases to this one.

Proof of Theorem 32.1 for a single difference. I deal with the more complicated range, $1 < \alpha < 2$. From the hypothesis that $f(n) - f(n-1)$ belongs to $\mathcal{L}^\alpha$, Theorem 31.1 and Lemma 3.2, the sums:

$$\sum_{\substack{q\le x \\ |f(q)-F\log q|\le y}} q^{-1}|f(q) - F\log q|^2 + \sum_{\substack{q\le x \\ |f(q)-F\log q|>y}} q^{-1}|f(q) - F\log q|^\alpha \tag{32.3}$$

are bounded uniformly in $x \ge 2$. Here $F = (f, \log)\|\log\|_2^{-2}$ in terms of the function f on $\mathbb{C}^s$, and $y = \|f - F\log\|$, the norm taken in $(L^2 \cap L^{\alpha'}(\mathbb{C}^s))'$.

Considering the contribution to (32.3) arising from $q = 2$, we see that the functions F are bounded uniformly in $x \ge 2$. There is an unbounded sequence x_j on which $D = \lim F$ exists. Likewise the norms $\|f - F\log\|$

are uniformly bounded. Without loss of generality we may assume that on the same sequence x_j, $z = \lim \|f - F\log\|$ exists.

Let $t > 0$, $0 < \varepsilon < 1$. Then

$$\sum_{\substack{q \le t \\ |f(q) - D\log q| \le z - 2\varepsilon}} q^{-1}|f(q) - D\log q|^2$$

$$\le 1 + \sum_{\substack{q \le t \\ |f(q) - F\log q| \le z - \varepsilon}} q^{-1}|f(q) - F\log q|^2$$

$$\le 1 + \sum_{\substack{q \le x_j \\ |f(q) - F\log q| \le y}} q^{-1}|f(q) - F\log q|^2$$

for all sufficiently large x_j. The initial sum of this chain is bounded above uniformly in ε, t. This gives the convergence of the first series at (32.2) with the condition $|f(q) - D\log q| \le 1$ replaced by a similar one with upper bound z in place of 1.

We may obtain a like result for the series involving α-powers, and then recover the convergence of the series at (32.2) themselves.

Since the convergence of the series at (32.2) guarantees $f - D\log$ membership in $\mathcal{L}^\alpha$, the converse proposition of Theorem 32.1 is in the present case evident.

The reduction of the general case of Theorem 32.1 to the special case employs shift operators algebraically. The hypothesis (32.1) of Theorem 32.1 makes sense for $\alpha > 0$, and the reduction argument works for that range, too.

A ring of operators

Consider the space of all doubly-infinite sequences of complex numbers

$$\ldots, b_{-2}, b_{-1}, b_0, b_1, b_2, \ldots$$

We regard an arithmetic function h as belonging to this space by setting $h(n) = 0$ when $n = 0, -1, -2, \ldots$

Let E be the shift operator defined by $Eb_n = b_{n+1}$ for all n. This definition is slightly at odds with that given in Chapters 23 and 30, where functions of finite support were considered. Note that $Eh(n)$ denotes the value of the function Eh at the integer n. If we define a new arithmetic function H by $H(n) = h(2n)$, then $EH(n) = h(2n+2) = E^2h(2n)$.

Both E^0 and I will denote the operator which takes b_n to b_n for all n.

Corresponding to each polynomial $F(x) = \sum_{j=1}^{r} \mu_j x^{d_j}$ in $\mathbb{C}[x]$ there is an operator

$$F(E) = \sum_{j=1}^{r} \mu_h E^{d_j}$$

with action given by

$$F(E)b_n = \sum_{j=1}^{r} \mu_j b_{n+d_j}.$$

For polynomials $F(x)$ and $G(x)$ in $\mathbb{C}[x]$, we define in an obvious manner operators $F(E) + G(E)$ and $F(E)G(E)$. The operators $F(E)$ form a ring that is isomorphic to $\mathbb{C}[x]$.

Let f be a complex-valued additive arithmetic function. We define a ring Γ of polynomials in $\mathbb{C}[x]$ by: $F(x)$ belongs to Γ if and only if $F(E)f$ belongs to $\mathcal{L}^\alpha$. Since $\mathcal{L}^\alpha$ is taken into itself under the action of E, Γ is not only a ring, but an ideal in $\mathbb{C}[x]$.

Hypothesis (32.1) of Theorem 32.1 asserts that $L(E)f$ belongs to $\mathcal{L}^\alpha$ for the polynomial

$$L(x) = \sum_{j=1}^{k} \lambda_j x^{a_j}.$$

Thus Γ contains a nontrivial polynomial.

Note that even without the hypothesis of Theorem 32.1, the ring Γ may depend upon f.

Lemma 32.2. *Let d be a positive integer. If $F(x^d)$ belongs to Γ, then so does $F(x)$.*

Proof. Let

$$F(x) = \sum_{j=1}^{m} \rho_j x^{\ell_j}.$$

We may assume that $0 = \ell_1 < \ell_2 < \cdots < \ell_m$, otherwise F is a constant and the desired result trivially valid. We may also assume that $\rho_1 \neq 0$, otherwise $x^{-d\ell_2} F(x^d)$ belongs to Γ. From the hypothesis of the lemma

$$\sum_{j=1}^{m} \rho_j f(n + d\ell_j)$$

belongs to $\mathcal{L}^\alpha$. Then considered as a function of n, so does

$$\sum_{j=1}^{m} \rho_j f(dn + d\ell_j).$$

We cannot immediately assert that $f(dn+d\ell_j) = f(d)+f(n+\ell_j)$ to conclude the proof, since the condition $(d, n+\ell_j) = 1$ may not be satisfied.

Assume for the moment that $d = q = p^\alpha$ is a power of the prime p. Set $b = \ell_m$, so that $b \geq 1$. If now n has the form $n_0 = p^u w + t$ for integers $u, w \geq 0$, and t in the range $0 < t \leq p^u - b - 1$, then none of the integers $n + \ell_j$, $1 \leq j \leq m$, is divisible by a high power of p. In fact, if p^β is the highest power of p which divides $n + \ell_j$, say, then $\beta \leq u - 1$. Hence

$$\begin{aligned} f(qn + q\ell_j) &= f(qp^\beta \cdot p^{-\beta}(n + \ell_j)) \\ &= f(qp^\beta) + f(p^{-\beta}(n+\ell_j)) \\ &= f(n+\ell_j) - f(p^\beta) + f(p^{\alpha+\beta}), \end{aligned}$$

so that

$$|f(qn + q\ell_j) - f(n+\ell_j)| \leq \sum_{s=1}^{\alpha+u-1} |f(p^s)|.$$

This bound is uniform in integers n_0, and

$$\sum_{n_0 \leq x} |F(E)f(n_0)|^\alpha \ll x, \qquad x \geq 1. \tag{32.4}$$

The implied constant depends upon u.

The polynomial $(F(x^d) - \rho_1)^{2b} - \rho_1^{2b}$ is a multiple of $F(x^d)$, and so belongs to Γ. There are integers $e_i \geq 2b$, $i = 1, \dots, J$, so that

$$f - \sum_{i=1}^{J} \mu_i E^{e_i} f$$

belongs to $\mathcal{L}^\alpha$. If now u is fixed at a sufficiently large value, and n has the form $p^u w$ or $p^u w + t$, $p^u - b \leq t < p^u - 1$, then every $n + e_j$ will be an n_0. The analogue of inequality (32.4) holds for those integers not of the form n_0. Lemma 32.2 holds when d is a prime power.

If in the general case of Lemma 32.2 $d = qv$, where q is a prime power and $(q, v) = 1$, then the argument so far allows us to assert that $F(x^v)$ belongs to Γ. We conclude the proof by stripping prime powers from d inductively.

Lemma 32.3. *If $L(E)f$ belongs to $\mathcal{L}^\alpha$, then there is a non-negative integer t so that $(E-I)^t f$ belongs to $\mathcal{L}^\alpha$.*

Proof. The ring $\mathbb{C}[x]$ has a euclidean algorithm using degrees. Every ideal, and in particular Γ, is principal. Let $\phi(x)$ be a monic polynomial of minimal degree whose polynomial multiples generate Γ. Let

$$\phi(x) = \prod_{j=1}^{r}(x-\alpha_j).$$

The restrictions upon $F(x)$ in the proof of Lemma 32.2 may also be made here. If $r=0$, then f belongs to $\mathcal{L}^\alpha$ and the assertion of Lemma 32.3 is valid with $t=0$. Suppose that $r \geq 1$. Since $\psi(E)f$ belongs to $\mathcal{L}^\alpha$ whenever $E\psi(E)f$ belongs, we may assume that no $\alpha_j = 0$.

For every positive integer m

$$\prod_{j=1}^{r}(x^m - \alpha_j^m)$$

is a multiple of $\phi(x)$, and belongs to Γ. According to Lemma 32.2, so therefore does the polynomial

$$\phi_m(x) = \prod_{j=1}^{r}(x-\alpha_j^m).$$

The polynomial $\phi(x)-\phi_m(x)$ is of degree at most $(r-1)$ and belongs to Γ. This contradicts the minimality of the degree r of ϕ, unless $\phi(x)=\phi_m(x)$ identically.

For example, $\alpha_1, \alpha_1^2, \ldots, \alpha_1^{r+1}$ lie amongst the roots of $\phi(x)$. At least two of them coincide, $\alpha_1^u = \alpha_1^v$ say. Hence α_1 must be an algebraic root of unity. Since this argument may be applied to each of the roots α_j, there is an integer h so that $\alpha_j^h = 1$, $j = 1, \ldots, r$. Then $\phi(x) = \phi_h(x) = (x-1)^r$. The proof of Lemma 32.3 is complete, with $t=r$.

If the value of t in Lemma 32.3 is minimal, then $(x-1)^t$ divides $L(x)$. In particular, if $\lambda_1 + \cdots + \lambda_k \neq 0$, then $L(1) \neq 0$, $t=0$ and f belongs to $\mathcal{L}^\alpha$.

In terms of a shift operator E, it is traditional to define a difference operator $\Delta = E - I$ and to extend it by $\Delta^k = \Delta(\Delta^{k-1})$ for $k \geq 2$.

It is convenient to further introduce $\Delta_q = E^q - I$ for a prime q, with again $\Delta_q^k = \Delta_q(\Delta_q^{k-1})$, $k \geq 2$. The operators Δ and Δ_q are connected by the following relation.

Lemma 32.4. *For all $n \geq 1$, $t \geq 1$,*

$$\Delta_q^t f(n) = \sum_{j=0}^{t(q-1)} \mu_{t,j} \Delta^t f(n+j),$$

where the sum of the non-negative coefficients $\mu_{t,j}$ is q^t.

Proof. We appeal to the identity

$$(x^q - 1)^t = (x-1)^t (x^{q-1} + x^{q-2} + \cdots + 1)^t.$$

Lemma 32.5. *In the notation of Lemma 32.4*

$$\Delta_q^t f(n) = \sum_{j=0}^{t(q-1)} \mu_{t,j} \Delta_q^t f(qn + qj)$$

provided $(n, q) = 1$.

Proof. Define the additive function $w(n) = f(qn) - f(q)$. Then $\Delta w(n) = \Delta_q f(qn)$. From Lemma 32.4 the right hand expression in Lemma 32.5 has the alternative representation

$$\sum_{j=0}^{t(q-1)} \mu_{t,j} \Delta^t w(n+j) = \Delta_q^t w(n).$$

When $(n, q) = 1$, $(n + qs, q) = 1$ for all integers s and $\Delta_q^t w(n) = \Delta_q^t f(n)$. The lemma is established.

Lemma 32.6. *If $(E - I)^t f$ belongs to $\mathcal{L}^\alpha$ for some $t \geq 2$, then $(E - I)^{t-1} f$ also belongs to $\mathcal{L}^\alpha$.*

Proof. The argument is in part an elaboration of that used to establish Lemma 31.2.

We begin by noting that

$$\Delta_q^{t-1} f(qn + qj) - \Delta_q^{t-1} f(qn) = \Delta_q^t (E^{q(j-1)} + \cdots + I) f(qn).$$

It follows from Lemma 32.5 that for $(n, q) = 1$

$$\Delta_q^{t-1} f(qn) - q^{-(t-1)} \Delta_q^{t-1} f(n) \ll \sum_{s=0}^{qt-q} |\Delta_q^t f(qn + s)|. \tag{32.5}$$

Given any integer $n(>2q)$ we can find a further integer w, $1 \le w \le 2q$, such that $n-w$ has the form qm where $(q,m)=1$. We have

$$\Delta_q^{t-1} f(n) - \Delta_q^{t-1} f(n-w) = (E^q - I)^{t-1}(E^w - I) f(n-w)$$

and the polynomial $(x^q-1)^{t-1}(x^w-1)$ is divisible by $(x-1)^t$. Hence

$$\Delta_q^{t-1} f(n) - \Delta_q^{t-1} f(n-w) \ll \sum_{v=-w}^{qt-q} |\Delta^t f(n+v)|.$$

Together with (32.5) this shows that for any $n(>2q)$ there is an n_1 not exceeding n/q such that

$$|\Delta_q^{t-1} f(n) - q^{-(t-1)}\Delta_q^{t-1} f(n_1)|^\alpha \ll \sum_{v=-2q}^{qt} |\Delta^t f(n+v)|^\alpha + \sum_{s=0}^{qt} |\Delta_q^t f(qn+s)|^\alpha,$$

$= \theta(n)$, say.

Arguing inductively we obtain

$$\begin{aligned} |\Delta_q^{t-1} f(n)| &\le \theta(n)^{1/\alpha} + q^{-(t-1)}\theta(n_1)^{1/\alpha} + \cdots \\ &\cdots + q^{-(r-1)(t-1)}\theta(n_{r-1})^{1/\alpha} + q^{-r(t-1)}|\Delta_q^{t-1} f(n_r)|, \end{aligned}$$

where $n_1 \le q^{-1}n$, $n_2 \le q^{-1}n_1, \ldots,$ and $1 \le n_r \le 2q$. If we define $\theta(n)$ to be a suitable (constant) value for $1 \le n \le 2q$, and to be zero for negative n, then for all n

$$|\Delta_q^{t-1} f(n)| \le \sum_{j=0}^{r} q^{-j(t-1)}\theta(n_j)^{1/\alpha},$$

with $n_i \le q^{-i}n$, $i = 1, \ldots, r$, $1 \le n_r \le 2q$.

Let $\sigma = q^{-\alpha(t-1)/2}$. If $\alpha > 1$, then by Hölder's inequality with $(\alpha')^{-1} + \alpha^{-1} = 1$,

$$(32.6)\quad |\Delta_q^{t-1} f(n)|^\alpha \le \left(\sum_{m=0}^{\infty} q^{-\alpha'(t-1)/2} \right)^{\alpha-1} \sum_{j=0}^{r} \sigma^j \theta(n_j) \ll \sum_{j=0}^{r} \sigma^j \theta(n_j).$$

If $0 < \alpha \le 1$, then from the inequality

$$(d_1 + \cdots + d_v)^\alpha \le d_1^\alpha + \cdots + d_v^\alpha$$

which holds for non-negative numbers d_i, $i = 1, \dots, v$, $v \geq 1$, we obtain an inequality similar to (32.6) with the middle step omitted, where it is possible to replace σ by $q^{-\alpha(t-1)}$.

We sum over the integers n not exceeding x:

$$\sum_{n \leq x} |\Delta_q^{t-1} f(n)|^\alpha \ll \sum_{n \leq x} \sum_{j=0}^{r} \sigma^j \theta(n_j),$$

and seek to invert the order of summation in the upper bound.

In the induction procedure n_j arose from some integer n after finitely many steps of the form $n \to (n-w)/q$, where $1 \leq w \leq 2q$. Consider an integer m which is arrived at after $\ell(\geq 1)$ such steps. We can recover n by the operations

$$m \to qm + w_1 \to q(qm + w_1) + w_2 \to \cdots,$$

each operation consisting of multiplying by q and adding an integer w in the range $1 \leq w \leq 2q$. In particular, $n \geq q^\ell m$.

The number of integers n which can be reached from m in ℓ (reverse) operations is at most $(2q)^\ell$, since there are $2q$ choices for each w_i. By counting those integers n which reach a given n_j in ℓ steps we see that

$$\sum_{n \leq x} \sum_{j=0}^{r} \sigma^j \theta(n_j) \leq \sum_{m \leq x} \theta(m) \sum_{\ell} \sigma^\ell (2q)^\ell,$$

where the inner sum runs over all non-negative integers consistent with the inequality $q^\ell m \leq x$.

Fixing q at a sufficiently large value ensures that

$$2\sigma q = 2q^{1-\alpha(t-1)/2} < q^{1-\delta}$$

for some δ depending only upon α and t, and lying in the range $0 < \delta < 1$. Let γ be the largest integer ℓ such that $q^\ell m \leq x$. Then typically

$$\sum_{\ell} (2\sigma q)^\ell \leq \sum_{\ell} q^{(1-\delta)\ell} \leq \frac{q^{(1-\delta)(\gamma+1)} - 1}{q^{1-\delta} - 1} < 2q^{(1-\delta)\gamma} \leq 2(m^{-1}x)^{1-\delta}.$$

Hence

$$\sum_{n \leq x} |\Delta^{t-1} f(n)|^\alpha \ll x^{1-\delta} \sum_{m \leq x} \theta(m) m^{-1+\delta}, \quad x \geq 1. \tag{32.7}$$

We turn aside for a moment to prove that

$$\sum_{m \le y} \theta(m) \ll y, \qquad y \ge 1. \tag{32.8}$$

It follows from the hypothesis of Lemma 32.6 that

$$\sum_{n \le y} \sum_{v=-2q}^{qt} |\Delta^t f(n+v)|^\alpha \le (qt + 2q + 1) \sum_{-2q \le n \le y+qt} |\Delta^t f(n)|^\alpha \ll y.$$

Moreover, we see from Lemma 32.4 that

$$\begin{aligned} \sum_{s=0}^{qt} |\Delta_q^t f(qn+s)|^\alpha &\ll \sum_{s=0}^{qt} \sum_{j=0}^{t(q-1)} \mu_{t,j}^\alpha |\Delta^t f(qn+s+j)|^\alpha \\ &\ll \sum_{u=0}^{(2q-1)t} |\Delta^t f(qn+u)|^\alpha. \end{aligned}$$

Hence

$$\sum_{n \le y} \sum_{s=0}^{qt} |\Delta_q^t f(qn+s)|^\alpha \ll \sum_{n \le qy+(2q-1)t} |\Delta^t f(n)|^\alpha \ll y,$$

and (32.8) is established.

From (32.8), using an integration by parts,

$$\sum_{m \le y} \theta(m) m^{-1+\delta} \ll y^\delta, \qquad y \ge 1.$$

Together with (32.7) this shows $\Delta_q^{t-1} f$ to belong to $\mathcal{L}^\alpha$.

Further application of Lemma 32.4, t replaced by $t-1$, gives

$$\Delta_q^{t-1} f(n) = \sum_{j=0}^{(t-1)(q-1)} \mu_{t-1,j} \Delta^{t-1} f(n) + O\left(\sum_{z=0}^{tq} |\Delta^t f(n+z)| \right).$$

The coefficient of $\Delta^{t-1} f(n)$ is q^{t-1}. The functions $\Delta_q^{t-1} f$, $E^z \Delta^t f$, $0 \le z \le tq$, all belong to $\mathcal{L}^\alpha$. So therefore do $q^{t-1}\Delta^{t-1} f$ and $\Delta^{t-1} f$.

Lemma 32.6 is proved.

Proof of Theorem 32.1 in the general case. Suppose that the function

$$\lambda_1 f(n + a_1) + \cdots + \lambda_k f(n + a_k)$$

at (32.1) belongs to the class $\mathcal{L}^\alpha$ for $\alpha > 1$. In the notation preceding Lemma 32.2, $L(E)f$ belongs to $\mathcal{L}^\alpha$. By Lemma 32.3, for some t, $(E-I)^t f$ also belongs to $\mathcal{L}^\alpha$. We may assume that $t \geq 2$. Then Lemma 32.6 applied inductively assures that $(E-I)f$ belongs.

From what I have already established concerning a single difference, there is a constant D so that $f - D\log$ is a member of $\mathcal{L}^\alpha$. Moreover, $L(E)(f - D\log)$ will then belong to $\mathcal{L}^\alpha$. This is consistent with the initial hypothesis of Theorem 32.1 if and only if $DL(1) = 0$.

That this last condition together with the convergence of the series at (32.2) ensures the membership of $L(E)f$ in $\mathcal{L}^\alpha$ follows readily from Theorem 8.1.

The proof of Theorem 32.1 is complete.

With minor changes, I gave this complete reduction by polynomial operators in [44], including a proof of Theorem 32.1 for the cases $\alpha = 2, \infty$.

Having looped into Theorem 8.1 we may set out again, this time to study linear forms of functions.

33

Exercises: Stability; Correlations of multiplicative functions

Disregarding for the moment the motivating philosophy, the overall method described in this volume succeeds in part because as the associated parameter varies, the operators $A_\alpha, A'_{\alpha'}$, underlying the Turán–Kubilius inequalities and their duals tend to preserve their form. As mentioned earlier, this is redolent of the notion of stability in dynamics.

Besides this diffused notion of stability, detailed exemplifications of the notion are employed. An example is the Ulam–Hyers result of Chapter 6: an approximate homomorphism is near to a genuine homomorphism. Another is the the spectral decomposition bound of Lemma 20.5: an approximate eigenvector of a self-adjoint operator is near to a genuine eigenvector. We may compare these particular results to that of Newton approximation, which asserts that under favourable circumstances, usually control on the derivative, an approximate root of a function is near to a genuine root. For linear maps A on Hilbert spaces, self-adjointness plays the rôle of the derivative. In the notation of Chapter 21,

$$\delta\|y - z\| \le \|Ay - \mu y\|$$

where $Az = \mu z$, and δ is the distance of μ from the remaining spectrum of A. We may appreciate the similarity of the Ulam–Hyers result by recasting it.

Let $L(G, B)$ be the linear space, over $\mathbb{C}$, of functions defined on an abelian group G, assuming values in a Banach space B. Likewise, $L(G^2, B)$ denotes the space of functions taking $G \times G$, the Cartesian product of G with itself, into B. The map $F : L(G, B) \to L(G^2, B)$ assigns to the function f, the function given by

$$(x, y) \mapsto \langle x + y, f\rangle - \langle x, f\rangle - \langle y, f\rangle = f(x + y) - f(x) - f(y).$$

Note that under the usual conventions, F is linear. We partially metrise $L(G, B)$ by

$$\rho_1(f_1, f_2) = \sup_{g\in G} \|\langle g, f_1 - f_2\rangle\|,$$

provided this is finite. Here the norm is that on B. Similarly, for w_1, w_2 in $L^2(G, B)$,

$$\rho_2(w_1, w_2) = \sup_{t \in G^2} \|\langle t, w_1 - w_2 \rangle\|,$$

again provided this is finite. The Ulam–Hyers result asserts that any approximate zero f of F is near to a genuine zero f_0. In fact

$$\rho_1(f, f_0) \leq \rho_2(F(f), F(0)).$$

In the parlance of functional analysis, we might view the maps $f \to \langle g, f \rangle$, for g in G, as characters on $L(B, G)$, and say that f_0 is a weak root of F.

I close this short discussion of stability by considering a continuous analogue of Theorem 32.1 in the preceding chapter. What can we say of a function $g : [0, \infty) \to \mathbb{C}$, which a differential operator with constant coefficients moves into $L^2(0, \infty)$?

In the following nine exercises, D will denote the differential operator $\frac{d}{dx}$.

1. Let $g : [0, \infty) \to \mathbb{C}$ be Lebesgue measurable, ω a complex number, $\operatorname{Re}(\omega) < 0$. Suppose that $h = (D + \omega)g$ belongs to $L^2(0, \infty)$. Prove that

$$A = \lim_{x \to \infty} e^{\omega x} g(x)$$

exists.

2. Prove that

$$A - e^{\omega x} g(x) = \int_x^\infty h(y) e^{\omega y} dy, \quad x \geq 0.$$

3. Prove that

$$|Ae^{-\omega x} - g(x)|^2 \leq \frac{e^{-\operatorname{Re} \omega x}}{|\operatorname{Re} \omega|} \int_x^\infty |h(y)|^2 e^{\operatorname{Re} \omega y} dy, \quad x \geq 0.$$

4. Prove that $Ae^{-\omega x} - g(x)$ belongs to $L^2(0, \infty)$.

5. With the hypotheses of exercise 1, save that $\operatorname{Re}(\omega) > 0$, prove that

$$g(x) - g(0)e^{-\omega x} = e^{-\omega x} \int_0^x h(y) e^{\omega y} dy, \quad x \geq 0.$$

6. Prove that $g(x) - g(0)e^{-\omega x}$ belongs to $L^2(0, \infty)$.

If we define g to be 1 on the intervals $(2n, 2n + 1]$, $n = 1, 2, \ldots$, and to be 2 otherwise in $[0, \infty)$, then for any purely imaginary ω, $k(x) = g(x)e^{-\omega x}$ fulfills the hypothesis $(D + \omega)k \in L^2(0, \infty)$. However, $k(x) - ce^{-\omega x}$ cannot belong to $L^2(0, \infty)$ for any c, since that would force $g - c$ to belong, too.

7. Let $\alpha > 1$. Let $L(D) = \sum_{j=0}^{r} \lambda_j D^j$ be a differential operator with complex coefficients λ_j, for which the accompanying equation $L(z) = 0$ has no purely imaginary roots. Prove that

$$\|g - v\|_\alpha \leq \prod_j |\mathrm{Re}\,\omega_j|^{-1} \|L(D)g\|_\alpha$$

with the understanding that if for the measurable $g : [0, \infty) \to \mathbb{C}$, $L(D)g$ belongs to $L^\alpha(0, \infty)$, then there is a solution v to $L(D)v = 0$ on $[0, \infty)$, for which the inequality is valid. Here ω_j runs through the roots of $L(z) = 0$, counted with multiplicity. Of course, v is a classical solution of the form

$$\sum_j R_j(x) e^{\omega_j x},$$

where the polynomials $R_j(x)$ have complex coefficients.

We may view this result as a measure of the stability under diffused forcing of solutions to linear differential equations with constant coefficients. Let $L(D)g = w$, and suppose that $L(D)g_1 = w + w_1$, where w_1 is small in an $L^2(0, \infty)$ sense. Then for some solution v to $L(D)v = 0$, $\|g - g_1 - v\|_2 \leq \gamma \|w_1\|_2$ with $\gamma = \Pi|\mathrm{Re}\,\omega_j|^{-1}$. Since v need not vanish identically, we cannot immediately conclude that g varies continuously with h_1. However, the forms of v are fixed and only its implicit coefficients may vary.

8. Prove that

$$\left(\int_0^\infty e^{\alpha\beta x} |g(x) - v(x)|^\alpha dx\right)^{1/\alpha} \leq \prod_j |\beta - \mathrm{Re}\,\omega_j|^{-1} \left(\int_0^\infty e^{\alpha\beta x} |(L(D)g)(x)|^\alpha dx\right)^{1/\alpha}$$

with understanding similar to that of exercise 7, save that β is permitted any value differing from the $\mathrm{Re}\,\omega_j$.

9. In the same notation, and with $b > 0$, prove that if $(L(D)g)(x) \ll e^{-ax}$ for each $a < b$, then there is a v, $L(D)v = 0$ on $[0, \infty)$, such that $g(x) - v(x) \ll e^{-ax}$ for each $a < b$.

Correlations of multiplicative functions

We begin with some general remarks.

10. Prove that if for some ε, $0 < \varepsilon < 1$, the unimodular complex numbers z_m, $m = 1, \ldots, n$, satisfy

$$n^{-1} \left| \sum_{m=1}^{n} z_n \right| > 1 - \varepsilon,$$

then there is a unimodular η such that

$$n^{-1}\sum_{m=1}^{n}|z_m-\eta|^2<4\varepsilon.$$

11. Establish the following converse: If for some $\varepsilon>0$, the unimodular η and z_m, $m=1,\dots,n$, satisfy

$$n^{-1}\sum_{m=1}^{n}|z_m-\eta|^2<\varepsilon,$$

then

$$n^{-1}\left|\sum_{m=1}^{n}z_m\right|>1-\varepsilon/2.$$

We adapt the shift operator E and its polynomial extensions defined in Chapter 32. Let q be a prime, F_q the finite field of residue classes $(\operatorname{mod} q)$. For each polynomial ψ in the ring $F_q[E]$ and homomorphism f from Q^* into the additive group of F_q, define

$$S(y,\psi)=y^{-1}\left|\sum_{n\le y}e_q((\psi(E)f)(n))\right|,$$

where $e_q(\alpha)$, for real α, denotes $\exp(2\pi i\alpha/q)$.

12. Let $0<\varepsilon<1$. Prove that (i) If $S(y,\psi_j)>1-\varepsilon$ for $j=1,2$, then $S(y,\psi_1+\psi_2)>1-8\varepsilon$.

(ii) If $y\ge 2r/\varepsilon$ for a positive integer r, then $|S(y,\psi)-S(y,E^r\psi)|<\varepsilon$.

(iii) If $S(y,\psi)>1-\varepsilon$ and d is a positive integer, then $S(y,d\psi)>1-2d^2\varepsilon$.

The combined effect of these three propositions is that the polynomials ψ for which $S(y,\psi)$ is near to 1 largely form an ideal in $F_q[E]$.

13. Prove that if for a positive integer d

$$y^{-1}\left|\sum_{n\le y}e_q((\psi(E^d)f)(n))\right|>1-\varepsilon,$$

then

$$(y/d)^{-1}\left|\sum_{n\le y/d}e_q((\psi(E)f)(n))\right|>1-2d\varepsilon.$$

14. Let $b_0, \dots, b_k$, be k integers whose sum is not divisible by the prime q, x real, $x \geq 2$. Prove that there are positive numbers c_1 (> 1) and c_2 so that any complex-valued completely multiplicative g which satisfies $g(p)^q = 1$ on the primes, and

$$y^{-1} \left| \sum_{n \leq y} g(n)^{b_0} \cdots g(n+k)^{b_k} \right| \geq 1 - \varepsilon > 0$$

for every y in $(x, c_1 x]$, also satisfies

$$w^{-1} \left| \sum_{n \leq w} g(n) \right| > 1 - c_2 \varepsilon$$

for some w in this same interval.

15. Let $b_0, \dots, b_k$, be k integers whose sum is not divisible by the prime q. Let g be a complex-valued completely multiplicative function satisfying $g(p)^q = 1$ on the primes, and for which the series $\sum p^{-1}(1 - \operatorname{Re} g(p))$ diverges. Prove that

$$\limsup_{x \to \infty} x^{-1} \left| \sum_{n \leq x} g(n)^{b_0} \cdots g(n+k)^{b_k} \right| \leq c < 1,$$

with a value of c not depending upon g.

The assertion of this exercise is almost certainly valid with the condition $(b_0 + \cdots + b_k, q) = 1$ replaced by the requirement that not all the b_j be divisible by q.

Liouville's function is defined by $\lambda(n) = (-1)^{\Omega(n)}$, where Ω counts the number of prime divisors of n with multiplicity. Thus $\lambda(12) = -1$. The prime number theorem amounts to the proposition

$$\lim_{x \to \infty} x^{-1} \sum_{n \leq x} \lambda(n) = 0.$$

No doubt the asymptotic mean value of $\lambda(n)\lambda(n+1)$ is zero, but we cannot currently establish the much weaker conjecture

$$\liminf_{x \to \infty} x^{-1} \left| \sum_{n \leq x} \lambda(n)\lambda(n+1) \right| < 1.$$

Even more hopeless seems the conjecture that $\lambda(n)\lambda(n+1)\lambda(n+2)$ also has asymptotic mean value zero. The following result therefore comes as a surprise.

16. Prove that

$$\limsup_{x\to\infty} x^{-1}\left|\sum_{n\le x}\lambda(n)\lambda(n+1)\lambda(n+2)\right| \le \frac{20}{21}.$$

The more complicated sum has the better estimate.

Given an additive real-valued function f and integers $b_0, \dots, b_k$ not all zero, let $F_x(z)$ denote the frequency amongst the integers n in the interval $[1, x]$ of those for which

$$b_0 f(n) + \cdots + b_k f(n+k) \le z.$$

What are the necessary and sufficient conditions that these frequencies converge weakly to a distribution function as $x \to \infty$?

For real t, define the multiplicative function $g(n) = \exp(itf(n))$. Then $F_x(z)$ has characteristic function

$$\phi_x(t) = [x]^{-1}\sum_{n\le x} g(n)^{b_0}\cdots g(n+k)^{b_k},$$

and converges weakly if and only if this characteristic function converges to some characteristic function $\phi(t)$, uniformly on compact sets of t-values. Since $\phi(t)$ is continuous and $\phi(0) = 1$, it will be necessary that

$$\limsup_{t\to 0}\limsup_{x\to\infty}|\phi_x(t) - 1| = 0.$$

We study the above question by introducing the set J of polynomials ψ in $Q[E]$ for which

$$\limsup_{t\to 0}\limsup_{x\to\infty}\left|x^{-1}\sum_{n\le x}\exp(it(\psi(E)f)(n)) - 1\right| = 0.$$

Here Q denotes the field of rational numbers, and E is again the shift operator.

17. Prove that the validity of this last condition is not affected by changing the values of f on the powers of primes greater than the first.

We may assume f to be completely additive from now on.

18. Prove that J is a principal ideal, with a generator of the form $(E-I)^r$ for an integer r, $0 \le r \le k$. One can obtain this same structural result even if the polynomials ψ are confined to the ring $\mathbb{Z}[E]$. Why?

If the sum $b_0 + \cdots + b_k$ is non-zero, then $r = 0$ so that

$$\limsup_{t\to 0} \limsup_{x\to\infty} \left| x^{-1} \sum_{n\le x} \exp(itf(n)) - 1 \right| = 0.$$

19. Employ this property to recover that for the weak convergence of the corresponding frequencies $F_x(z)$, the condition from the classical Erdős–Wintner theorem (cf. Chapter 16) is again necessary and sufficient.

If $\sum_{j=0}^{k} b_j$ vanishes but $\sum_{j=1}^{k} jb_j$ does not, then $r = 1$, and the present reduction argument yields

$$\limsup_{t\to 0} \limsup_{x\to\infty} \left| x^{-1} \sum_{n\le x} \exp(it\{f(n+1) - f(n)\}) - 1 \right| = 0.$$

This condition, too, may be employed to characterise the corresponding frequencies $F_x(z)$ which possess a limiting distribution, Hildebrand, [104], Elliott, [70]. It is necessary and sufficient that for some constant α the series

$$\sum_{|f(p)-\alpha \log p|>1} \frac{1}{p}, \qquad \sum_{|f(p)-\alpha \log p|\le 1} \frac{(f(p) - \alpha \log p)^2}{p}$$

converge.

Doubtless this same criterion is appropriate to the cases $r \ge 2$ as well, but we cannot prove so. The convergence of these two series for some α will guarantee that the frequencies

$$[x]^{-1} \sum_{\substack{n\le x \\ f(n+2)-2f(n+1)+f(n)\le z}} 1$$

possess a weak limiting distribution as $x \to \infty$, but is is not currently known whether the converse is true.

The results of this section elaborate part of a colloquium that I gave at the University of Chile, Santiago, whilst visiting the Pontificia Universidad Catolica de Chile in June and July of 1991. Cf. 'On the correlation of multiplicative functions', *Notas Soc. Mat. Chile* **11** (1992), no. 1, 1–11.

34

Further readings

This chapter contains a selection of results and exercises largely pertaining to the method of the stable dual. Unlike practice in the previous chapters, theorems will be stated without proof.

Norms of compositions of arithmetic operators

The following results generalise Theorem 2.1.

Theorem 34.1. *Let $w(n)$ be a non-negative real-valued arithmetic function which satisfies $w(q) \ll 1$, $w(qm) \ll w(q)w(m)$ uniformly for prime powers q, and positive integers m, $(q,m)=1$. Suppose that for some $\beta > 1$ the sums*

$$y^{-1}\sum_{n\le y} w(n)^{\beta}, \qquad y \ge 1$$

are uniformly bounded.

If $\alpha \ge 2$, then

$$\left(\frac{1}{x}\sum_{n\le x} w(n)\left|f(n)-\sum_{q\le x}\frac{f(q)}{q}\right|^{\alpha}\right)^{1/\alpha} \ll \left(\sum_{q\le x}\frac{|f(q)|^2}{q}\right)^{1/2} + \left(\sum_{q\le x}\frac{|f(q)|^{\alpha}}{q}\right)^{1/\alpha}$$

uniformly for complex-valued additive functions f, and for $x \ge 1$.

If $0 < \alpha < 2$, then a similar inequality holds with the upper bound replaced by

$$\inf_{\eta>0}\left\{\left(\sum_{\substack{q\le x\\ |f(q)|\le\eta}}\frac{|f(q)|^2}{q}\right)^{1/2} + \left(\sum_{\substack{q\le x\\ |f(q)|>\eta}}\frac{|f(q)|^{\alpha}}{q}\right)^{1/\alpha}\right\}.$$

Theorem 34.2. *Let $w(n)$ be a non-negative real-valued arithmetic function which satisfies $w(qm) \le w(q)w(m)$ for prime powers q, and positive integers m, $(q,m)=1$. Suppose, further, that the sums*

$$\sum_{q \le y} \frac{w(q)-1}{q}, \qquad \frac{1}{y}\sum_{q \le y} w(q)\log q,$$

are bounded above uniformly in $y \ge 2$. Then inequalities of the type in Theorem 34.1 hold, but with every weight q^{-1} replaced by $q^{-1}w(q)$.

The assertions of Theorem 34.1 remain valid if in the sum $\sum q^{-1}f(q)$, $q \le x$, the weights q^{-1} are replaced by $q^{-1}(1-q_0^{-1})$.

These results are established in [61]. The proof of Theorem 2.1 in the present volume is a simplified version of the argument given there.

A version of Theorem 34.1 with $\alpha = 2$ is obtained in [56] Chapter 1. It is applied in Chapter 3 of that same reference with $w(n) = (n/\phi(n))^2$, and ϕ Euler's function. A particular aim of that volume is the following inequality, there labelled Theorem 10.1.

Theorem 34.3. *Let $a > 0$, b, $A > 0$, B be integers for which the determinant $\Delta = aB - Ab$ is non-zero. Then there are parameters x_0, c, depending at most upon the four integers, such that*

$$\sum_{\substack{q \le x \\ (q,aA\Delta)=1}} \frac{1}{q}\,|f(q) - F\log q|^2 \ll \sup_{x \le y \le x^c} \frac{1}{y} \sum_{x < n \le y} |f(an+b) - f(An+B)|^2$$

with

$$F = \sum_{\substack{x^{1/2} < q \le x \\ (q,aA\Delta)=1}} q^{-1}f(q) \Big/ \sum_{\substack{x^{1/2} < q \le x \\ (q,aA\Delta)=1}} q^{-1}\log q$$

holds uniformly for all complex valued additive functions f, and real $x \ge x_0$.

Theorem 34.3, introduced in the notes concluding Chapter 6, may be compared with the case $\alpha = 2$ of Theorem 31.1. I constructed a proof of Theorem 34.3 in three parts. The first part derived, from a bound

$$\sum_{n \le x} |f(an+b) - f(An+B)|^2 \ll x(\log x)^{c_1},$$

a bound for f itself:

$$\sum_{\substack{n \le x \\ (n,aA\Delta)=1}} |f(n)|^2 \ll x(\log x)^{c_2}.$$

Besides the generalised Turán–Kubilius inequality mentioned above, I applied an estimate for Kloosterman trigonometric sums in order to guarantee a reasonable distribution to the solutions of a certain diophantine equation. This first part of the argument I had already made precise by the Fall of 1979. Moreover, I knew from an earlier work, [40], that if $c_2 = 1$ were obtained, then f could be determined with accuracy. My hope was to employ the method of the stable dual to strip powers of $\log x$ from the upper bound $x(\log x)^{c_2}$, using the fact that the differences $f(an+b) - f(An+B)$ were uniformly bounded in mean square: $c_1 = 0$. I was successful in the Spring of 1980, as I mentioned earlier. A coarse version of the proof methodology may be found in my earlier work [53], [54], in which I settle a conjecture of Kátai. What amounts to an L^∞ analogue of Theorem 34.3 had been conjectured by Ruzsa, but I was not aware of it.

To establish a generalisation of Theorem 31.1 that treats $f(an+b) - f(An+B)$ rather than $f(n) - f(n-1)$, we need an analogue of Lemma 31.2. This may be obtained by modifying the argument given in [56] for the case $\alpha = 2$, employing an appropriate version of Theorem 34.1. Indeed, one may treat $f_1(an+b) + f_2(An+B)$, with the additive functions f_1 and f_2 not necessarily related.

Applications of Theorem 34.3 to the representation of rationals by products of integers may be found in [56] Chapters 15 and following.

Various remarks concerning the application of Theorems 34.1 and 34.2 to number theory may be found in [61]. I note here only that

$$\sum_{n \le x} \frac{\tau(n)^2}{n^{11}} \left| f(n) - \sum_{q \le x} \frac{\tau(q)^2 f(q)}{q^{12}} \right|^2 \ll x \sum_{q \le x} \frac{\tau(q)^2 |f(q)|^2}{q^{12}},$$

with $\tau(n)$ Ramanujan's function, holds uniformly for complex additive f and real $x \ge 2$. In the study of the value distribution of $\tau(n)$, we may confine ourselves to those integers n which have about $\log\log n$ distinct prime factors.

An interesting generalisation of the mean square Turán–Kubilius inequality and its dual to integers without large prime factors is established in Alladi [2].

Multiplicative functions and almost periodicity

The Fourier–Bohr spectrum of a function in the space B^α was defined in Chapter 12. More generally, of a function h in $\mathcal{L}^\alpha$, $\alpha \ge 1$, it will comprise the real β for which

$$\limsup_{x \to \infty} x^{-1} \left| \sum_{n \le x} h(n) e^{-2\pi i \beta x} \right| > 0.$$

Using the dual of the Turán–Kubilius inequality arithmetically, Daboussi proved that the Fourier–Bohr spectrum of a multiplicative function with values in the complex unit disc contains no irrational points. There was an announcement in his joint paper with Delange, [16]. Delange sketches the original proof (and this is the first I saw of it), together with some generalisations, in his paper [22].

A natural way to approach generalisations is through the generalised Turán–Kubilius duals given as Theorem 3.1 of the present volume.

I adapt the argument of Daboussi in exercises 12–15 of Chapter 12. In place of a bound by (what is essentially) a power norm, I employ an approximate representation in terms of a bilinear form. Delange later told me that he had, himself, drawn Daboussi's attention to my 1972 St. Louis paper [32]. See also, [31].

The result of Daboussi went far. It provoked a sharper bound by Montgomery and Vaughan, [127], which they applied to estimate the coefficients of cyclotomic polynomials, [128]. In turn, their study was advanced to a largely finished form by Bachmann, [4].

In their book [154], Schwarz and Spilker study almost periodic arithmetic functions in terms of their expansions by Ramanujan sums, employing explicit and implicit Gelfand representations. Most suitable for the characterisation of such functions, their methodology extends to the treatment of (otherwise arbitrary) multiplicative functions in $\mathcal{L}^\alpha$ with a non-zero (asymptotic) mean value only if $\alpha \geq 2$.

In Chapter III of their book two multiplicative functions f and g are defined to be *related* if the series $\sum p^{-1}|f(p) - g(p)|$, taken over the prime numbers, converges. In that same chapter they set out to axiomatise the notion, familiar to experts but not commonly expressed in the literature, that subject to certain conditions related multiplicative functions have similar asymptotic behaviour. In particular, for the function h defined by Dirichlet convolution: $f = h * g$, the series $\sum |h(n)|n^{-1}$, taken over the positive integers, then converges. To this end they employ the analogue for Dirichlet series, and established by Hewitt and Williamson using a Gelfand representation, of Wiener's celebrated theorem on the invertibility of non-vanishing functions with an L^1 Fourier series. The several examples of such a convolution relation that arise in the present text I deal with by direct computation of Euler products.

I encountered the book of Schwarz and Spilker after I had already completed the text and many of the exercises of the present volume. These two works overlap very little and may with advantage be read alongside each other.

A central object in the study of multiplicative arithmetic functions is the dual group of Q^*, but that is for another day.

Approximate functional equations

Literature concerning the stability of structural homomorphisms is now more widespread, as may be seen from the survey article of Hyers and Rassias, [107]. The streamlining of these studies into a discipline is perhaps hampered by the fact that stability is a notion imported from dynamics. As with many notions adopted into mathematics from the physical world, it is suggestive but imprecise. In their paper Hyers and Rassias demonstrate several viable interpretations for the stability of equations.

Distribution of primes

Since he was interested in factorials, sums of logarithms arise naturally in Chebyshev's work on the primes, but there is no sign that he thought of using a logarithm to weight an arbitrary arithmetic function. I mention, in passing, that one of Riemann's integrals for the analytic continuation of $\zeta(s)$ was already considered at length by Chebyshev, but only for a real variable s.

Besides avoidance of the theory of complex variables, there seems to be a feeling that an elementary proof of the prime number theorem should provide an estimate for $x^{-1}\pi(x)\log x - 1$, or $x^{-1}\sum \mu(n)$, $n \le x$, without excursions to infinity. A discrete proof of the prime number theorem which meets such a demand and in part employs ideas related to those of the present work is given by Hildebrand, [101].

Another proof of the prime number theorem, of an elementary nature and due to Daboussi, may be found in the aforementioned book of Schwarz and Spilker.

Maximal inequalities

The inequalities of Chapter 29 are of interest for their form as much as for their precision. Individual bounds need not be best. For example, an inequality of Hunt, developed following Carlson's work on the Fourier expansion of functions in $L^2(0,1)$, and so not immediate, allows the factor $\log \delta^{-1}$ in exercise 3 to be removed, Montgomery [125], Elliott, [73]. In general it is not clear to what extent the introduction of a uniformity into a norm inequality might degrade an upper bound.

Arithmetic functions on short intervals

There is a considerable literature concerning the value distribution of general arithmetic functions, but much less when the functions are considered over intervals short in comparison with their distance from the origin.

The following appear as questions 13, 14 and 15 on p. 420 of [56].

Let $0 < \beta < 1$. What are the necessary and sufficient conditions that

$$\lim_{x\to\infty} x^{-\beta} \sum_{x<n\le x+x^\beta} |f(n)|^2$$

exist and be finite for a real additive function?

When does the (approximate) frequency

$$x^{-\beta} \sum_{\substack{x<n\le x+x^\beta \\ f(n)\le z}} 1$$

converge weakly as $x \to \infty$?

What are the necessary and sufficient conditions to be satisfied by a multiplicative function g, with values in the complex unit disc, in order that

$$\lim_{x\to\infty} x^{-\beta} \sum_{x<n\le x+x^\beta} g(n)$$

exist?

As indicated there, these questions were paradigms. I was aware of analogues of the Turán–Kubilius inequality which considered additive functions f over an interval $(x-y, x]$ where y could be small in comparison with x. It suffices that the support of f on the prime powers last only up to a certain fixed power of y. The inequalities could be obtained directly.

Applying the dual of such an inequality to the study of (say) an additive function imposed restrictions upon the values which f could take on the prime powers. Naturally, some care would be needed to simplify the resulting functional equations, since the interval $(x - y, x]$ was to be short in comparison with x, say $y = x^\beta$, with β in (0,1) and regarded as fixed. Varying x, y would give better control on the equations but require an hypothesis concerning a patch of intervals $(x_1 - y_1, x_1]$, with x_1 near to x and y_1 near to y, in some sense. The complications attendant to this approach were offset by there being no other generally applicable argument in analytic number theory which took the direction *integers to primes*, and set out from hypotheses concerning the values of functions on intervals of the form $(x - x^\beta, x]$.

An immediate difficulty is that the values of an additive (or multiplicative) function on a single interval $(x - y, x]$ with y small in comparison with x, need not determine the function on the larger interval $(1, x]$.

Several versions of a Turán–Kubilius inequality for intervals are given in [62]. The following appears there as Theorem 2, and is typical.

For $1 \le y \le x$ define

$$\Delta(f) = y^{-1} \sum_{x-y<n\le x} \left| f(n) - \frac{1}{y} \sum_{x-y<m\le x} f(m) \right|^2 .$$

Theorem 34.4. *Let c be a positive constant. Then*

$$\Delta(f) \ll |\lambda|^2 \Delta(\log) + \sum_{q \le y} \frac{|g(q)|^2}{q} + \frac{1}{y} \sum_{y < q \le x}{}' |g(q)|^2$$

uniformly for $y > x^c$, $x \ge 2$, complex λ and complex additive functions f. Here $g = f - \lambda \log$, q denotes a prime power, and $'$ that summation is restricted to those prime powers which exactly-divide at least one integer in the interval $(x - y, x]$.

It is shown as Theorem 3 of [62], that for y not too small in comparison with x this version of the Turán–Kubilius inequality is an approximate isometry.

Let ψ be the function log, scaled to have norm 1 in the Hilbert space $L^2(\mathbb{C}^s)$.

Theorem 34.5. *With $F = (f, \psi)$ the inequality*

$$|F|^2 \Delta(\psi) + \|f - F\psi\|_2^2 + \frac{1}{y} \sum_{y < q \le x}{}' |f(q) - F\psi(q)|^2 \ll \Delta(f)$$

holds uniformly for all complex additive functions f, $x(\log x)^{-1/8} \log\log x \le y < x$, and x absolutely large.

The restriction upon y is no doubt unnecessarily strong, but allows application of the function analytic ideas described in Chapter 20 of the present work. Extensive remarks, concerning the application of the method of the stable dual to the study of inequalities of Turán–Kubilius type may be found in Chapter 10 of [56].

An upper bound for non-negative multiplicative functions g on an interval $(x - y, x]$ is given by Shiu, [160].

1. Let $0 < \alpha < 1/2$, $0 < \beta < 1/2$, $B > 0$. Check that the argument of Shiu gives

$$\sum_{\substack{x - y < n \le x \\ n \equiv a (\operatorname{mod} k)}} g(n) \ll \frac{y}{\phi(k) \log x} \exp\left(\sum_{\substack{p \le x \\ (p,k)=1}} \frac{g(p)}{p} \right)$$

uniformly for $(a, k) = 1$, $1 \le k \le y^{1-\alpha}$, $x^\beta < y \le x$, $x \ge 2$ and all multiplicative functions g which satisfy $0 \le g(q) \le B$ on the prime powers. The implied constant depends at most upon α, β and B.

Note that (2.13) on p. 163 of Shiu's account needs $\delta > 1/2$ rather than $\delta > 0$. Since only the cases $\delta \ge 3/4$ are applied, no damage is done.

2. Let $0 < c < 1$. Prove that Ramanujan's function satisfies

$$\sum_{x-y<n\le x} |\tau(n)| n^{-11/2} \ll y(\log x)^{-1/18}$$

uniformly for $2 \le x^c \le y \le x$.

To treat multiplicative functions g with values in the complex unit disc, over short intervals $(x-y, x]$, Hildebrand, [103], applies the method of the stable dual. I denote by $M(g(m), x)$ the average $x^{-1}\sum g(m)$ taken over the positive integers m not exceeding x.

Theorem 34.6. *In the above notation*

$$y^{-1} \sum_{x-y<n\le x} g(n) = M(g(m)m^{i\alpha}, x) y^{-1} \int_{x-y}^{x} t^{-i\alpha} dt + O(R(x,y))$$

with

$$R(x,y) = \left(\log\left(\frac{2\log x}{\log 2x/y}\right)\right)^{-1/4}$$

and α any real number, $|\alpha| \le x$, for which $|M(g(m)m^{i\alpha}, x)|$ is maximal, holds uniformly for $3 \le y \le x$. The implied constant is absolute.

The error term $R(x,y)$ will not be small unless y falls below a (small) power of x; and to ensure that $R(x,y) \to \infty$ as $x \to \infty$, we need $\log y/\log x$ to approach 1.

The parameter α may depend upon g and x. In applications of Theorem 34.6 an effort is usually made to better localise α.

A commentary on key steps in the establishment of Theorem 34.6 may be given by adapting the proofs of Theorems 13.1 and 13.2. The details may differ from those of Hildebrand.

3. Let $c > 0$. Establish a suitable version of the Turán–Kubilius inequality and deduce that

$$\sum_{p\le(y\log y)^{1/2}} p \left| \sum_{\substack{x-y<n\le x\\ n\equiv 0 \pmod p}} a_n - \frac{1}{p}\sum_{x-y<n\le x} a_n \right|^2 \ll y \sum_{x-y<n\le x} |a_n|^2$$

holds uniformly for $2 \le y \le x$ and all complex numbers a_n.

Let g be a multiplicative function with values in the complex unit disc. For $1 \le y \le x$ define

$$m(x,y) = y^{-1} \sum_{x-y<n\le x} g(n).$$

4. Prove that

$$\sum_{P<p\le P^K} \frac{1}{p}\left|g(p)m\left(\frac{x}{p},\frac{y}{p}\right) - m(x,y)\right| \ll (\log K)^{1/2}$$

uniformly for g, and real $1 \le y \le x$, $P \ge 2$, $K \ge 2$, $P^K \le y^{1/2}$.

5. Prove that

$$\sum_{p\le y} p^{-1}|1-|g(p)||^2 \left|m\left(\frac{x}{p},\frac{y}{p}\right)\right|^2 \ll 1.$$

Show that $g(p)$ in the inequality of exercise 4 may be replaced by $g(p)/|g(p)|$ when $g(p) \ne 0$, and by 1 otherwise.

6. Prove that

$$\left|m\left(\frac{x}{z},\frac{y}{z}\right) - m\left(\frac{x}{w},\frac{y}{w}\right)\right| \le (w-z)\left(\frac{1}{z}+\frac{2x}{zy}\right)+\frac{3w}{y}$$

holds uniformly for $1 \le y \le x$, $0 < z \le w$.

This coarse continuity condition will ultimately only legitimise a useful result when $\log y/\log x$ is near to 1. It also gives only a weak bound on the size of the parameter α in Theorem 34.6.

7. In the notation of exercise 4 prove that if $P^{-1/3}xy^{-1} \ll (\log K)^{-1/2}$, then for some real λ

$$m\left(\frac{x}{w},\frac{y}{w}\right) = w^{i\lambda}m(x,y) + O((\log K)^{-1/2}),$$

uniformly for $1 \le w \le P^{\sqrt{K}}$. Moreover, λ may be taken to be zero when g is real valued.

8. Deduce that

$$\left(\frac{|\lambda|^2y^2}{x^2} + \sum_{p\le T^{c_1}} \frac{1}{p}|g(p)p^{i\lambda}-1|^2\right)|m(x,y)|^2 \ll 1$$

with

$$T = \min(\exp(\sqrt{\log x}), \exp(\log x(\log x/y)^{-1})),$$

c_1 and the implied constant absolute.

Note that $\lambda \ll xy^{-1}$ improves $|\lambda| \le x$, but allows λ to be large when y is small in comparison with x, a case of great interest.

The following generalisation of Theorem 13.1 of Wirsing was first obtained by Hildebrand.

9. Let g be a real valued multiplicative function which satisfies $|g(q)| \le 1$ on the prime powers, and for which the series $\sum p^{-1}(1 - g(p))$, taken over the primes, diverges. Prove that

$$y^{-1} \sum_{x-y<n\le x} g(n) \to 0, \quad x \to \infty,$$

provided $\log y/\log x \to 1$.

A slightly more elaborate application of the slowly oscillating estimate obtained in exercise 7 leads to a bound on $m(x, y)$ sharper than that of exercise 8.

10. After exercise 7, multiply the estimate for $m(xw^{-1}, yw^{-1})$ by $w^{-i\lambda-2}$ and integrate with respect to w over the range $1 \le w \le P^{\sqrt{K}}$ to obtain

$$m(x, y) = M(g(m)m^{i\lambda}, x)y^{-1} \int_{x-y}^{x} t^{-i\lambda} dt + O(yx^{-1} + (\log K)^{-1/2}).$$

Let

$$\rho(x, T) = \exp\left(-\min_{|\tau|\le T} \sum \frac{1}{p}(1 - \operatorname{Re} g(p)p^{i\tau})\right) + T^{-1}.$$

11. Check that [41], Volume 1, Lemma 6.10 gives

$$\sum_{n\le z} g(n) \ll x\rho(x, T)^{1/5}$$

uniformly for $1 \le z \le x$, $T > 0$. The exponent $1/5$ may be improved.

12. Deduce from exercises 10 and 11 that

$$m(x, y) \ll \rho(x, T + 2|\lambda|)^{1/10} + \left\{\log\left(\frac{\log x}{\log 2x/y}\right)\right\}^{-1/2}$$

uniformly for $x^{1/2} < y \le x$, $T > 0$.

Again, the exponent $1/10$ may be improved.

If g is real valued, then

$$\sum_{n\le y} g(n) \ll y \exp\left(-\frac{1}{40}\sum_{p\le y} p^{-1}(1 - g(p))\right)$$

holds uniformly for $y \ge 2$; [72] Theorem 3, p. 163. It is shown in Tenenbaum, [162] Théorème 7, p. 382, that $1/40$ may be replaced by $1/8$.

13. Prove that

$$y^{-1} \sum_{x-y<n\le x} g(n) \ll \exp\left(-\frac{1}{16} \sum_{p\le x} p^{-1}(1-g(p)) \right) + \left\{ \log\left(\frac{\log x}{\log 2x/y} \right) \right\}^{\frac{-1}{2}}$$

uniformly for $2 \le x^{1/2} \le y \le x$, and real multiplicative functions g which are at most 1 in absolute value.

For an individual multiplicative function the estimate vouchsafed by Theorem 34.6 is non-trivial, but may be weak in comparison with that afforded using local information. Theorem 34.6 shows to advantage when applied to distributional problems concerning functions whose local behaviour is largely unknown, problems for which the method of the stable dual was initially constructed. I give some examples.

For $1 \le y \le x$ and a real valued arithmetic function h, define the frequencies

$$\nu_{x,y}(n; h(n) \le z) = \frac{1}{[x]-[x-y]} \sum_{\substack{x-y<n\le x \\ h(n)\le z}} 1$$

which count integers n.

Theorem 34.7. *Let $c > 1$. Let the numbers M_j, N_j satisfy $N_{j+1} \le N_j^c$, $N_j \to \infty$, $\log M_j / \log N_j \to 1$, as $j \to \infty$. Then the frequencies*

$$\nu_{N_j,M_j}(n; f(n) \le z), \qquad j \to \infty,$$

converge weakly to a limiting distribution, for the additive function f, if and only if the three series

$$\sum_{|f(p)|>1} \frac{1}{p}, \quad \sum_{|f(p)|\le 1} \frac{f(p)}{p}, \quad \sum_{|f(p)|\le 1} \frac{f(p)^2}{p},$$

taken over the prime numbers, converge.

This result is proved in [66]. It represents a far reaching localised version of the classical result of Erdős and Wintner considered in Chapter 16. The example $N_j = 3^{3^j}$, $M_j = N_j \exp(-(\log N_j)^{1/2})$, shows that the information concerning $f(n)$ may be given on short intervals very widely spaced. I have the feeling that if the interpolation points N_j are allowed much farther apart, then the three series condition may no longer be appropriate.

More generally let $F_x(z)$ denote the distribution function $y^{-1}\{\min(x, e^z) - \min(x-y, e^z)\}$, $0 < y \le x$. Let H be a distribution function whose characteristic function does not vanish on the real line.

Theorem 34.8. *Assume that $1 \le y \le x$, $\log y/\log x \to 1$, as $x \to \infty$ in some manner. Then there are parameters $\gamma(x,y)$ so that for these x-values*

$$\nu_{x,y}(n; f_x(n) - \gamma \le z) \Rightarrow H(z), \quad x \to \infty,$$

if and only if there are further parameters $\lambda = \lambda(x)$, $|\lambda| \le x^{1/2}$, so that

$$F_x(\lambda^{-1}(z+\gamma)) * \nu_{x,x}(n; f_x(n) - \lambda \log n \le z) \Rightarrow H, \quad x \Rightarrow \infty.$$

Here $\Rightarrow$ denotes (the) weak convergence (of measures), and $*$ the convolution of distribution functions. The additive function $f_x(n)$ is defined anew for each value of x.

The class of possible limit laws H is very wide. It includes the infinitely divisible laws, such as the Normal, the Poisson or the Cauchy.

Theorem 34.8 is established in [67]. In a 1988 lecture at Bordeaux, [68], concerned with the distribution of additive functions defined on short intervals, I made the following comment:

"A possibility that I raised in a lecture at Cambridge, England, fifteen years ago, is here made manifest: we need to know about convolution decompositions of the limit law. This news is received with mixed feelings. In spite of the work of the mighty Russian school, amongst others, the decomposition of probability laws, within the theory of probability proper, is still not well understood."

This comment still stands. The difficulties attending the decomposition of probability laws are in part of an aesthetic nature. Owing to their wide occurrence in the study of the physical world, it is natural to single out for attention the infinitely divisible laws. However, their class is too narrow to include limit laws which arise naturally in the theory of arithmetic functions. Illustrative examples of this phenomenon are furnished in Chapter 18 of [41], and in Elliott and Erdős, [79].

A beautiful conjecture of Lévy, established by Cramer, asserts that any (convolution) factor of a Normal law must again be a Normal law. Cramer's theorem allows a simplification of an important case of Theorem 34.8; [67] Theorem 2.

Theorem 34.9. *Assume that $1 \le y \le x$, $\log y/\log x \to 1$, as $x \to \infty$ in some manner. Then there are parameters $\gamma = \gamma(x,y)$ so that for the same x*

$$\nu_{x,y}(n; f_x(n) - \gamma \le z) \Rightarrow \frac{1}{\sqrt{2\pi}} \int_{-\infty}^{z} e^{-u^2/2} du$$

if and only if there are further parameters $\lambda = \lambda(x)$, *for those* x *values, so that* $\lambda y/x \to 0$ *and*

$$\nu_{x,x}(n; f_x(n) - \lambda \log n + \lambda \log x - \gamma \le z) \Rightarrow \frac{1}{\sqrt{2\pi}} \int_{-\infty}^{z} e^{-u^2/2} du.$$

To investigate the convergence to the normal law over intervals $(x - y, x]$ with $\log y / \log x \to 1$, it suffices to consider the same problem over the whole interval $[1, x]$. Paradoxically, this apparently simpler case is still not solved in complete generality.

The following result, of Levin and Timofeev, and Elliott, in part established by the method of the stable dual, [34], [38], see also [41], Volume 2, Chapters 13, 14, 16, typifies what is currently known concerning the value distribution of additive functions on initial intervals. We confine attention to $f_x(n)$ of the form $f(n)/\beta(x)$, where the positive function β satisfies $\beta(x) \to \infty$, and for each $t > 0$, $\beta(x^t)/\beta(x) \to 1$, as $x \to \infty$.

Theorem 34.10.

$$\nu_{x,x}(n; f(n) - \alpha(x) \le z\beta(x)) \Rightarrow F(z), \quad x \to \infty$$

if and only if there is a constant A *so that the independent random variables* Y_p, *one defined for each prime* p *not exceeding* x, *and distributed according to*

$$Y_p = \begin{cases} (f(p) - A \log p) & \text{with probability } p^{-1}, \\ 0 & \text{with probability } 1 - p^{-1}, \end{cases}$$

satisfy

$$P\left(\sum_{p \le x} Y_p - \alpha(x) + A \log x \le z\beta(x) \right) \Rightarrow F(z), \quad x \to \infty.$$

Here x *runs through all positive real values exceeding* 2.

In general, only the convergence of the frequencies $\nu_{x,y}(n; f_x(n) - \gamma \le z)$ to the improper law is untrammelled by side conditions, and so in a more satisfactory state; [67] Theorem 1.

Theorem 34.11. *Assume that* $1 \le y \le x$, $\log y/\log x \to 1$ *as* $x \to \infty$. *Then there are parameters* $\gamma = \gamma(x, y)$ *so that*

$$\nu_{x,y}(n; f_x(n) - \gamma \le z) \to \begin{cases} 1 & \text{if} \quad z > 0, \\ 0 & \text{if} \quad z < 0, \end{cases}$$

as $x \to \infty$, *if and only if there are further parameters* λ *so that* $\lambda y/x \to 0$, *and*

$$\min_{\mu} \left(\mu + \sum_{q \le x} \frac{1}{q} \min(1, |f(q) - (\lambda + \mu) \log q|^2) \right) \to 0$$

as $x \to \infty$.

As usual, q denotes a prime power. Note that in these last two theorems x runs through all reals exceeding 2 (say), and not through a chosen subsequence, as was the case in Theorems 34.8 and 34.9.

The correlation of multiplicative functions and sums of additive functions

A number of conjectures concerning the correlations of multiplicative functions are made in [71], [77]. Note that the first minimum in Conjecture III of [77] should be taken over Dirichlet characters with moduli not exceeding T. The following is typical.

Conjecture. *Let* $a_j > 0$, b_j, $j = 1, 2$, *be integers for which* $a_1 b_2 \ne a_2 b_1$. *Let* g_j, $j = 1, 2$, *be multiplicative functions with values in the complex unit disc, and satisfying*

$$\limsup_{x \to \infty} x^{-1} \left| \sum_{n \le x} g_1(a_1 n + b_1) g_2(a_1 n + b_2) \right| > 0.$$

Then there are reals τ_j *and Dirichlet characters* χ_j *so that the series*

$$\sum p^{-1}(1 - Re\, \chi_j(p) g_j(p) p^{-i\tau_j}), \qquad j = 1, 2,$$

taken over the prime numbers, converge.

A corollary of this conjecture would be that

$$x^{-1} \sum_{n \le x} \mu(n) g(n+1) \to 0, \qquad x \to \infty,$$

for every multiplicative g with values in the complex unit disc.

If f_j, $j = 1, 2$, are additive functions, then the characteristic function of the frequency

$$\nu_x(n; f_1(a_1 n + b_1) + f_2(a_1 n + b_2)) \le z$$

has the form

$$\phi_x(t) = [x]^{-1} \sum_{n \le x} g_1(a_1 n + b) g_2(a_2 n + b)$$

with $g_j(n) = \exp(itf_j(n))$ and t real. The study of the value distribution of sums of additive functions is equivalent to that of the correlations of a class of multiplicative functions. Thus $f_1(a_1 n + b_1) + f_2(a_2 n + b_2)$ belongs to $\mathcal{L}^2$ if and only if $\partial^2 \phi_x(t)/\partial t^2$ at $t = 0$ is bounded uniformly for $x \ge 2$. An example of such a study is given in [56]. Further developments may be found in Hildebrand, [104]; Elliott [70], [77]. All these results are established using variants of the method of the stable dual. The following comes from the last of these references.

Theorem 34.12. *Let $x \ge 2$, $(\log x)^{-1/100} \le \delta \le 1/3$. Let g, g_y, $x^\delta \le y \le x$, be complex-valued multiplicative functions, of absolute value at most 1, which satisfy*

$$y^{-1} \left| \sum_{n \le y} g(a_1 n + b_1) g_y(a_2 n + b_2) \right| \ge 1 - \delta^2$$

uniformly for $x^\delta \le y \le x$.

If there is a character χ mod $a_1(a_1, b_1)^{-1}$, and a real τ, $|\tau| \le x^\delta (\log x)^{-2}$ for which

$$\sum_{p \le x} \frac{1}{p} |1 - g(p)\chi(p)p^{i\tau}|^2 \le \frac{1}{3} \log \frac{1}{\delta},$$

then there exists a real μ, $|\mu| \le \min(1, (\log x)^{-9/10})$ so that

$$Re \sum_{p \le x} \frac{1}{p} \left(1 - g(p)\chi(p)p^{i(\tau + \mu)}\right) \le c_0$$

where the constant c_0 depends at most upon the four integers a_j, b_j.

A similar acceleration process plays an important rôle in earlier of my studies of sums and differences of additive functions. Note that in Theorem 34.12 the function g_y may be defined anew for each y. This generality is very useful in applications. Theorem 34.12 represents a preliminary application of the method of the stable dual to the study of the correlation of multiplicative functions. In turn Theorem 34.12 has the following important application, [77], otherwise currently unobtainable.

For $j = 1, 2$, let $\beta_j(x)$ be positive real measurable functions, defined for $x \ge 2$, which satisfy $\beta_j(x) \to \infty$, and $\beta_j(x^y)/\beta_j(x) \to 1$ for each $y > 0$, as $x \to \infty$.

Theorem 34.13. *In order that for suitably chosen real $\alpha_j(x)$ the frequencies*

$$\nu_x\left(n; \sum_{j=1}^{2} \frac{f_j(a_j n + b_j) - \alpha_j(x)}{\beta_j(x)} \le z\right)$$

should converge weakly to a distribution function as $x \to \infty$, it is both necessary and sufficient that there exist constants λ_j, $j = 1, 2$, and a real $\gamma(x)$ so that if the independent random variables Y_p, one for each prime p not exceeding x, are distributed according to

$$Y_p = \begin{cases} (f_1(p) - \lambda_1 \log p)/\beta_1(x) & \text{with probability } \dfrac{1}{p}, \\ (f_2(p) - \lambda_2 \log p)/\beta_2(x) & \text{with probability } \dfrac{1}{p}, \\ 0 & \text{with probability } 1 - \dfrac{2}{p}, \end{cases}$$

then the

$$P\left(\sum_{p \le x} Y_p - \gamma(x) \le z\right)$$

converge to the same distribution function.

Theorem 34.13 may be compared with Theorem 34.10, which it includes.

Rings of shift operators

Variant forms of the algebraic treatment of shift operators given in Chapter 32 may be applied in wide circumstances. Examples are: the study of congruences [47]; the representation of rationals by products of specified rationals [48], [51], [52], [56]; the correlation of multiplicative functions as in Chapter 33; the factorisation of polynomials [64]. The following result is taken from the last of these references.

Let $K(x), P_i$ denote typical polynomials in $\mathbb{Z}[x]$ with positive leading coefficients. Let $Q(x)^*$ be the multiplicative group generated by the rational functions P_1/P_2. For a given rational function $S(x)$ in $Q(x)^*$ let $\Delta(S(x))$ be the subgroup of $Q(x)^*$ which is generated by the $S(K(x))$ for varying $K(x)$. Define the quotient group $H(S(x)) = Q(x)^*/\Delta(S(x))$.

Theorem 34.14. *The group $H(x(x^2+a))$ with $a \neq 0$ is cyclic of order 3, generated by the image of x.*

As a corollary, every positive rational number r has a representation

$$r = \prod_{j=1}^{m} (n_j(n_j^2+a))^{\varepsilon_j},$$

with positive integers n_j, and each ε_j having the value 1 or -1. Moreover, the representation may be largely parameterised by polynomials. Representations of this type bear upon the value distribution of $\chi(n(n^2+c))$ for Dirichlet characters χ, [64].

The product representation of integers by integers of a particular form plays an important rôle in the rapid factorisation of large numbers by computers; Buhler, Lenstra and Pomerance, [11].

Abstract semigroups

The method of the stable dual is very flexible. As presented here it rests upon two structural properties: the positive integers are a freely generated semigroup possessing an appropriate (approximate) homomorphism into the non-negative reals; the group of positive reals is locally compact, and so has a Haar measure.

A positive real-valued function $a \mapsto Na$, on a freely generated commutative semigroup, is a norm if it satisfies $Nab = NaNb$ for all a, b in the semigroup, $Np \geq c > 1$ uniformly for all generators (primes). The notation continues in force for the next five questions.

The notions of divisibility and of (complex valued) additive and multiplicative functions, concerned initially with the integers, extend naturally to such semigroups.

Suppose that

$$\sum_{Np \leq y} \log Np \ll y, \quad y \geq 2.$$

14. Let $\sigma = 1 + (\log x)^{-1}$. Prove that

$$\sum_{Np \leq x} \left(\frac{1}{Np} - \frac{1}{(Np)^\sigma} \right) \ll 1, \quad \sum_{Np > x} \frac{1}{(Np)^\sigma} \ll 1,$$

uniformly for positive x.

Suppose, further, that

$$\sum_{Na \leq y} 1 \ll y, \qquad y \geq 2.$$

15. Prove that

$$\sum_{a}(Na)^{-w} \ll (w-1)^{-1}$$

uniformly for real $w > 1$.

16. Prove that

$$\sum_{Np\le x} \frac{1}{Np} \le \log\log x + O(1), \quad x \ge e^2.$$

17. Prove that any non-negative real multiplicative function g on the semigroup, at most M on the generator-powers q, satisfies

$$\sum_{Na\le x} g(a) \ll Mx \exp\left(\sum_{Nq\le x} \frac{g(q)-1}{Nq}\right), \quad x \ge 2.$$

For a complex valued additive function f on the semigroup, define

$$E(x) = \sum_{Nq\le x} \frac{f(q)}{Nq}\left(1 - \frac{1}{Nq_0}\right),$$

$$D(x) = \left(\sum_{Nq\le x} \frac{|f(q)|^2}{Nq}\right)^{\frac{1}{2}} \ge 0.$$

As usual in this volume, q_0 denotes the generator of which q is a power.

18. (Turán–Kubilius inequality for normed semigroups) Prove that, subject to the above hypotheses on the norm,

$$\sum_{Na\le x} |f(a) - E(x)|^2 \ll xD(x)^2, \quad x > 0.$$

Analogues of the other inequalities in Chapter 2 may be likewise established. Elements of the semigroup are not explicitly constrained to grow in norm. Moreover, the upper bound hypotheses on the norm may be appropriately varied. Within this generality the classical argument of Turán (and of Kubilius) is not applicable. Remarks concerning the present argument and the application of inequalities of Turán–Kubilius type to the study of abstract number theoretical functions may be found in [65] p. 207.

Examples of semigroups to which the method of the stable dual might be applied are: the integral ideals in the ring of algebraic integers of a finite algebraic extension of the rationals; the ring of polynomials in one or more variables over a field. These examples and their connection with a study of product representations are discussed in the concluding chapter of my memoir [77].

35

Rückblick

(after the manner of Johannes Brahms)

In the notation of Chapter 23, the Fourier transform

$$F : L^2(G) \to L^2(\widehat{G})$$

is an isometry. It has as inverse its adjoint:

$$f = F^*Ff.$$

On the classical groups $\mathbb{R}$ and the multiplicative positive reals, this procedure changes ends: Knowledge of a function near to the identity of the group is equivalent to knowledge of its transform far from the identity of the (dual) group.

In the present volume I have constructed an arithmetic analogue of this procedure. Given a function f defined on the prime powers up to x, in the notation of Chapter 3 a point in $\mathbb{C}^s$, we transform it to an additive arithmetic function using the map

$$A : L^2(\mathbb{C}^s) \to M^2(\mathbb{C}^t), \quad t = [x],$$

of Chapter 3. Although one-to-one, A does not cover $\mathbb{C}^t$. Rather than inverting A directly, we operate upon $\mathbb{C}^t$ with the adjoint map A^*. Then $(I-S)f = A^*Af$, where $S = I - A^*A$ largely coincides with the operator T of Chapter 20. In particular, we can estimate the spectrum of S somewhat accurately. The largest eigenvalue λ_1 of S is close to 1. The corresponding (unit) eigenvector ψ is essentially the classical logarithmic function, suitably renormalised. Removing this eigenvector, we obtain an operator $S_1 = S - \lambda_1(\ ,\psi)\psi$ with spectral radius close to 1/2. Since $(I-S_1)(f-(f,\psi)\psi) = (I-S)f$, we can invert our map according to the scheme

$$f - (f,\psi)\psi = (I-S_1)^{-1}A^*Af = (I + S_1 + S_1^2 + \cdots)A^*Af.$$

Note that the operator $T : \mathbb{C}^s \to \mathbb{C}^s$ also changes ends. The values of Tf on the prime powers up to w ($\le x$) are determined by the values of f on the prime powers in $[x/w, x]$.

The whole procedure may be carried out in terms of other norms.

Within this structure, the method of the stable dual studies the stability of the various operators as the parameter x varies.

References

1. Aczél, J. *Lectures on Functional Equations and their Applications*, Academic Press, New York, London, 1966.
2. Alladi, K. The Turán–Kubilius inequality for integers without large prime factors, *J. für die reine und angew. Math.* **335** (1982), 180–196.
3. Atkinson, F.V. and Lord Cherwell. The mean values of arithmetical functions, *Quart. J. Math. Oxford Ser.* **20** (1949), 65–79.
4. Bachmann, G. *On the Coefficients of Cyclotomic Polynomials*, Amer. Math. Soc. Memoirs **106** (1993), no. 510.
5. Bergh, J. and Löfström, J. *Interpolation Spaces*, Grundlehren **223**, Springer–Verlag, Berlin, Heidelberg, New York, 1976.
6. Bombieri, E. On the large sieve, *Mathematika* **12** (1965), 201–225.
7. Bombieri, E. and Davenport, H. On the large sieve method, *Abh. aus Zahlentheorie und Analysis zur Erinnerung an Edmund Landau*, Deut. Verlag. Wiss., Berlin, 1968, 11–22.
8. Bombieri, E., Friedlander, J.B. and Iwaniec, I. Primes in arithmetic progressions to large moduli, III., *J. Amer. Math. Soc.* **2** (1989), 215–224.
9. Bruggeman, R.W. Fourier coefficients of cusp forms, *Invent. Math.* **45** (1978), 1–18.
10. Bruggeman, R.W. *Fourier Coefficients of Automorphic Forms*, Lecture Notes in Math. **865**, Springer–Verlag, Berlin, Heidelberg, New York, 1981.
11. Buhler, J.P., Lenstra, H.W. Jr. and Pomerance, C. Factoring integers with the number field sieve, in *The Development of the Number Field Sieve*, eds. A.K. Lenstra and H.W. Lenstra, Jr., Lecture Notes in Math. **1554**, Springer–Verlag, Berlin, Heidelberg, 1993, 50–94.
12. Chudakov, N.G. Theory of the characters of number semigroups, *J. Indian Math. Soc.* **20** (1956), 11–15.

13. Conway, J.B. *Functions of One Complex Variable*, Springer–Verlag Graduate Texts in Math. **11**, second edition, New York, Berlin, Heidelberg, Tokyo, 1973, 1978, 1986.

14. Daboussi, H. Caractérisation des fonctions multiplicatives p.p. B^λ à spectre non vide, *Ann. Inst. Fourier Grenoble* **30** (1980), 141–166.

15. __________. Sur les fonctions multiplicatives ayant une valeur moyenne non nulle, *Bull. Soc. Math. France* **109** (1981), 183–205.

16. Daboussi, H. and Delange, H. Quelques propriétés des fonctions multiplicatives de module au plus égàl à 1, *C. R. Acad. Sci. Paris Sér. A*, **278** (1974), 657–660.

17. __________. On a theorem of P.D.T.A. Elliott on multiplicative functions, *J. London Math. Soc.*, **14** (1976), 345–356.

18. __________. On multiplicative arithmetical functions whose modulus does not exceed one, *J. London Math. Soc.* **26** (1982), 245–264.

19. Davenport, H. *Multiplicative Number Theory*, Markham, Chicago, 1967; second edition, revised by H. L. Montgomery, Graduate Texts in Math. **74**, Springer–Verlag, New York, Heidelberg, Berlin, 1980.

20. Davenport, H. and Halberstam, H. The values of a trigonometrical polynomial at well spaced points, *Mathematika* **13** (1966), 91–96; Corrigendum and addendum, *Mathematika* **14** (1967), 229–232.

21. Delange, H. Sur les fonctions arithmétiques multiplicatives et ses applications, *Ann. Scient. Éc. Norm. Sup.* 3e *série* t. **78** (1961), 273–304.

22. __________. Generalisation of Daboussi's Theorem, in *Colloquia Mathematica Societatis János Bolyai* **34**, *Topics in Classical Number Theory* (Budapest 1981), published 1984, 305–318.

23. Deligne, P. La conjecture de Weil, I., *Publ. Math. I.H.E.S.* **43** (1974), 273–307.

24. Duke, W. and Iwaniec, H. Estimates for coefficients of L-functions, II., *Proc. Amalfi Conf. on Analytic Number Theory (Maiori, 1989)*, 71–82, Univ. Salerno, Salerno, 1992.

25. Elliott, P.D.T.A. On certain additive functions, *Acta Arith.* **12** (1967), 365–384.

26. __________. On certain additive functions, II., *Acta Arith.* **14** (1968), 51–64.

27. __________. The Turán–Kubilius inequality and a limitation theorem for the large sieve, *Amer. J. Math.* **92** (1970), 293–300.

28. __________. On the mean value of $f(p)$, *Proc. London Math. Soc.*, Davenport Memorial Volume, **21** (1970), 28–96.

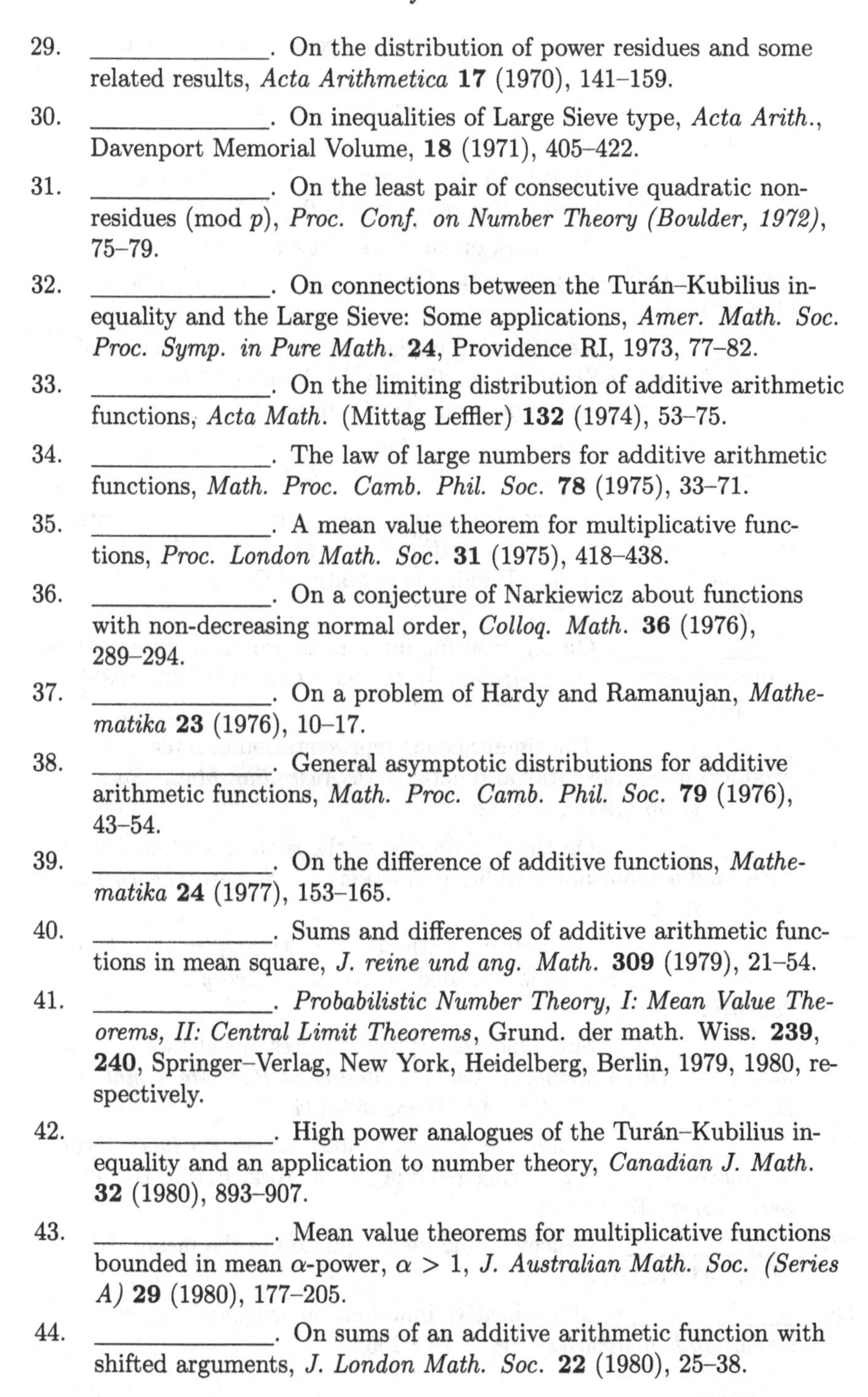

29. ______________. On the distribution of power residues and some related results, *Acta Arithmetica* **17** (1970), 141–159.

30. ______________. On inequalities of Large Sieve type, *Acta Arith.*, Davenport Memorial Volume, **18** (1971), 405–422.

31. ______________. On the least pair of consecutive quadratic non-residues (mod p), *Proc. Conf. on Number Theory (Boulder, 1972)*, 75–79.

32. ______________. On connections between the Turán–Kubilius inequality and the Large Sieve: Some applications, *Amer. Math. Soc. Proc. Symp. in Pure Math.* **24**, Providence RI, 1973, 77–82.

33. ______________. On the limiting distribution of additive arithmetic functions, *Acta Math.* (Mittag Leffler) **132** (1974), 53–75.

34. ______________. The law of large numbers for additive arithmetic functions, *Math. Proc. Camb. Phil. Soc.* **78** (1975), 33–71.

35. ______________. A mean value theorem for multiplicative functions, *Proc. London Math. Soc.* **31** (1975), 418–438.

36. ______________. On a conjecture of Narkiewicz about functions with non-decreasing normal order, *Colloq. Math.* **36** (1976), 289–294.

37. ______________. On a problem of Hardy and Ramanujan, *Mathematika* **23** (1976), 10–17.

38. ______________. General asymptotic distributions for additive arithmetic functions, *Math. Proc. Camb. Phil. Soc.* **79** (1976), 43–54.

39. ______________. On the difference of additive functions, *Mathematika* **24** (1977), 153–165.

40. ______________. Sums and differences of additive arithmetic functions in mean square, *J. reine und ang. Math.* **309** (1979), 21–54.

41. ______________. *Probabilistic Number Theory, I: Mean Value Theorems, II: Central Limit Theorems*, Grund. der math. Wiss. **239**, **240**, Springer–Verlag, New York, Heidelberg, Berlin, 1979, 1980, respectively.

42. ______________. High power analogues of the Turán–Kubilius inequality and an application to number theory, *Canadian J. Math.* **32** (1980), 893–907.

43. ______________. Mean value theorems for multiplicative functions bounded in mean α-power, $\alpha > 1$, *J. Australian Math. Soc. (Series A)* **29** (1980), 177–205.

44. ______________. On sums of an additive arithmetic function with shifted arguments, *J. London Math. Soc.* **22** (1980), 25–38.

45. ______________. Recent results in the theory of arithmetic functions, in *Journées de Théorie Analytique et Elémentaire de Nombres (Orsay, Juin 1980)*, Publications Mathématiques d'Orsay, Paris (1981), 19–27.

46. ______________. Multiplicative functions and Ramanujan's τ-function, *J. Australian Math. Soc. (Series A)* **30** (1981), 461–468.

47. ______________. A remark on the Dirichlet character values of a completely reducible polynomial (mod p), *J. Number Theory*, **13** (1981), 12–17.

48. ______________. Representing integers as products of integers of a specified type, in *Séminaire de Théorie des Nombres (Université de Bordeaux I, 1981–82)*, exposé no. **38**, 11 Juin 1982.

49. ______________. The exponential function characterised by an approximate functional equation, *Illinois J. Math.* **26** (1982), 503–518.

50. ______________. A new inequality in the theory of additive arithmetic functions, in *Journées Arithmétiques (S.M.F.) Colloque Hubert Delange (7,8 Juin 1982)*, Publications Math. d'Orsay, Paris (83.04), 53–58.

51. ______________. On representing integers as products of integers of a prescribed type, *J. Australian Math. Soc. (Series A)* **35** (1983), 143–161.

52. ______________. The simultaneous representation of integers by products of certain rational functions, *J. Australian Math. Soc. (Series A)* **35** (1983), 404–420.

53. ______________. On the distribution of the roots of certain congruences and a problem for additive functions, *J. Number Theory* **16** (1983), 267–282.

54. ______________. On additive arithmetic functions $f(n)$ for which $f(an+b) - f(cn+d)$ is bounded, *J. Number Theory* **16** (1983), 285–310.

55. ______________. Subsequences of primes in residue classes to prime moduli, in *Turán Memorial Volume, Studies in Pure Mathematics*, Birkhäuser, Basel, Boston MA, 1983, 157–164.

56. ______________. *Arithmetic Functions and Integer Products*, Grund. der math. Wiss. **272**, Springer–Verlag, New York, Berlin, Heidelberg, Tokyo, 1985.

57. ______________. Cauchy's functional equation in the mean, *Adv. Math.* **51** (1985), 253–257.

58. ______________. Multiplicative functions on arithmetic progressions, *Mathematika* **34** (1987), 199–206.

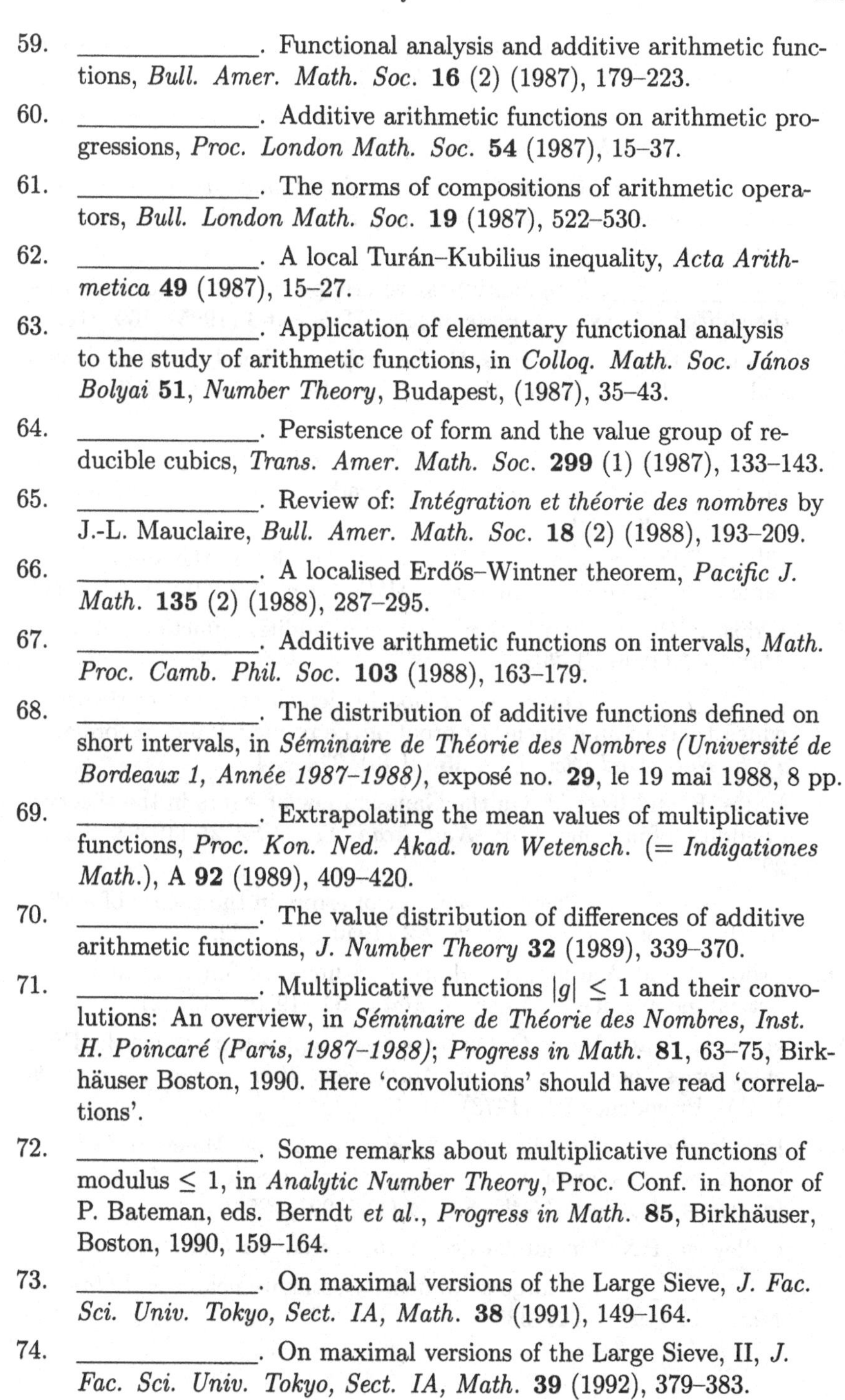

59. ______________. Functional analysis and additive arithmetic functions, *Bull. Amer. Math. Soc.* **16** (2) (1987), 179–223.

60. ______________. Additive arithmetic functions on arithmetic progressions, *Proc. London Math. Soc.* **54** (1987), 15–37.

61. ______________. The norms of compositions of arithmetic operators, *Bull. London Math. Soc.* **19** (1987), 522–530.

62. ______________. A local Turán–Kubilius inequality, *Acta Arithmetica* **49** (1987), 15–27.

63. ______________. Application of elementary functional analysis to the study of arithmetic functions, in *Colloq. Math. Soc. János Bolyai* **51**, *Number Theory*, Budapest, (1987), 35–43.

64. ______________. Persistence of form and the value group of reducible cubics, *Trans. Amer. Math. Soc.* **299** (1) (1987), 133–143.

65. ______________. Review of: *Intégration et théorie des nombres* by J.-L. Mauclaire, *Bull. Amer. Math. Soc.* **18** (2) (1988), 193–209.

66. ______________. A localised Erdős–Wintner theorem, *Pacific J. Math.* **135** (2) (1988), 287–295.

67. ______________. Additive arithmetic functions on intervals, *Math. Proc. Camb. Phil. Soc.* **103** (1988), 163–179.

68. ______________. The distribution of additive functions defined on short intervals, in *Séminaire de Théorie des Nombres (Université de Bordeaux 1, Année 1987–1988)*, exposé no. **29**, le 19 mai 1988, 8 pp.

69. ______________. Extrapolating the mean values of multiplicative functions, *Proc. Kon. Ned. Akad. van Wetensch.* (= *Indigationes Math.*), A **92** (1989), 409–420.

70. ______________. The value distribution of differences of additive arithmetic functions, *J. Number Theory* **32** (1989), 339–370.

71. ______________. Multiplicative functions $|g| \leq 1$ and their convolutions: An overview, in *Séminaire de Théorie des Nombres, Inst. H. Poincaré (Paris, 1987–1988)*; *Progress in Math.* **81**, 63–75, Birkhäuser Boston, 1990. Here 'convolutions' should have read 'correlations'.

72. ______________. Some remarks about multiplicative functions of modulus ≤ 1, in *Analytic Number Theory*, Proc. Conf. in honor of P. Bateman, eds. Berndt *et al.*, *Progress in Math.* **85**, Birkhäuser, Boston, 1990, 159–164.

73. ______________. On maximal versions of the Large Sieve, *J. Fac. Sci. Univ. Tokyo, Sect. IA, Math.* **38** (1991), 149–164.

74. ______________. On maximal versions of the Large Sieve, II, *J. Fac. Sci. Univ. Tokyo, Sect. IA, Math.* **39** (1992), 379–383.

75. ______________. Multiplicative functions on arithmetic progressions VI: More middle moduli, *J. Number Theory* **44** (2) (1993), 178–208.

76. ______________. From a large sieve to the orthogonality of operators, *J. London Math. Soc.* **48** (1993), 427–440.

77. ______________. *On the Correlation of Multiplicative and the Sum of Additive Arithmetic Functions*, Amer. Math. Soc. Memoirs **112** (1994), no. 538.

78. ______________. The multiplicative group of rationals generated by the shifted primes, I, *J. reine angew. Math.* **463** (1995), 169–216.

79. Elliott, P.D.T.A. and Erdős, P. The tails of infinitely divisible laws and a problem of number theory, *J. Number Theory* **11** (1979), 542–551.

80. Elliott, P.D.T.A. and Halberstam, H. A conjecture in prime number theory, *Symposia Mathematica*, IV, Academic Press, London and New York, 1970, 59–72.

81. Elliott, P.D.T.A., Moreno, C.J. and Shahidi, F. On the absolute value of Ramanujan's τ function, *Math. Ann.* **266** (1984), 507–511.

82. Erdős, P. On the distribution function of additive functions, *Ann. of Math.* **47** (1946), 1–20.

83. ______________. On a new method in elementary number theory which leads to an elementary proof of the prime number theorem, *Proc. Nat. Acad. Sci. USA* **35** (1949), 374–384.

84. Erdős, P. and Kac, M. On the Gaussian law of errors in the theory of additive functions, *Proc. Nat. Acad. Sci. USA* **25** (1939), 206–207.

85. ______________. The Gaussian law of errors in the theory of additive functions, *Amer. J. Math.* **62** (1940), 738–742.

86. Erdős, P. and Wintner, A. Additive arithmetical functions and statistical independence, *Amer. J. Math.* **61** (1939), 713–721.

87. Forti, M.C. and Viola, C. On large sieve type estimates for the Dirichlet series operator, in *Amer. Math. Soc. Proc. Symp. Pure Math.* XXIV, Providence RI, (1973), 31–49.

88. Friedlander, J., Granville, A., Hildebrand, A. and Maier, H. Oscillation theorems for primes in arithmetic progressions and for sifting functions, *J. Amer. Math. Soc.* **4** (1) (1991), 25–86.

89. Gallagher, P.X. The large sieve, *Mathematika* **14** (1967), 14–20.

90. ______________. A large sieve density estimate near $\sigma = 1$, *Invent. Math.* **11** (1970), 329–339.

91. Hadamard, J. *The Psychology of Invention in the Mathematical Field*, Princeton Univ. Press, 1945, enlarged 1949, reprinted by Dover, 1954.

92. Halász, G. Über die Mittelwerte multiplikativer zahlentheoretischer Funktionen, *Acta Math. Acad. Sci. Hung.* **19** (1968), 365–403.

93. Halberstam, H. and Richert, H.-E. *Sieve Methods*, Academic Press, London, New York, 1974.

94. Hardy, G.H. Note on Ramanujan's arithmetical function $\tau(n)$, *Proc. Camb. Phil. Soc.* **23** (1927), 675–680.

95. ______________. *Ramanujan. Twelve Lectures on Subjects suggested by his Life and Work*, Cambridge Univ. Press, 1940.

96. Hardy, G.H. and Wright, E.M. *An Introduction to the Theory of Numbers*, fourth edition, Clarendon Press, Oxford, 1960.

97. Heath-Brown, D.R. Zero-free regions for Dirichlet L-functions, and the least prime in an arithmetic progression, *Proc. London Math. Soc.* **64** (2) (1992), 265–338.

98. Hildebrand, A. Sur les moments d'une fonction additive, *Ann. Inst. Fourier* **33** (1983), 1–22.

99. ______________. An asymptotic formula for the variance of an additive function, *Math. Zeit.* **183** (2) (1983), 145–170.

100. ______________. Additive functions on arithmetic progressions, *J. London Math. Soc.* **34** (1986), 394–402.

101. ______________. The prime number theorem via the large sieve, *Mathematika* **33** (1986), 23–30.

102. ______________. Additive functions at consecutive integers, *J. London Math. Soc.* **35** (1987), 217–232.

103. ______________. Multiplicative functions in short intervals, *Canadian J. Math.* **39** (3) (1987), 646–672.

104. ______________. An Erdős–Wintner theorem for differences of additive functions, *Trans. Amer. Math. Soc.* **310** (1) (1988), 257–276.

105. Hildebrand, A. and Spilker, J. Charakterisierung der additiven fastgeraden Funktionen, *Manuscripta Math.* **32** (1980), 213–230.

106. Hyers, D.H. On the stability of the linear functional equation, *Proc. Nat. Acad. Sci. USA* **27** (1941), 222–224.

107. Hyers, D.H. and Rassias, T.M. Approximate homomorphisms, *Aequationes Mathematicae* **44** (1992), 125–153.

108. Kátai, I. Some results and problems in the theory of additive functions, *Acta Sc. Math. Szeged.* **30** (1969), 305–311.

109. Kubilius, J. Probabilistic methods in the theory of numbers, *Uspekhi Mat. Nauk. (N.S.)* **11** (1956), 2 (68), 31–66; = *Amer. Math. Soc. Translations* **19** (1962), 47–85.

110. ______. On the estimation of the second central moment for strongly additive arithmetic functions, *Liet. Mat. Rinkinys*, = *Litovsk Mat. Sbornik*, **23** (1) (1983), 122–133.

111. ______. On the estimate of the second central moment for arbitrary additive arithmetic functions, *Liet. Mat. Rinkinys*, = *Litovsk Mat. Sbornik*, **23** (2) (1983), 110–117.

112. ______. Improvement of the estimation of the second central moment for additive arithmetical functions, *Liet. Mat. Rinkinys*, = *Litovsk Mat. Sbornik*, **25** (3) (1985), 104–110.

113. Kuznetsov, N.V. Petersson's conjecture for cusp forms of weight zero and Linnik's conjecture. Sums of Kloosterman sums, *Mat. Sbornik, SSSR* **111** (153) (1980), 334–383. This paper was preceded by an influential Complex Science Research Institute of Khabarovsk preprint, 1977. (I have not seen it.)

114. Lakatos, I. *Proofs and Refutations*, Cambridge Univ. Press, 1976.

115. Landau, E. Neuer Beweis des Primzahlsatzes und Beweis des Primidealsatzes, *Math. Ann.* **56** (1903), 645–670; *Collected Works, Vol. 1*, Thales Verlag, Essen, 1985, 327–352.

116. ______. Über die Äquivalenz zweier Hauptsätze der analytischen Zahlentheorie, *Wiener Sitzungsberichte* **120** (1911), 973–988; *Collected Works, Vol. 5*, Thales Verlag, Essen, 1985, 46–61.

117. Lee, J. *On the constant in the Turán–Kubilius inequality*, Ph.D. dissertation, Univ. of Michigan, 1989.

118. Lehmer, D.H. Ramanujan's function $\tau(n)$, *Duke Math. J.* **10** (1943), 483–492.

119. Levin, B.V. and Timofeev, N.M. An analytic method in the probabilistic theory of numbers, *Uch. Zap. Vladimir gos. ped. inst. mat.*, **57** (2) (1971), 57–150.

120. Linnik, Yu.V. The Large Sieve, *Dokl. Akad. Nauk. SSSR* **30** (1941), 292–294.

121. Maass, H. Über eine neue Art von nichtanalytischen automorphen Funktionen und die Bestimmung Dirichletscher Reihen durch Funktionalgleichung, *Math. Ann.* **121** (1949), 141–183.

122. Manstavičius, E. Inequalities for the pth-moment, $0 < p < 2$, of a sum of independent random variables, *Liet. Mat. Rinkinys*, = *Litovsk Mat. Sbornik*, **22** (1982), 112–116.

123. Mauclaire, J.-L. *Intégration et théorie des nombres*, Hermann, Paris, 1986.

124. Montgomery, H.L. *Topics in Multiplicative Number Theory*, Lecture Notes in Math. **227**, Springer–Verlag, Berlin, 1971.

125. ____________. Maximal variants of the large sieve, *J. Fac. Sci. Univ. Tokyo, Sect. IA, Math.* **28** (1981), 805–812.

126. Montgomery, H.L. and Vaughan, R.C. On the large sieve, *Mathematika* **20** (1973), 119–134.

127. ____________. Exponential sums with multiplicative coefficients, *Invent. Math.* **43** (1977), 69–82.

128. ____________. The order of magnitude of the mth coefficients of cyclotomic polynomials, *Glasgow Math. J.* **27** (1985), 143–159.

129. Mordell, L.J. On Mr. Ramanujan's empirical expansions of modular functions, *Proc. Camb. Phil. Soc.* **19** (1917), 117–124.

130. Murty, M. Ram. Some Ω-results for Ramanujan's τ function, *Proc. Third Mat. Science Conf. (Mysore, 1981)*, Lecture Notes in Math. **938** 1982), Springer–Verlag, 122–137.

131. ____________. Oscillations of Fourier coefficients of modular forms, *Math. Ann.* **262** (1983), 431–446.

132. ____________. The Ramanujan τ function, in *Ramanujan Revisited*, Proc. Centenary Conf., (Urbana–Champaign, IL, 1987), eds. George E. Andrews, *et al.*, Academic Press, London, New York, Tokyo, 1988, 269–288.

133. Nachbin, L. *The Haar Integral*, Van Nostrand, Princeton, Toronto, New York, London, 1965.

134. Prachar, K. *Primzahlverteilung*, Springer–Verlag, Berlin, 1957.

135. Rainville, Earl D. *Special Functions*, Macmillan, New York, 1960.

136. Ramanujan, S. On certain arithmetical functions, *Trans. Camb. Phil. Soc.* **22** (1916), 1159–184.

137. Rankin, R.A. Contributions to the theory of Ramanujan's function $\tau(n)$ and similar arithmetical functions, II, *Proc. Camb. Phil. Soc.* **35** (1939), 357–372.

138. ____________. Sums of powers of cusp form coefficients, *Math. Ann.* **263** (1983), 227–236.

139. ____________. Sums of powers of cusp form coefficients, II, *Math. Ann.* **272** (1985), 593–600.

140. ____________. Fourier coefficients of cusp forms, *Math. Proc. Camb. Phil. Soc.* **100** (1986), 5–29.

141. Rényi, A. On the representation of an even number as the sum of a prime and an almost prime, *Izv. Akad. Nauk. SSSR Ser. Mat.*

12 (1948), 57–78; = *Amer. Math. Soc. Translations*, (2) **19** (1962), 299–321.

142. ______________. Un nouveau théorème concernant les fonctions indépendentes et ses applications à la théorie des nombres, *J. math pures appl.* **28** (1949), 137–149.

143. ______________. A new proof of a theorem of Delange, *Publ. Math. Debrecen* **12** (1965), 323–329.

144. Rosenthal, H.P. On the subspaces of L^p $(p > 2)$ spanned by sequences of independent random variables, *Israel J. Math.* **8** (1970), 273–303.

145. Roth, K.F. On the Large Sieves of Linnik and Rényi, *Mathematika* **12** (1965), 1–9.

146. Rudin, W. *Fourier Analysis on Groups*, Interscience Publishers, New York NY, 1962.

147. Ruzsa, I.Z. The law of large numbers for additive functions, *Stud. Sci. Math. Hung.* **14** (1979), 247–253.

148. ______________. Additive functions with bounded difference, *Periodica Math. Hung.* **10** (1979), 67–70.

149. ______________. On the concentration of additive functions, *Acta Math. Acad. Sci. Hungar.* **36** (1980), 215–232.

150. ______________. On the variance of additive functions, in *Turán Memorial Volume of the J. Bolyai Math. Soc.*, eds. L. Alpár, G. Halász, A. Sárközy, Birkhäuser Verlag, Basel, Boston, Stuttgart, 1983, 577–586.

151. Sarnak, P. *Some applications of modular forms*, Cambridge Tracts in Math. **99**, Cambridge Univ. Press, 1990.

152. Schwarz, W. Fourier–Ramanujan Entwicklungen zahlentheoretischer Funktionen und Anwendungen, in *Festschrift der wissenschaftlichen Gesellschaft an der Johann Wolfgang Goethe Universität*, Frankfurt am Main.

153. Schwarz, W. and Spilker, J. Remarks on Elliott's theorem on mean value of multiplicative functions, *Recent Progress in Analytic Number Theory*, Proc. Conf. Number Theory (Durham, 1979), 325–339, Academic Press, London, 1981.

154. ______________. *Arithmetical Functions*, London Math. Soc. Lecture Note Series **184**, Cambridge Univ. Press, 1994.

155. Selberg, A. An elementary proof of the prime number theorem, *Ann. of Math.* **50** (1949), 305–313.

156. ______________. Harmonic analysis and discontinuous groups in weakly symmetric Riemannian spaces with applications to Dirichlet series, *J. Indian Math. Soc.* **20** (1956), 47–87.

157. ______________. On the estimation of Fourier coefficients of modular forms, *Amer. Math. Soc. Proc. Symp. in Pure Math.* **8**, Providence RI, 1965, 1–15.

158. Serre, J.-P. *Abelian ℓ-adic Representations and Elliptic Curves*, Benjamin, New York, 1968.

159. ______________. Quelques applications du théorème de densité de Chebotarev, *Publ. Math. I.H.E.S.* **54** (1982), 123–201.

160. Shiu, P. A Brun–Titchmarsh theorem for multiplicative functions, *J. reine und angew. Math.* **313** (1980), 161–170.

161. Siebert, H. and Wolke, D. Über einige Analoga zum Bombierischen Primzahlsatz, *Math. Zeit.* **27** (1977), 327–341.

162. Tenenbaum, G. *Introduction à la théorie analytique et probabiliste des nombres*, Institut Elie Cartan, 13, Université de Nancy I, 1990. English translation *Introduction to Analytic and Probabilistic Number Theory*, Cambridge University Press, 1994. Second edition of French edition published by SMF, 1996.

163. Titchmarsh, E.C. *The Theory of Functions*, Oxford Univ. Press, second edition, 1939.

164. ______________. *Introduction to the Theory of Fourier Integrals*, Oxford Univ. Press, second edition, 1948.

165. ______________. *The Theory of the Riemann Zeta-Function*, Oxford Univ. Press, 1951; new edition (D. R. Heath-Brown, ed.), 1986.

166. Vaughan, R.C. *The Hardy–Littlewood Method*, Cambridge Univ. Press, 1981.

167. Venkov, A.B. *Spectral Theory of Automorphic Functions and its Applications*, Kluwer Academic Publishers, Dordrecht, 1990.

168. Wiener, N. Tauberian theorems, *Ann. of Math.* **33** (1932), 1–100.

169. ______________. *The Fourier Integral and Certain of its Applications*, Cambridge Univ. Press, 1933, reprinted by Dover, New York.

170. Wirsing, E. Das asymptotische Verhalten von Summen über multiplikativer Funktionen, II, *Acta Math. Acad. Sci. Hung.* **18** (1967), 411–467.

171. ______________. A characterisation of $\log n$ as an additive arithmetic function, in *Symposia Mathematica*, IV, 45–57, Academic Press, London and New York, 1970.

172. ______________. Additive and completely additive functions with restricted growth, in *Recent Progress in Analytic Number Theory*, Proc. Conf. Number Theory (Durham, 1979), Vol. 2, 231–280, Academic Press, London, 1981.

173. Wolke, D. Über die mittlere Verteilung der Werte zahlentheoretischer Funktionen auf Restklassen, I, *Math. Ann.* **202** (1973), 1–25.

174. ______________. Über die mittlere Verteilung der Werte zahlentheoretischer Funktionen auf Restklassen, II, *Math. Ann.* **204** (1973), 145–153.

175. Yosida, K. *Functional Analysis*, Grund. der math. Wiss. **123**, Springer–Verlag, New York, 1971.

Author Index

Subject Index